Gas Hydrate

Gas Hydrate

Environmental and Climate Impacts

Special Issue Editors

Umberta Tinivella
Michela Giustiniani
Ivan De La Cruz Vargas Cordero
Atanas Vasilev

MDPI • Basel • Beijing • Wuhan • Barcelona • Belgrade

MDPI

Special Issue Editors

Umberta Tinivella
Istituto Nazionale di
Oceanografia e di Geofisica
Sperimentale
Italy

Michela Giustiniani
Istituto Nazionale di
Oceanografia e di Geofisica
Sperimentale
Italy

Ivan De La Cruz Vargas Cordero
Universidad Andrés Bello
Chile

Atanas Vasilev
Bulgarian Academy of Sciences
Bulgaria

Editorial Office
MDPI
St. Alban-Anlage 66
4052 Basel, Switzerland

This is a reprint of articles from the Special Issue published online in the open access journal *Geosciences* (ISSN 2076-3263) in 2019 (available at: https://www.mdpi.com/journal/geosciences/special_issues/gas_hydrate_climate).

For citation purposes, cite each article independently as indicated on the article page online and as indicated below:

LastName, A.A.; LastName, B.B.; LastName, C.C. Article Title. *Journal Name* **Year**, *Article Number*, Page Range.

ISBN 978-3-03921-844-8 (Pbk)
ISBN 978-3-03921-845-5 (PDF)

Cover image courtesy of Umberta Tinivella.

Contents

About the Special Issue Editors

Umberta Tinivella, PhD, Senior Researcher, Vice-Director Geophysical Department. In 1998, she received the Cagniard Award at the EAGE Conference. Since 1996, she has been developing a procedure to estimate the concentrations of gas hydrate and free gas from seismic and well data, as well as to predict overpressure zones from the analysis of seismic, log, and laboratory data by using elastic theories. She has performed simulation of acoustic wave propagation along the drill string. In 2005, she received the Best Poster Presentation at the Near Surface Conference. Her present work addresses different on gas hydrate topics, considering standard and non-conventional processing of seismic data such as wave equation datuming, amplitude versus offset, and modeling and theoretical models to describe the physical properties of gas phase-bearing sediments. She is the leader of several past and ongoing gas hydrate projects that are supported by oil companies and the EU. She is the author of many technical reports and papers on the topic of gas hydrates, and actively disseminates.

Michela Giustiniani, PhD. She received her degree (Final Grade 110/110) in Geological Sciences from University of Rome "La Sapienza" in 1999. She then specialized in flow and transport modeling at Lancaster University, obtaining a scholarship from University of Rome. In 2005, she obtained her PhD in Applied Geophysics at the University of Trieste. During her PhD, she focused on high-resolution seismic analysis as applied to environmental problems such as landslides and aquifers. She has worked as a Researcher at OGS since May 2006. In OGS, she was experienced in the acquisition, processing, inversion, and interpretation of both 2D and 3D crustal and high-resolution seismic data reaching a high level of experience in applied geophysics. She has been working on gas hydrates since 2007, in particular, on the gas reservoir located in Antarctica.

Ivan de la Cruz Vargas Cordero, PhD, Associate Professor. He graduated in Geological Sciences at the University of Concepción in 1999. After this degree, he received a grant from the Italian Government for his doctoral studies at University of Trieste (IT) and National Institute of Oceanography and Experimental Geophysics (OGS). During this three-year period, his main topic was to study gas hydrate and free gas occurrences along the Chilean margin. These studies were focused on standard and advanced processing of marine multichannel seismic data. In particular, prestack depth migration was applied in order to determine accurate velocity models using Seismic Unix and scripts and codes that were created ad hoc. In 2009, he relocated to Barcelona, Spain, to continue his postdoctoral research at the University of Barcelona. Additionally, in the framework of national and international projects, he has participated in oceanographic cruises and geophysical field works. In 2014, he relocated to Valparaíso, Chile, at the Universidad Andres Bello, where he was recently appointed as Associated Professor of Geology.

Atanas Vasilev is a marine geophysicist with 40 years' experience in marine seismic, magnetic, and heat flow data acquisition, processing, and interpretation. His research interests include scientific and applied studies of complex processes and geo-objects: gas hydrates, mud volcanoes, oil and gas, geohazards and coastal infrastructure. He is author of 162 academic publications and 95 research reports, a participant in 48 research cruises onboard of 30 research vessels, and also involved in the main European projects for the study of marine gas hydrates: cesum-blacksea, assemblage, euroseismic, gashydat, seiscanex, crimea, geo-methane, geo-hydrate, pergamon, marine geohazards, sugar and migrate. The study of Black Sea gas hydrates is the main scientific interest of Dr Vasilev over the last 20 years. His last project KP-06-OPR04/7 GEOHydrate: "Geothermal Evolution of Marine Gas Hydrate Deposits", 2019–2021 (Bulgarian Science Fund) is aimed at reconstruction of the 4D formation of marine gas hydrate deposits from heat flow data.

Preface to "Gas Hydrate"

In recent decades, gas hydrates have been considered a possible reservoir of natural gas, even if the actual global estimate is very rough. The growing interest in the scientific and industrial communities for gas hydrates is focused on: (1) the assessment of methane hydrate as a new "clean" energy source, (2) the relationship between gas hydrate and global climate change, (3) the geological hazards connected to the gas hydrate and, recently, (4) a wide range of industrial applications based on the specifics of the processes of gas hydrate formation and dissociation. Gas hydrates can be related to environmental risks because their dissociation could affect seafloor stability and release methane and associated gases into the water column. Also well known is the role of methane as an important greenhouse gas, and any methane release into the atmosphere would have an impact on climate change.

Gas hydrates also have an influence on geopolitics. In fact, the biggest natural gas importers as China, India, and Japan have significant hydrate reserves and have started challenging and generously funded programs for marine gas hydrate production. On the other hand, other countries, such as in Europe, have reduced financial resources dedicated to this topic. Gas hydrate deposits are investigated using geophysical methods. Significant progress in the development of deep-water high-resolution geophysical tools and technology during the 21st century is due mainly to the acceleration of gas hydrate studies. The seismic technique, which is used mostly for gas hydrate investigations, allows for detection of a clear indicator of the boundary between hydrate and free gas accumulation, known as bottom-simulating reflector (BSR). Moreover, the seismic data provide information about the geometry of the main geological structures, allowing for possible explanations of the presence of gas hydrate. In the last few years, the integration of geophysical (mainly seismic and electromagnetic data), geochemical, and heat-flow data have allowed for detecting and characterizing gas hydrate and free gas volumes and their distribution in the sediments. Thus, reviews of extensive geophysical surveys and direct measurements combined with geological interpretation and theoretical modeling will increase our understanding of the occurrence, distribution, and concentration of gas hydrate and the underlying free gas beneath the ocean bottom and in the permafrost regions.

This Special Issue offers the scientific community an opportunity to illustrate the multidisciplinary research developed in parts of the world such as Arctic and offshore Chile, where the interest in gas hydrates is from an energy and environmental point of view.

Umberta Tinivella, Michela Giustiniani, Ivan De La Cruz Vargas Cordero, Atanas Vasilev
Special Issue Editors

geosciences

MDPI

Editorial

Gas Hydrate: Environmental and Climate Impacts

Umberta Tinivella [1,*], Michela Giustiniani [1], Ivan de la Cruz Vargas Cordero [2] and Atanas Vasilev [3]

1   Geophysical Department, OGS (Istituto Nazionale di Oceanografia e di Geofisica Sperimentale), Borgo Grotta Gigante 42/C, 34010 Sgonico, Italy; mgiustiniani@inogs.it
2   Facultad de Ingeniería, Universidad Andrés Bello, Quillota 980, Viña del Mar 2531015, Chile; ivan.vargas@unab.cl
3   Institute of Oceanology, Bulgarian Academy of Sciences, 9000 Varna, Bulgaria; gasberg@io-bas.bg
*   Correspondence: utinivella@inogs.it; Tel.: +39-040-2140-219

Received: 10 October 2019; Accepted: 14 October 2019; Published: 18 October 2019

check for updates

Abstract: This Special Issue reports research spanning from the analysis of indirect data, modelling, laboratory and geological data confirming the intrinsic multidisciplinarity of the gas hydrate studies. The study areas are (1) Arctic, (2) Brazil, (3) Chile and (4) the Mediterranean region. The results furnished an important tessera of the knowledge about the relationship of a gas hydrate system with other complex natural phenomena such as climate change, slope stability and earthquakes, and human activities.

Keywords: natural gas hydrate; methane cycle; global change; ecosystem; geohazards; risk assessment; environmental impact; multidisciplinary; blue growth

1. Introduction

In recent decades, gas hydrates have been considered a possible reservoir of natural gas, even if the actual global estimate is very rough [1–4]. The growing interest in gas hydrate of the scientific and industrial communities is focused on: (1) the assessment of methane hydrate as a new "clean" energy source, (2) the relationship between gas hydrate and global climate change, (3) the geological hazards connected to the gas hydrate, and, recently, (4) a wide range of industrial applications based on the specifics of the processes of gas hydrates formation and dissociation. Gas hydrates can be related to environmental risks because their dissociation could affect seafloor stability and release methane (and associated gases) into the water column. Also well known, methane is an important greenhouse gas and any release of it into the atmosphere would have an impact on climate change [1,4–9].

Gas hydrates could have an influence on geopolitics. In fact, the biggest natural gas importers as China, India and Japan have significant hydrate reserves and started challenging and generous funded programs for marine gas hydrate production, i.e., [10–14]. On the other hand, other countries, such as Europe, have reduced financial resources dedicated to this topic.

Generally, gas hydrate deposits are investigated using geophysical methods, i.e., [15,16]. A significant progress/improvement in the twenty-first century of the deep water high resolution geophysical tools and technology is due mainly to acceleration of gas hydrate studies. The seismic technique, which is used mostly for gas hydrate investigation, allows for detecting a clear indicator of the boundary between hydrate and free gas accumulation, known as bottom simulating reflector (BSR), i.e., [17]. Moreover, the seismic data provide information about the geometry of the main geological structures, allowing for possible explanations of the presence/absence of gas hydrate [18,19]. In the last few years, the integration of geophysical (mainly seismic and electromagnetic data), geochemical, and heat-flow data have allowed for detecting and characterizing gas hydrate and free gas volumes and their distribution in the sediments, i.e., [20–23]. Thus, reviews of extensive geophysical surveys and direct measurements combined with geological interpretation and theoretical modelling will increase our understanding of the occurrence, distribution, and concentration of gas hydrate and the underlying free gas beneath the ocean bottom and in the permafrost regions, i.e., [24–29].

This Special Issue has offered to the scientific community an opportunity to illustrate multidisciplinary research developed in part of the word, such as Arctic and offshore Chile, where the interest about gas hydrate is from an energy and environmental point of view.

2. An Overview of the Special Issue

The Special Issue is composed by 9 scientific articles and 1 review paper, spanning from analysis of indirect data, modelling, laboratory and geological data confirming the intrinsic multidisciplinarity of the gas hydrate studies. The papers are grouped based on the study areas that are (1) Arctic, (2) Brazil, (3) Chile and (4) Mediterranean region.

2.1. Arctic

Natural gas hydrates are discovered for the first time in a permafrost region in Russia in 1976 [30]. Then, the number of studies was increased year by year, mainly due to the rapid increase of the surface temperature in this region in order to understand the relationship between gas hydrate stability and global warming, i.e., [31–34].

Chuvilin et al. [35,36] modeled the role of salt migration and warming in the destabilization of intra permafrost hydrates in order to understand if the destabilization of intrapermafrost gas hydrate could be related to methane emission on the Arctic shelf. The intrapermafrost hydrate could be present at a shallow depth and transform into a relict state. In the paper [35], the authors' studies of the interaction of frozen sandy sediments containing relict methane hydrates with salt solutions of different concentrations at negative temperatures to assess the conditions of intrapermafrost gas hydrates dissociation. The results of the experiments are that the migration of salts into frozen hydrate-containing sediments activates the decomposition of pore gas hydrates and increases the methane emission. Moreover, in the paper [36], the authors analyzed the effect of temperature increase on frozen sand and silt containing metastable pore methane hydrate in order to reconstruct the conditions for intrapermafrost gas hydrate dissociation. The experiments showed that the dissociation process in hydrate-bearing frozen sediments exposed to warming begins and ends before the onset of pore ice melting. The critical temperature sufficient for gas hydrate dissociation varies from −3.0 °C to −0.3 °C and depends on lithology (particle size) and salinity of the host frozen sediments. Considering an almost gradientless temperature distribution during degradation of subsea permafrost, even minor temperature increases can be expected to trigger large-scale dissociation of intrapermafrost hydrates. So, References [35,36] have furnished an important piece of the knowledge about the mechanism of massive methane release from bottom sediments of the East Siberian Arctic shelf.

Many studies have demonstrated the coexistence of subaqueous permafrost, gas hydrate and the effect of the subaqueous on their formation/dissociation, i.e., [37]. Nevertheless, before Reference [38], an empirical method, which allows for an easy initial estimation of the conditions sufficient to have the stability of hydrate below subaqueous permafrost in absence of direct geological or geophysical data, was missed. In this Special Issue, for the first time a quick-look method that allows estimating the steady-state conditions for gas hydrate stability in the presence of subaqueous permafrost is presented. Different thermodynamic conditions typical of subaqueous permafrost in shallow waters both in marine and lacustrine environments are considered. The approach is derived for pressure, temperature, and salinity conditions typical of subaqueous permafrost in marine (brine) and lacustrine (freshwater) environments and it can be easily and reliably applied to assess if the sufficient conditions to have hydrate stability are satisfied.

2.2. Brazil

In this area the gas hydrate is explored only recently. This Special Issue reported the review of the evidences of venting from gas hydrate provinces along Brazil's continental margin in [39]. In literature, only indirect indications of the presence of gas hydrate were reported analyzing seismic data in two deep-water depocenters: the Rio Grande cone in the Pelotas Basin and the Amazon deep-sea fan in the Foz do Amazonas basin. Recently, direct data, such as seafloor sampling of gas venting, confirmed gas hydrate presence. The modeling of the hydrate stability zone confirmed that the hydrate is stable for water depth greater than about 500–700 m. Moreover, the identified gas venting is located along the feather edge of the stability zone, suggesting gas hydrate dissociation or upward fluid flow through the stability zone facilitated by tectonic structures recording the gravitational collapse of depocenters.

Reference [40] focused their attention on the Amazon deep-sea fan and adjacent continental slope, investigating the molecular stable isotope compositions of hydrate bound and dissolved gases in sediments. A dominant microbial origin of methane via carbon dioxide reduction was detected; however, a possible mixture of thermogenic and microbial gases are recovered in sites located in the adjacent continental slope.

Finally, Reference [41] analyzed the deep structures related to the high concentrations of CO_2 detected along the southeastern Brazilian Margin by using a multidisciplinary approach. Gravimetric and magnetic potential methods were used to identify major intrusive bodies, crustal thinning and other geotectonic elements of the southeastern Brazilian Margin. Modeling based on magnetic, gravity and seismic data suggests a major intrusive magmatic body just below the reservoir where a high CO_2 accumulation was found. Small faults connecting this magmatic body with the sedimentary section could be the fairway for the magmatic sourced gas rise to reservoirs, confirming that mapping and understanding the crustal structure of sedimentary basins are important steps for "de-risking" in the exploration process.

To conclude, these three papers indicated that it is important to model the quantities of gas that may be transferred from sediments to the oceans offshore Brazil. Considering the possible existence of gas hydrate provinces in other basins along the Brazilian margin, further investigations are necessary.

2.3. Chile

In the last decade, the studies about gas hydrate presence along the Chilean Margin are increased rapidly, furnishing information about distribution and quantification of gas hydrate and free gas from seismic data analysis in several zones of the Chilean Margin, i.e., [42,43]. Here, Reference [44] presented an analysis of the spatial distribution, concentration, estimate of gas-phases (gas hydrate and free gas) and geothermal gradients in the accretionary prism, and forearc sediments offshore Taitao at the Chile Triple Junction. Seismic data analysis indicated high gas hydrate concentration and extremely high geothermal gradients. The large amount of hydrate and free gas estimated, the high seismicity, the mechanically unstable nature of the sediments, and the anomalous conditions of the geothermal gradient set the stage for potentially massive releases of methane to the ocean, mainly through hydrate dissociation and/or migration directly to the seabed through faults. So, the Chile Triple Junction is an important methane seepage area and should be the focus of novel geological, oceanographic, and ecological research.

In order to extrapolate information about potential hydrate distribution along the whole Chilean margin, Reference [45] modeled the gas hydrate stability zone using a steady state approach to evaluate the effects of climate change on gas hydrate stability. Present day conditions were modelled using published literature and compared with available measurements. Then, the effects of climate change on gas hydrate stability in 50 and 100 years on the basis of Intergovernmental Panel on Climate Change and National Aeronautics and Space Administration forecasts are modeled. An increase in temperature might cause the dissociation of gas hydrate that could strongly affect gas hydrate stability. Moreover, it is important to consider that the high seismicity of this area could have a strong effect on gas hydrate stability.

The results of these two papers confirm that the Chilean margin should be considered as a natural laboratory for understanding the relationship between gas hydrate systems and complex natural phenomena, such as climate change, slope stability and earthquakes.

2.4. Mediterranean Region

In the Mediterranean Sea, evidences of the hydrate presence are unclear from indirect data analysis. Nerveless, the Eastern Mediterranean Sea is expected to host a significant amount of hydrate because large areas of the seabed are located within the hydrate stability zone [46]. Multiple observations indicate the availability of gas, required for the formation of hydrate, across the seafloor. In particular, numerous mud volcanoes are present, primarily along the accretionary complex and to a lesser degree in the Nile fan [47]. The scope of known seepage is continuously expanding as new data become available, providing further evidence for the potential for hydrate formation. To date, hydrate has been sampled only in several mud volcanoes of the accretionary complex, starting in the Anaximander Seamount region, i.e., [48,49]. In addition, a recent 3D dataset acquired in the Levan Basin, southeastern Mediterranean Sea, suggested that this region could be promising in regards to gas hydrate [50]. Reference [50] estimated the potential inventory of natural gas hydrate in the Levant Basin correlating the gas hydrate stability zone with seismic indicators of gas and providing a potentiality of carbon in this area.

Another key point to understand is whether or not the Mediterranean region hosted hydrate in the past. Compared to the abundant literature on present-day gas hydrates, only few studies deal with their past occurrence or with fossil seep-carbonates recording the dissociation of gas hydrates, i.e., [51]. In fossil sediments, the paleo-occurrence of gas hydrate is particularly challenging to assess, due to the lack of well-established proxies and to the uncertainties on the reconstruction of paleoenvironmental conditions (pressure, temperature, depth) controlling the hydrate stability field. Clathrate-like structures have been reported in fossil deposits and can be used as an indication of past gas hydrate destabilization, i.e., [52]. Additional evidences can be yielded by geochemical signatures, the large dimensions of seep-carbonate deposits (several hundred meters in lateral extent and tens of meters in thickness) and the association with sedimentary instability (soft-sediment deformations) in hosting sediments [53]. Reference [54] could be considered pioneer in this background. In fact, they combined multiple field and geochemical indicators for paleo-gas hydrate occurrence based on present-day analogues to investigate fossil seeps located in the northern Apennines. They recognized clathrate-like structures, such as thin-layered, spongy and muggy textures and microbreccias. Non-gravitational cementation fabrics and pinch-out terminations in cavities within the seep-carbonate deposits are ascribed to irregularly oriented dissociation of gas hydrates. Additional evidences for paleo-gas hydrates are provided by the large dimensions of seep-carbonate masses and by the association with sedimentary instability in the host sediments. Moreover, heavy oxygen isotopic values in the examined seep-carbonates indicated a contribution of isotopically heavier fluids released by gas hydrate decomposition. Their result agrees with the calculation of the stability field of methane hydrates for the northern Apennine wedge-foredeep system during the Miocene indicating the potential occurrence of shallow gas hydrates in the upper few tens of meters of sedimentary column.

So, References [50,54] suggest that the Mediterranean region should be investigated in order to understand the reason of the past-presence and the quite-absence of gas hydrate by using a multidisciplinary approach spanning from field data to modeling.

3. Key Message for Future Research

This Special Issue points out that more studies are necessary to better understand the complexity of the natural gas hydrate system around the world. More efforts should be devoted to correctly quantify the global amount of carbon stored in hydrate form and their relationship with other complex natural phenomena, such as climate change, slope stability and earthquakes, and human activities. Therefore, we hope that new research will be started in order to acquire new data by using innovative technologies to refine the existing theories or define new theoretical models that cover all aspects of this complex phenomena.

Author Contributions: This Editorial is the result of the collaboration of all authors. U.T. and M.G. created the main text; I.d.l.C.V.C. and A.V. made a minor edition and added a part of the Introduction.

Funding: A.V. contribution is in the frame of the Project KP-06-OPR04/7 GEOHydrate (Bulgarian Science Fund).

Acknowledgments: The Guest Editors thank all the authors, Geosciences' editors, and reviewers for their great contributions and commitment to this Special Issue. A special thank goes to Richard Li, Geosciences' Managing Editor, for his dedication to this project and his valuable collaboration in the design and setup of the Special Issue.

Conflicts of Interest: The authors declare no conflict of interest.

References

1. Kvenvolden, K.A. Gas hydrates—Geological perspective and global change. *Rev. Geophys.* **1993**, *31*, 173–187. [CrossRef]

2. Milkov, A.; Sassen, R. Economic Geology of offshore gas hydrate accumulations and provinces. *Mar. Pet. Geol.* **2002**, *19*, 1–11. [CrossRef]

3. Makagon, Y.F. Natural gas hydrate—A promising source of energy. *J. Nat. Gas Sci. Eng.* **2010**, *2*, 49–59. [CrossRef]

4. Boswell, R.; Collett, T.S. Current perspectives on gas hydrate resources. *Energy Environ. Sci.* **2011**, *4*, 1206–1215. [CrossRef]

5. Henriet, J.-P.; Mienert, J. (Eds.) *Gas Hydrates. Relevance to World Margin Stability and Climatic Change*; Geological Society Special Publication No. 137; Geological Society of London: London, UK, 1998; 338p.

6. Kvenvolden, K.A. Potential effects of gas hydrate on human welfare. *Proc. Natl. Acad. Sci. USA* **1999**, *96*, 3420–3426. [CrossRef]

7. de Garidel-Thoron, T.; Beafort, L.; Bassinot, F.; Hensy, P. Evidence for large methane releases to the atmosphere from deep-sea gas-hydrate dissociation during the last glacial episode. *Proc. Natl. Acad. Sci. USA* **2004**, *101*, 9187–9192. [CrossRef]

8. Waite, W.F.; Santamarina, J.C.; Cortes, D.D.; Dugan, B.; Espinoza, D.N.; Germaine, J.; Jang, J.; Jung, J.W.; Kneafsey, T.J.; Shin, H.; et al. Physical properties of hydrate-bearing sediments. *Rev. Geophys.* **2009**, *47*. [CrossRef]

9. Ruppel, C.D.; Kessler, J.D. The interaction of climate change and methane hydrates. *Rev. Geophys.* **2017**, *55*, 126–168. [CrossRef]

10. Dallimore, S.R.; Wright, J.F.; Nixon, F.M.; Kurihara, M.; Yamamoto, K.; Fujii, T.; Fujii, K.; Numasawa, M.; Yasuda, M.; Imasato, Y. Geologic and porous media factors affecting the 2007 production response characteristics of the JOGMEC/NRCAN/AURORA Mallik gas hydrate production research well. In Proceedings of the 6th International Conference on Gas Hydrates, Vancouver, BC, Canada, 6–10 July 2008; p. 10.

11. Dallimore, S.R.; Wright, J.F.; Yamamoto, K. Appendix D: Update on Mallik. In *Energy from Gas Hydrates: Assessing the Opportunities and Challenges for Canada*; Council of Canadian Academies: Ottawa, ON, Canada, 2008; pp. 196–200.

12. Gabitto, J.F.; Tsouris, C. Physical properties of gas hydrates: A review. *J. Thermodyn.* **2010**, *2010*, 271291. [CrossRef]

13. Song, Y.; Yang, L.; Zhao, J.; Liu, W.; Yang, M.; Li, Y.; Liu, Y.; Li, Q. The status of natural gas hydrate research in China: A review. *Renew. Sustain. Energy Rev.* **2014**, *31*, 778–791. [CrossRef]

14. Yamamoto, K.; Kanno, T.; Wang, X.-X.; Tamaki, M.; Fujii, T.; Chee, S.-S.; Wang, X.-W.; Pimenov, V.; Shako, V. Thermal responses of a gas hydrate-bearing sediment to a depressurization operation. *R. Soc. Chem.* **2017**, *7*, 5554–5577. [CrossRef]

15. Tinivella, U.; Accaino, F.; Della Vedova, B. Gas hydrates and active mud volcanism on the South Shetland continental margin, Antarctic Peninsula. *Geo-Mar. Lett.* **2008**, *28*, 97–106. [CrossRef]

16. Vargas-Cordero, I.; Tinivella, U.; Accaino, F.; Loreto, M.F.; Fanucci, F. Thermal state and concentration of gas hydrate and free gas of Coyhaique, Chilean Margin (44 30′ S). *Mar. Pet. Geol.* **2010**, *27*, 1148–1156. [CrossRef]

17. Vargas-Cordero, I.; Tinivella, U.; Accaino, F.; Loreto, M.F.; Fanucci, F.; Reichert, C. Analyses of bottom simulating reflections offshore Arauco and Coyhaique (Chile). *Geo-Mar. Lett.* **2010**, *30*, 271–281. [CrossRef]
18. Villar-Muñoz, L.; Bento, J.P.; Klaeschen, D.; Tinivella, U.; Vargas-Cordero, I.; Behrmann, J.H. A first estimation of gas hydrates offshore Patagonia (Chile). *Mar. Pet. Geol.* **2018**, *96*, 232–239. [CrossRef]
19. Song, S.; Tinivella, U.; Giustiniani, M.; Singhroha, S.; Bünz, S.; Cassiani, G. OBS data analysis to quantify gas hydrate and free gas in the South Shetland margin (Antarctica). *Energies* **2018**, *11*, 3290. [CrossRef]
20. Coren, F.; Volpi, V.; Tinivella, U. Gas hydrate physical properties imaging by multi-attribute analysis—Blake Ridge BSR case history. *Mar. Geol.* **2001**, *178*, 197–210. [CrossRef]
21. Loreto, M.F.; Tinivella, U.; Accaino, F.; Giustiniani, M. Offshore Antarctic Peninsula gas hydrate reservoir characterization by geophysical data analysis. *Energies* **2011**, *4*, 39–56. [CrossRef]
22. Loreto, M.F.; Tinivella, U. Gas hydrate versus geological features: The South Shetland case study. *Mar. Pet. Geol.* **2012**, *36*, 164–171. [CrossRef]
23. Tinivella, U.; Giustiniani, M. Numerical simulation of coupled waves in borehole drilling through a BSR. *Mar. Pet. Geol.* **2013**, *44*, 34–40. [CrossRef]
24. Tinivella, U. A method for estimating gas hydrate and free gas concentrations in marine sediments. *Boll. Geofis. Teor. Appl.* **1999**, *40*, 19–30.
25. Tinivella, U. The seismic response to overpressure versus gas hydrate and free gas concentration. *J. Seism. Explor.* **2002**, *11*, 283–305.
26. Chand, S.; Minshull, T.A.; Gei, D.; Carcione, J.M. Elastic velocity models for gas-hydrate-bearing sediments—A comparison. *Geophys. J. Int.* **2004**, *159*, 573–590. [CrossRef]
27. Kumar, D.; Sen, M.K.; Bangs, N.L. Gas hydrate concentration and characteristics within Hydrate Ridge inferred from multicomponent seismic reflection data. *J. Geophys. Res. Solid Earth* **2007**, *112*. [CrossRef]
28. Vargas-Cordero, I.; Tinivella, U.; Villar-Muñoz, L.; Giustiniani, M. Gas hydrate and free gas estimation from seismic analysis offshore Chiloé island (Chile). *Andean Geol.* **2016**, *43*, 263–274. [CrossRef]
29. Vargas-Cordero, I.; Tinivella, U.; Villar-Muñoz, L.; Bento, J.P. High Gas Hydrate and Free Gas Concentrations: An Explanation for Seeps Offshore South Mocha Island. *Energies* **2018**, *11*, 3062. [CrossRef]
30. Makogon, Y. *Hydrates of Hydrocarbon*; Penn Well Publisher: Tulsa, OK, USA, 1997.
31. Tinivella, U.; Giustiniani, M. Variations in BSR depth due to gas hydrate stability versus pore pressure. *Glob. Planet. Chang.* **2013**, *100*, 119–128. [CrossRef]
32. Marin-Moreno, H.; Giustiniani, M.; Tinivella, U. The Potential Response of the Hydrate Reservoir in the South Shetland Margin, Antarctic Peninsula, to Ocean Warming over the 21st Century. *Polar Res.* **2015**, *34*, 27443. [CrossRef]
33. Marín-Moreno, H.; Giustiniani, M.; Tinivella, U.; Piñero, E. The challenges of quantifying the carbon stored in Arctic marine gas hydrate. *Mar. Pet. Geol.* **2016**, *71*, 76–82. [CrossRef]
34. Giustiniani, M.; Tinivella, U.; Sauli, C.; Della Vedova, B. Distribution of the gas hydrate stability zone in the Ross Sea, Antarctica [Distribución de la zona de estabilidad de hidratos de metano en el mar de Ross, Antártica]. *Andean Geol.* **2018**, *45*, 78–86. [CrossRef]
35. Chuvilin, E.; Ekimova, V.; Bukhanov, B.; Grebenkin, S.; Shakhova, N.; Semiletov, I. Role of Salt Migration in Destabilization of Intra Permafrost Hydrates in the Arctic Shelf: Experimental Modeling. *Geosciences* **2019**, *9*, 188. [CrossRef]
36. Chuvilin, E.; Davletshina, D.; Ekimova, V.; Bukhanov, B.; Shakhova, N.; Semiletov, I. Role of Warming in Destabilization of Intrapermafrost Gas Hydrates in the Arctic Shelf: Experimental Modeling. *Geosciences* **2019**, *9*, 407. [CrossRef]
37. Tinivella, U.; Giustiniani, M. Gas hydrate stability zone in shallow Arctic Ocean in presence of sub-sea permafrost. *Rend. Lincei* **2016**, *27*, 163–171. [CrossRef]
38. Tinivella, U.; Giustiniani, M.; MarÃn-Moreno, H. A Quick-Look Method for Initial Evaluation of Gas Hydrate Stability below Subaqueous Permafrost. *Geosciences* **2019**, *9*, 329. [CrossRef]
39. Ketzer, M.; Praeg, D.; Pivel, M.; Augustin, A.; Rodrigues, L.; Viana, A.; Cupertino, J. Gas Seeps at the Edge of the Gas Hydrate Stability Zone on Brazilia's Continental Margin. *Geosciences* **2019**, *9*, 193. [CrossRef]
40. Rodrigues, L.; Ketzer, J.; Oliveira, R.; dos Santos, V.; Augustin, A.; Cupertino, J.; Viana, A.; Leonel, B.; Dorle, W. Molecular and Isotopic Composition of Hydrate-Bound, Dissolved and Free Gases in the Amazon Deep-Sea Fan and Slope Sediments, Brazil. *Geosciences* **2019**, *9*, 73. [CrossRef]

41. Gamboa, L.; Ferraz, A.; Baptista, R.; Neto, E. Geotectonic Controls on CO2 Formation and Distribution Processes in the Brazilian Pre-Salt Basins. *Geosciences* **2019**, *9*, 252. [CrossRef]

42. Vargas-Cordero, I.; Tinivella, U.; Accaino, F.; Fanucci, F.; Loreto, M.F.; Lascano, M.E.; Reichert, C. Basal and frontal accretion processes versus BSR characteristics along the Chilean margin. *J. Geol. Res.* **2011**, *2011*, 846101. [CrossRef]

43. Vargas-Cordero, I.; Tinivella, U.; Villar-Muñoz, L. Gas Hydrate and Free Gas Concentrations in Two Sites inside the Chilean Margin (Itata and Valdivia Offshores). *Energies* **2017**, *10*, 2154. [CrossRef]

44. Villar-Munoz, L.; Vargas-Cordero, I.; Bento, J.; Tinivella, U.; Fernandoy, F.; Giustiniani, M.; Behrmann, J.; Calderon-Diaz, S. Gas Hydrate Estimate in an Area of Deformation and High Heat Flow at the Chile Triple Junction. *Geosciences* **2019**, *9*, 28. [CrossRef]

45. Alessandrini, G.; Tinivella, U.; Giustiniani, M.; de la Cruz Vargas-Cordero, I.; Castellaro, S. Potential Instability of Gas Hydrates along the Chilean Margin Due to Ocean Warming. *Geosciences* **2019**, *9*, 234. [CrossRef]

46. Merey, S.; Longinos, S.N. Does the Mediterranean Sea have potential for producing gas hydrates? *J. Nat. Gas Sci. Eng.* **2018**, *55*, 113–134. [CrossRef]

47. Minshull, T.A.; Marín-Moreno, H.; Betlem, P.; Bialas, J.; Bünz, S.; Burwicz, E.; Cameselle, A.L.; Cifci, G.; Giustiniani, M.; Hillman, J.I.T.; et al. Hydrate occurrence in Europe: A review of available evidence. *Mar. Pet. Geol.* **2020**, *111*, 735–764. [CrossRef]

48. Zitter, T.A.C.; Huguen, C.; Woodside, J.M. Geology of mud volcanoes in the eastern Mediterranean from combined sidescan sonar and submersible surveys. *Deep-Sea Res. Part I. Oceanogr. Res.* **2005**, *52*, 457–475. [CrossRef]

49. Mascle, J.; Mary, F.; Praeg, D.; Brosolo, L.; Camera, L.; Ceramicola, S.; Dupre, S. Distribution and geological control of mud volcanoes and other fluid/free gas seepage features in the Mediterranean Sea and nearby Gulf of Cadiz. *Geo Mar. Lett.* **2014**, *34*, 89–110. [CrossRef]

50. Tayber, Z.; Meilijson, A.; Ben-Avraham, Z.; Makovsky, Y. Methane Hydrate Stability and Potential Resource in the Levant Basin, Southeastern Mediterranean Sea. *Geosciences* **2019**, *9*, 306. [CrossRef]

51. Accaino, F.; Bratus, A.; Conti, S.; Fontana, D.; Tinivella, U. Fluid seepage in mud volcanoes of the northern Apennines: An integrated geophysical and geological study. *J. Appl. Geophys.* **2007**, *63*, 90–101. [CrossRef]

52. Dela Pierre, F.; Martire, L.; Natalicchio, M.; Clari, P.; Petrea, C. Authigenic carbonates in Upper Miocene sediments of the Tertiary Piedmont Basin (NW Italy): Vestiges of an ancient gas hydrate stability zone? *GSA Bull.* **2010**, *122*, 994–1010. [CrossRef]

53. Conti, S.; Fontana, D.; Lucente, C.C.; Pini, G.A. Relationships between seep-carbonates, mud volcanism and basin geometry in the Late Miocene of the northern Apennines of Italy: The Montardone mélange. *Int. J. Earth Sci.* **2014**, *103*, 281–295. [CrossRef]

54. Argentino, C.; Conti, S.; Fioroni, C.; Fontana, D. Evidences for Paleo-Gas Hydrate Occurrence: What We Can Infer for the Miocene of the Northern Apennines (Italy). *Geosciences* **2019**, *9*, 134. [CrossRef]

geosciences

MDPI

Article

Role of Salt Migration in Destabilization of Intra Permafrost Hydrates in the Arctic Shelf: Experimental Modeling

Evgeny Chuvilin [1,*], Valentina Ekimova [1], Boris Bukhanov [1], Sergey Grebenkin [1], Natalia Shakhova [2,3] and Igor Semiletov [2,3,4]

1   Skolkovo Institute of Science and Technology (Skoltech), 3, Nobel st., Innovation Center Skolkovo, Moscow 121205, Russia; Valentina.Ekimova@skoltech.ru (V.E.); b.bukhanov@skoltech.ru (B.B.); S.Grebenkin@skoltech.ru (S.G.)
2   National Research Tomsk Polytechnic University, Tomsk Polytechnic University (TPU), 30, Lenin Avenue, Tomsk 634050, Russia; shahova@tpu.ru (N.S.); ipsemiletov@alaska.edu (I.S.)
3   International Arctic Research Center, University Alaska Fairbanks, 903 Koyukuk Drive, Fairbanks, AK 99775, USA
4   Pacific Oceanological Institute, Far Eastern Branch of Russian Academy of Sciences, 43, Baltiiskaya st., Vladivostok 690041, Russia
*   Correspondence: e.chuvilin@skoltech.ru

Received: 4 April 2019; Accepted: 20 April 2019; Published: 23 April 2019

check for
updates

Abstract: Destabilization of intrapermafrost gas hydrate is one possible reason for methane emission on the Arctic shelf. The formation of these intrapermafrost gas hydrates could occur almost simultaneously with the permafrost sediments due to the occurrence of a hydrate stability zone after sea regression and the subsequent deep cooling and freezing of sediments. The top of the gas hydrate stability zone could exist not only at depths of 200–250 m, but also higher due to local pressure increase in gas-saturated horizons during freezing. Formed at a shallow depth, intrapermafrost gas hydrates could later be preserved and transform into a metastable (relict) state. Under the conditions of submarine permafrost degradation, exactly relict hydrates located above the modern gas hydrate stability zone will, first of all, be involved in the decomposition process caused by negative temperature rising, permafrost thawing, and sediment salinity increasing. That's why special experiments were conducted on the interaction of frozen sandy sediments containing relict methane hydrates with salt solutions of different concentrations at negative temperatures to assess the conditions of intrapermafrost gas hydrates dissociation. Experiments showed that the migration of salts into frozen hydrate-containing sediments activates the decomposition of pore gas hydrates and increase the methane emission. These results allowed for an understanding of the mechanism of massive methane release from bottom sediments of the East Siberian Arctic shelf.

Keywords: Arctic shelf; permafrost; gas hydrate; salt migration; thawing; hydrate dissociation; methane emission; environmental impact; geohazards

---

1. Introduction

The Arctic shelf is the most promising hydrocarbon production area. However, its development is associated with the solution of a number of problems that are associated with the conditions of relict permafrost and the presence of gas hydrates [1–4]. Of particular interest is the assessment of methane emissions during the shelf permafrost degradation and the hydrate decomposition due to heat and mass transfer processes. According to many researchers, dissociation of gas hydrate formations in bottom sediments makes the largest contribution to methane emissions on the Arctic shelf [5–12].

Gas hydrates are crystalline clathrate compounds that are formed from gas (mainly methane in natural conditions) and water under certain temperature and pressure conditions [13,14]. An important characteristic of gas hydrates is a huge accumulation of gas in the clathrate structure-up to 160 volumes of gas in one volume of hydrate. As it is well known, methane is one of the most active greenhouse gases. In this regard, the dissociation of Arctic gas hydrates, accompanied by the active emission of methane into the atmosphere, can have a significant greenhouse effect and cause climate change [15,16].

Under natural conditions, gas hydrates are formed and exist in the bottom sediments of the seas and oceans, as well as in the areas of permafrost distribution where there exist favorable temperatures, pressurization, and geochemical conditions. On the Arctic shelf, gas hydrates can be expected at depths of the sea of 250–300 m, as well as at shallower depths in the presence of underwater permafrost. Considering that the thickness of the submarine permafrost can reach several hundred meters, gas hydrate formations can be located both in the sub-permafrost and intrapermafrost horizons [17–20]. The possible existence of hydrates within the gas hydrate stability zone (GHSZ) in the East Siberian Arctic shelf (ESAS) was predicted in the late 1970s, when it was understood that the high-latitude, shallow ESAS has been alternately subaerial and inundated with seawater during glacial and interglacial periods, respectively. Submarine conditions foster the formation of permafrost and associated underlying hydrate deposits, whereas inundation with relatively warm seawater destabilizes the permafrost and hydrates [21,22]. Later, hydrates were found at shallower depths (around 20 m), and their existence within the entire permafrost body was attributed to a so-called "self-preservation phenomenon" [23–25]. These "relict" gas hydrate formations in permafrost soils could have formed earlier, when there were favorable thermobaric conditions. Subsequently, when thermobaric conditions changed, hydrate transferred to the metastable state due to the manifestation of the self-preservation effect [26–29]. Transition from the last glacial period to the current warm Holocene, accompanied by sea level rise that inundated the previously-exposed shelf area, started 5–12 years ago. According to modeling results, sufficient time has elapsed since inundation to cause permafrost/hydrate system destabilization, which is manifested by formation of taliks (areas of completely thawed sediments within a permafrost) in a certain fraction of the ESAS affected by fault zones, runoff from large rivers, and thermokarst [11,18,21,30,31].

ESAS sediments have not been considered a CH_4 source to hydrosphere or atmosphere because submarine permafrost, which underlies most of the ESAS, was considered to be continuous and to act as an impermeable lid [18], preventing CH_4 escape through the seabed. However, multi-year data (2000–2018) showed extreme CH_4 super-saturation of surface waters (up to 10,000% saturation), implying that about 90% of the total ESAS area serves as a source of CH_4 to the atmosphere [12] and high air-to-sea bubble fluxes occur at numerous seepage sites [11,31]. Conservative estimation of CH_4 ebullition from the coastal ESAS areas yields an annual contribution of at least 9 Tg-CH_4, which increases annual atmospheric flux from the ESAS to 17 Tg-CH_4, on par with flux from the entire Arctic tundra [11]. That estimate does not include the non-gradual CH_4 release discovered recently on the outer ESAS. Sustained CH_4 release to the atmosphere from thawing Arctic subsea permafrost and dissociating hydrates were suggested to be positive and likely to be significant feedbacks to climate warming [32,33].

Since most hydrate deposits in the Arctic are permafrost-controlled, stability of permafrost is a key to whether hydrates are stable [31]. According to the permafrost thermobaric conditions of the Laptev Sea shelf, there are gas hydrates in the underwater permafrost. However, due to the absence of deep drilling, direct data on the presence of gas hydrate formations is not observed. Nevertheless, by indirect evidence, a number of researchers associate active gas shows with dissociation of gas hydrate formations [12,34–36].

The reason for the hydrate destabilization on the Arctic shelf can be both permafrost degradation as a result of the temperature increase, and the processes of penetration of seawater and salt ions contained in it into the layer of hydrate saturated frozen sediments. Today, the issues of salt transfer as a result of the interaction of cooled sea water with frozen hydrate-containing sediments are not sufficiently considered. In this regard, the experimental studying of the mechanism and hydrate dissociation parameters in frozen sediments as a result of salt migration is of particular interest for assessing the role of salt transfer in destabilization of intra-permafrost gas hydrate formations and methane emission on the Arctic shelf.

2. Methods

Experimental modeling of gas hydrate dissociation in frozen sediments as a result of salt migration included the following steps:

1) Preparation of hydrate containing sediment samples, using a pressure cell for artificial hydrate saturation;

2) Freezing of hydrate saturated samples in the pressure cell and transfer pore hydrates to a metastable state by reducing gas pressure to atmospheric at a negative temperature;

3) Extraction of frozen hydrate-saturated samples from the pressure chamber and their contact with a cooled NaCl solution at constant negative temperature and atmospheric pressure.

The objects of study were sandy samples:

- Quarts fine sand (sand 1)
- Silty sand (predominantly quartz composition) sampling during drilling operations on the Laptev Sea shelf (Buor-Khaya Bay in the area of the Muostakh island) (sand 2) (Table 1), where active gas emission was registered, and according to some indirect data there is a probability of the existence of natural hydrate formations in submarine permafrost [11,16,31,37,38].

Table 1. Particle size distribution and mineral composition of investigated sediments.

Sample	Sampling Site	Particle Size Distribution, %						Mineralogy
		1–0.5	0.5–0.25	0.25–0.1	0.1–0.05	0.05–0.001	<0.001	
Sand 1	-	6.5	6.5	79.6	2.2	3.1	2.1	>90% quartz
Sand 2	Laptev Sea shelf (well 1D-11, 40–46 m)	1	9	52	20	16	2	54% quartz 41% microcline + albite 4% illite

The listed mineral phases have percentages >1%.

The initial values of salinity of sandy samples are given in Table 2.

Table 2. Salinity and chemical composition of water extracts from investigated sediments.

Sample	Anions, mEq/100 g				Cations, mEq/100 g			Salinity, %
	pH	HCO_3^-	Cl^-	SO_4^{2-}	Ca^{2+}	Mg^{2+}	$Na^+ + K^+$	
Sand 1	7.1	0.075	0.025	0.06	0.025	-	0.135	0.01
Sand 2	8.4	0.89	5.00	-	0.4	4.9	0.6	0.4

The method of obtaining frozen hydrate-containing samples included sediment samples (twins) preparation of a cylindrical shape (about 3 cm in diameter and 6–9 cm long) with a given moisture content (14–16%), and placing them in a pressure cell, sealing and vacuuming the pressure cell with samples, filling the pressure cell by hydrate-forming gas (CH_4—99.98%), and creation of conditions for uniform saturation of the sediment pore space with gas hydrate [23,39]. Several samples were prepared (about 5–6 pcs.) at the same time, which had similar values of water content, density, and hydrate saturation. After hydrate saturation, which lasted at least 1 month, hydrate saturated sediment samples were frozen at the temperature of −7 ± 1 °C. As a result, the residual pore moisture in the samples that did not transfer to the hydrate became frozen out. Subsequently, the gas pressure in the cell, which was at a negative temperature, was dropped to 0.1 MPa, converting the frozen pore hydrate to a metastable state. The frozen hydrate-saturated samples were then taken out. The samples had a massive ice–hydrate texture with pore hydrate contents uniformly distributed over the sample height [23,40].

For the obtained frozen hydrate-containing samples, prior to their contact with the salt solution, the initial physical parameters (moisture content, density, hydrate content) were determined. Then, the hydrate-containing samples were brought into contact and cooled up to the experiment temperature NaCl solution at a temperature below zero. The experiments were carried out at temperatures from −2.5 °C to −4 °C and concentrations of salt solution from 0.1 to 0.4 N (accordingly to the natural Laptev Sea water concentration, which equals 10–34‰ [41]). The duration of the tests depends on the experimental conditions and ranged from several tens of minutes to several days. For control, one of the samples was stored at a given negative temperature, but was not brought into contact with the salt solution. These control data showed that during the experiment the change in the hydrate content of sediment samples under self-preservation conditions was insignificant, and the main decrease in hydrate saturation of other samples was caused by salt migration.

During the experiments, the dynamics of the interaction of frozen hydrate-containing samples with a salt solution over time, as well as the effect of the concentration of the NaCl contact solution and the ambient temperature on the hydrate dissociation processes in the porous media of sediments were studied. In experimental modeling, hydrate-containing samples were removed from contact solution at certain time intervals. A number of parameters were determined for each sample, characterizing the content of moisture, salts and gas hydrate along the sample length separately in nine-ten 7–10-mm-thick layers of the samples after interaction with the salt solution. This made it possible to trace the dynamics of saline front migration, thawing, and dissociation of gas hydrate formations in frozen sediment samples.

Gas contents were estimated by measuring the volume of gas released (with 2–3 times repeatability) as the samples were thawing in a saturated NaCl solution. Samples with residual hydrate content were put in a special glass tube filled with brine (saline solution), and as a result of gas hydrate dissociation and methane release volume of the brine in the glass tube was changed. And by the volume difference, it was possible to estimate gas content in studied samples. The obtained values were used to estimate hydrate content and hydrate coefficient [23,40], assuming a hydrate number of 5.9 for methane hydrate [19,39,42]. Specific gas content (G, cm^3/g) was found as

$$G = \frac{(V_2 - V_1) * T}{m_s},$$

where: ($V_2 - V_1$)—change in the volume of liquid in the gas collector tube (cm^3); T—temperature correction; m_s—the mass of the sediment sample (g).

The weight gas hydrate content (H, wt.% of sample weight) was determinate for each interval as

$$H = m_g * 7.64 * 100\%, \qquad (2)$$

where m_g is a specific gravity of methane in gas hydrate form (g/g i.e., grams of gas in per gram of sediment) calculated from the specific gas content (G) for pure methane.

The fraction of water converted to hydrate or the hydrate coefficient (K_h, u.f.) is given by

$$K_h = \frac{W_h}{W}, \qquad (3)$$

where W_h is the percentage of water in a hydrate form (wt.% of sample weight) and W is the total amount of water or moisture content (wt.%) [19,43].

The samples on which the moisture content was determined were further used for the interval determination of the content of salt ions that migrated from the contact salt solution. First of all, water extractions of salt were made from samples, and then analysis of the Na^+ ion concentration was carried out by the method of water extracts on a flame photometer PFP-7 (Jenway).

3. Results

Experimental modeling showed that the interaction of frozen hydrate saturated sediments with a cooled salt solution leads to an active diffusion of salt ions into the sediment sample, which results in sample salinization. Comparison of experimental data on the accumulation of salt ions in frozen hydrate-containing samples with frozen non-hydrate-containing samples under the same conditions indicates that migration of salt ions in the hydrate-containing samples occurs more intensively. So, in a frozen hydrate-containing sample, 46 h after the start of interaction with a 0.2 N solution of NaCl, ions (Na^+) penetrated 7.3 cm deep into the sample, and in a frozen non-hydrate-containing sample, only 5.6 cm deep with higher salt accumulation (Figure 1).

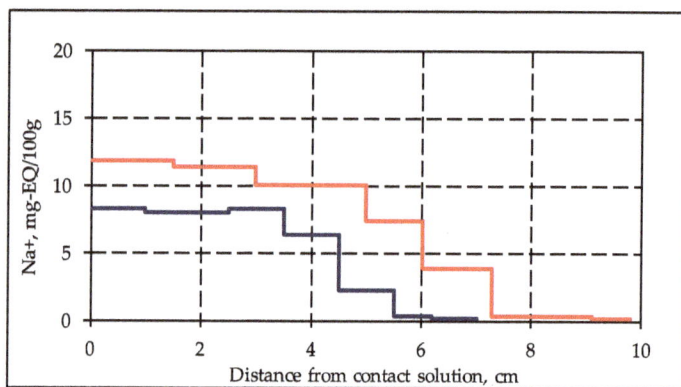

Figure 1. Salt ion accumulations (Na^+) in frozen sand samples (sand 1, W = 14%) containing (red line) and not containing (blue line) hydrate in porous media after 46 h of contact with 0.2N NaCl solution under the temperature −4 °C.

The process of migration of salt ions in the frozen hydrate-containing sample was accompanied by an increase in the liquid phase content in the sample and hydrate dissociation in the pore space.

Migrating salt ions cause melting of ice in porous media, including ice on the surface of conserved gas hydrates in the pore space of the frozen sample, and thereby activate the gas hydrate decomposition. Experimental studies allowed us to trace the movement in time of the salinity front of the frozen hydrate-containing sample under negative temperature conditions (−4 °C) (Figure 2). After 4 h of frozen hydrate-containing sample contact with a salt solution, the salt ions (Na +) penetrated on 2.5 cm, the salt ion content in the contact zone increased up to 0.7 mg EQ/ 100 g, and a day later (29 h after the start of the experiment) salts penetrated to a depth of about 5 cm. The maximum accumulation of salt ions in the sample reached 1.6 mg EQ/ 100 g.

The accumulation of salt ions in the studied sand samples affected the hydrate content. In the initial state (before contact with the solution) for sand samples (sand 1) about 60% of the pore moisture was in the hydrate form, while the hydrate saturation of the samples was about 40%. In the process of unilateral salinization, the proportion of pore moisture in the hydrate form (K_h) decreased, and the front of the complete decomposition of the gas hydrate appeared (Figure 3).

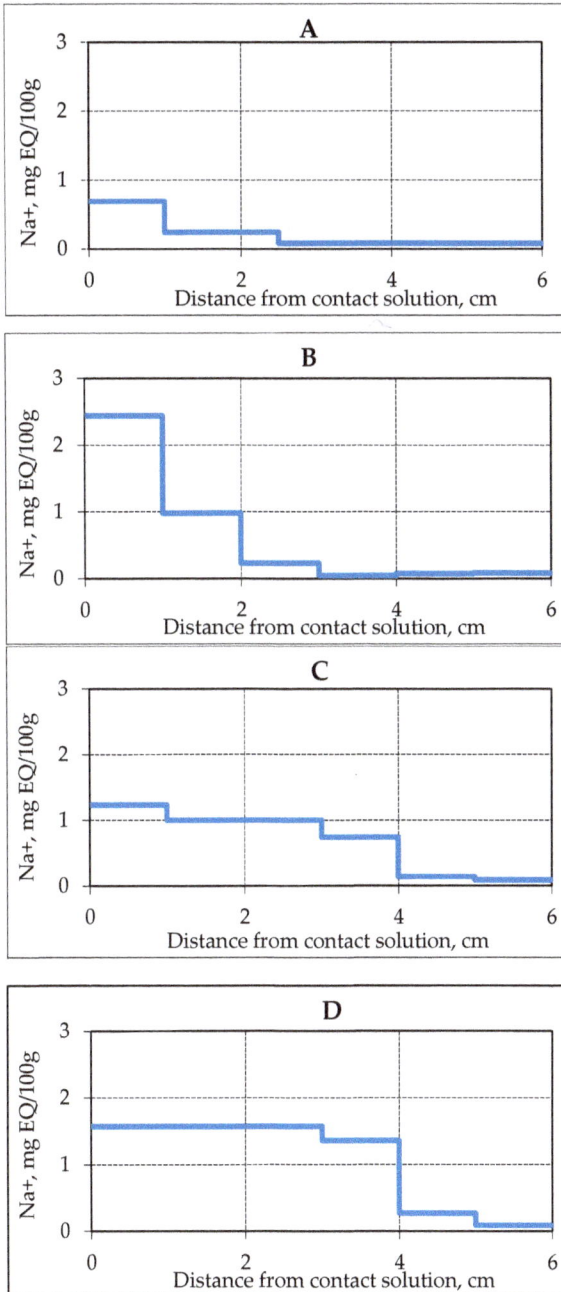

Figure 2. Accumulation of Na$^+$ ions in samples of frozen hydrate saturated sand (sand 1, W = 14%) in time when interacting with 0.2 N NaCl solution at a temperature of −4 °C. (**A–D**) is the time of interaction with the salt solution, respectively, 3.9, 17.5, 25.5 and 28.8 h.

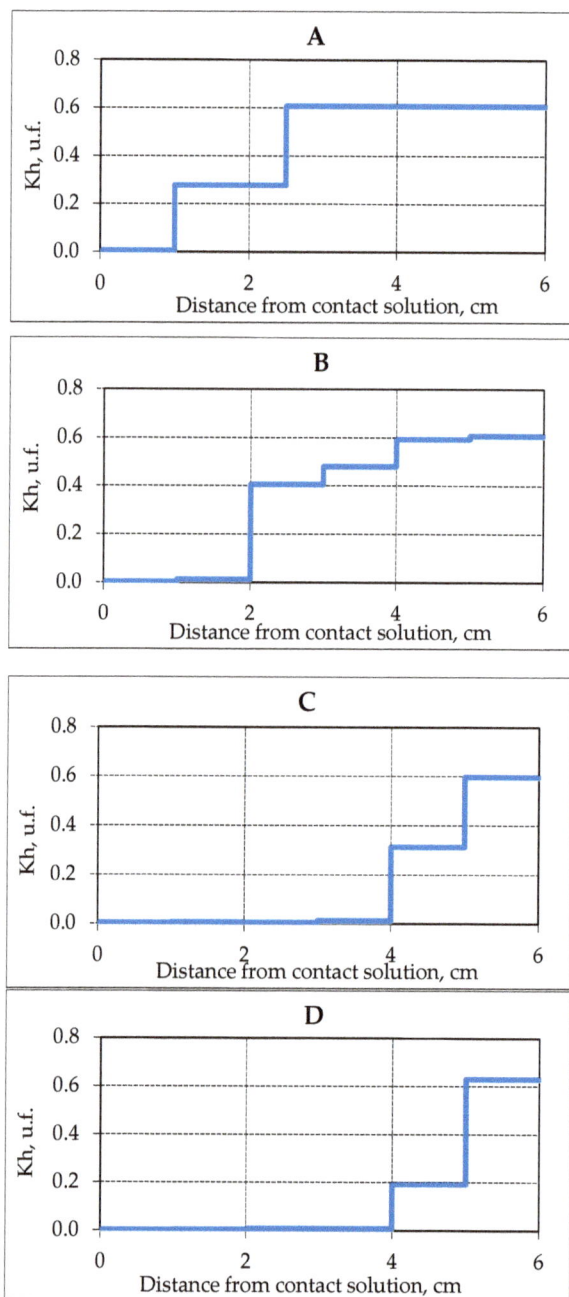

Figure 3. The change in the hydration coefficient (K_h) in frozen hydrate saturated samples of sand (sand 1, W = 14%) in time when interacting with 0.2 N NaCl salt solution at a temperature of −4 °C. (**A–D**) is the time of interaction with the salt solution, respectively, 3.9, 17.5, 25.5 and 28.8 h.

A joint analysis of the distribution of salt ions (Figure 2) and the hydration coefficient (K_h) over the height of the sample (Figure 3) shows that the amount of salinity determines the residual content of pore gas hydrate. In this case, it is possible to determine a certain critical content of salt ions in sandy samples, which causes the complete decomposition of gas hydrate formations in the sediment. Under specified conditions (temperature −4 °C and the concentration of a contact solution of NaCl 0.2 N) in sediment samples (sand 1), the critical salt accumulation was about 0.7–0.8 mg EQ/ 100 g.

In the process of salts migration in the frozen hydrate-containing sample, in addition to the hydrate decomposition front in porous media, a thawing front may occur when the accumulated salt ions completely transform pore ice into water. The movement of two fronts can be traced on Figure 4: hydrate decomposition front and thawing of pore ice.

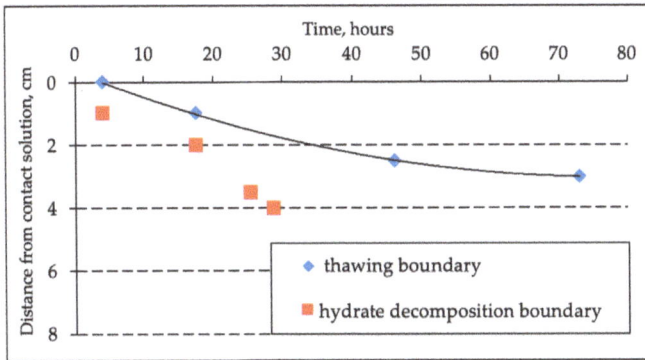

Figure 4. Experimental assessment of the gas hydrate decomposition front in porous media and the thawing front in frozen hydrate-containing sandy samples (sand-1, W = 14%) when interacting with 0.2 N NaCl salt solution at the temperature −4 °C.

The hydrate decomposition front is ahead of the thawing front since a higher value of salt accumulation in the frozen sample is needed for thawing. The thawing front in the sample is well marked by the characteristic change in the color of the sample (Figure 5). Additionally, it was evaluated using a special needle probe.

Figure 5. Frozen sandy samples (sand 2, W = 15%) before (**A**) and after interaction (**B**) with 0.2 N NaCl solution at −3 °C.

With an increase in the ambient temperature, the processes of salinity migration of the frozen hydrate-containing sediment, and, consequently, the gas hydrate decomposition rate in pore space increase. The study of the interaction dynamics of frozen hydrate saturated sediment (sand 2) with 0.2 N NaCl solution at the conditions of negative temperature increase (up to −2.5 °C) showed that the penetration rate of salt ions rises significantly. At a higher negative temperature, the NaCl ions penetrated deeper in a shorter time (Figure 6).

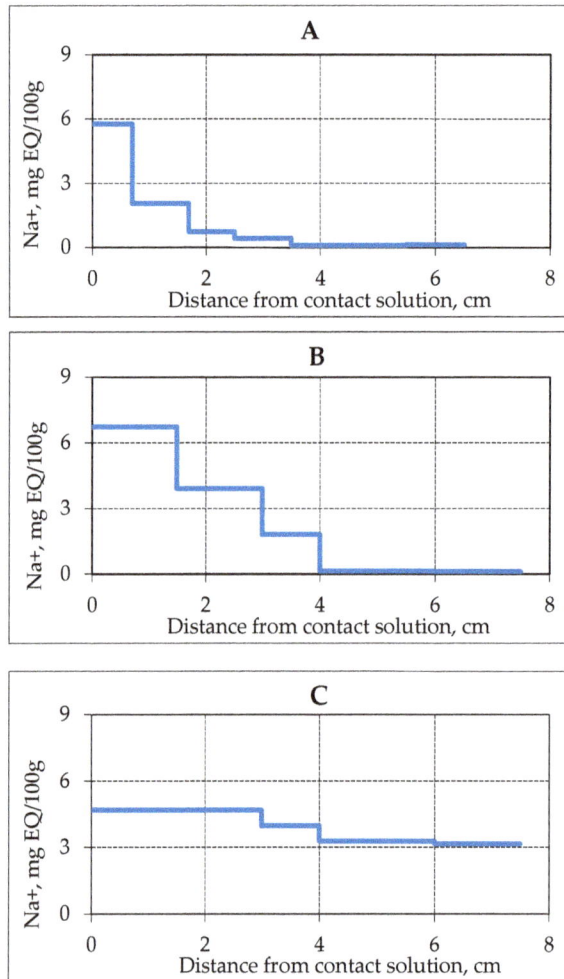

Figure 6. Accumulation of Na$^+$ ions in samples of frozen hydrate saturated sand (sand 2, W = 16%) over time when interacting with 0.2 N NaCl solution at a temperature −2.5 °C. (**A–C**)—the time of interaction with salt solution, respectively, 0.3, 0.9 and 2.6 h.

So, already 1 h after the experiment start, salt ions penetrated on a distance of 4 cm from the salt solution contact zone, and after 2.6 h, the migrating salt ions were registered along the entire length of the sample (about 7 cm). An increase in the moisture content of the sample due to the migrating solution was also observed. At the beginning of the experiment (0.3 h), the increase in water content was only in the contact zone, and at the end of the experiment (2.6 h) along the entire length of the sample. The overall increase in moisture content occurred from 16 to 30%, and in the contact area up to 50%. The concentration of ions (Na$^+$) was about 5 mg EQ/ 100 g in the contact area at the end of the experiment, which is more than 3 times the maximum concentration of salt ions in the experiment at a lower negative temperature (−4 °C). More intensive migration of NaCl solution in frozen hydrate saturated sediments at a temperature of −2.5 °C cause more active gas hydrate decomposition in pore space (Figure 7).

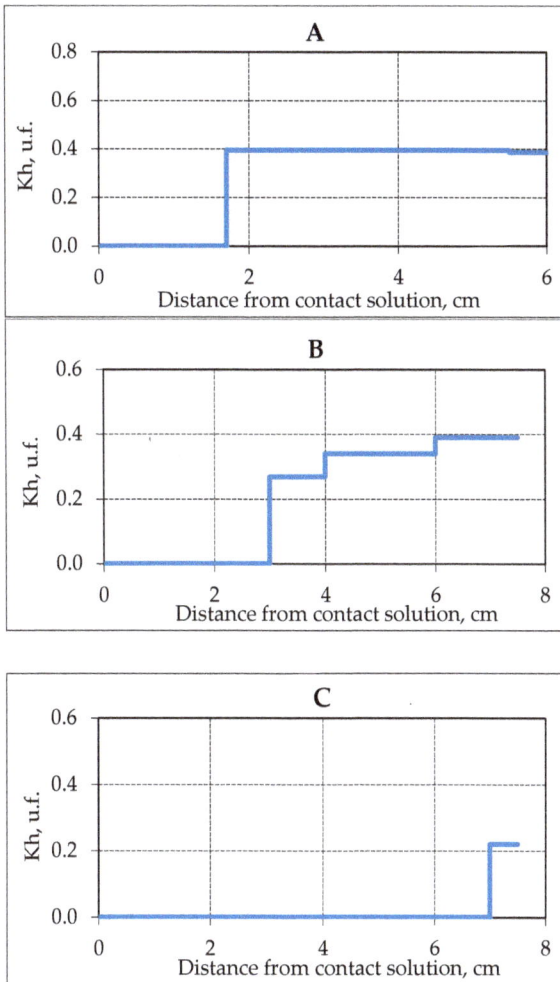

Figure 7. The change in the hydration coefficient (K_h) in samples of frozen hydrate saturated sand (sand 2, W = 16%) in time when interacting with 0.2 N NaCl solution at a temperature of −2.5 °C. (**A–C**) is the time of interaction with the salt solution, respectively, 0.3; 0.9; 2.6 h.

Before the contact of the frozen hydrate-containing sample (sand 2) with a salt solution, about 40% of the pore moisture was in the gas hydrate form. After 0.3 h, from the beginning of the contact of frozen hydrate-containing sample with the salt solution at −2.5 °C, the pore hydrate completely decomposed at a distance of 1.8 cm from the contact area, after 0.9 h—at a distance of 3 cm, and after 2.6 h–7 cm. At the time of the experiment end, the residual values of K_h (about 25%) were registered only in a narrow zone of the sample at the sample opposite end from the contact.

An increase in the concentration of a saline solution in contact with a frozen hydrate-containing sample at a fixed negative temperature also leads to more intense gas hydrate dissociation in porous media, an increase in the rate of movement of the salinity fronts and gas hydrate decomposition. For example, at a fixed interaction time (3 h) of a frozen hydrate saturated sample (sand 2) at a temperature of −3 °C with a saline solution 0.1 N salt ions (Na +) migrated to a depth of 1.8 cm, and at a concentration of contacting solution 0.4 N salt ions migrated almost over the entire length of the sample. The content of salt ions in the contact area of the sample with an increase in the concentration of the contacting solution (from 0.1N to 0.4 N) increased 6 times in 3 h of the experiment (Figure 8).

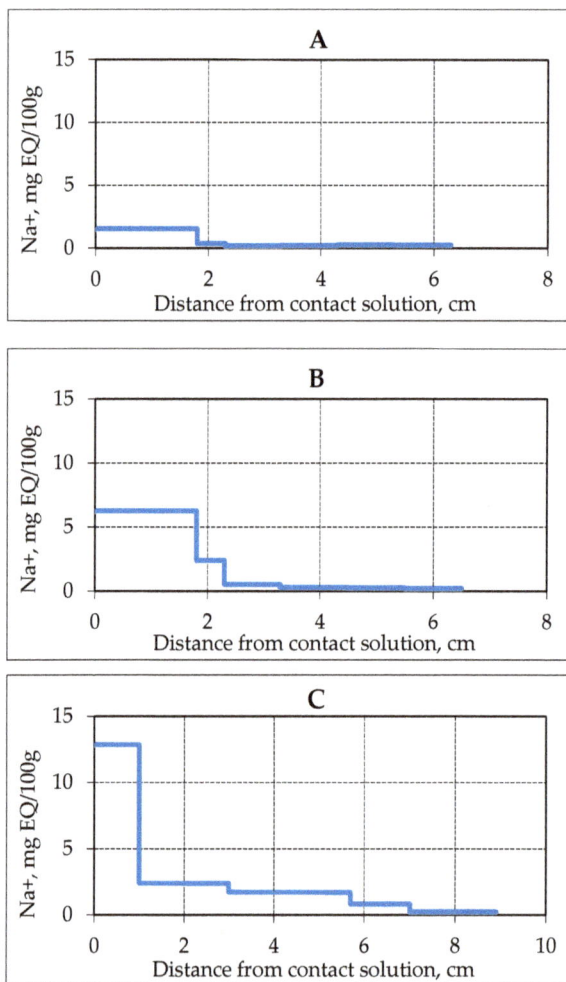

Figure 8. Accumulation of Na$^+$ ions in samples of frozen hydrate saturated sand (sand 2, W = 16%) after 3 h of contact with NaCl solution of various concentrations at a temperature of −3 °C. (**A–C**)—solution concentration, respectively, 0.1, 0.2 and 0.4 N.

At the same time, an increase in the moisture content of the sample from 16% to 60% was recorded in the contact zone.

An increase in the concentration of the contact solution accelerated movement of the gas hydrate dissociation front in the porous space of the frozen hydrate saturated sample. Thus, at a concentration of a contact solution of 0.1 N NaCl, the complete decomposition of a pore gas hydrate in a sediment sample occurred at a distance of 1.9 cm, and at a concentration of 0.4 N, the pore gas hydrate completely dissociated at a distance of 3 cm from the contact area. The maximum value of K_h in the samples decreased with an increase in the concentration of the contact solution from 26% to a residual value of 17% after the experiment (Figure 9).

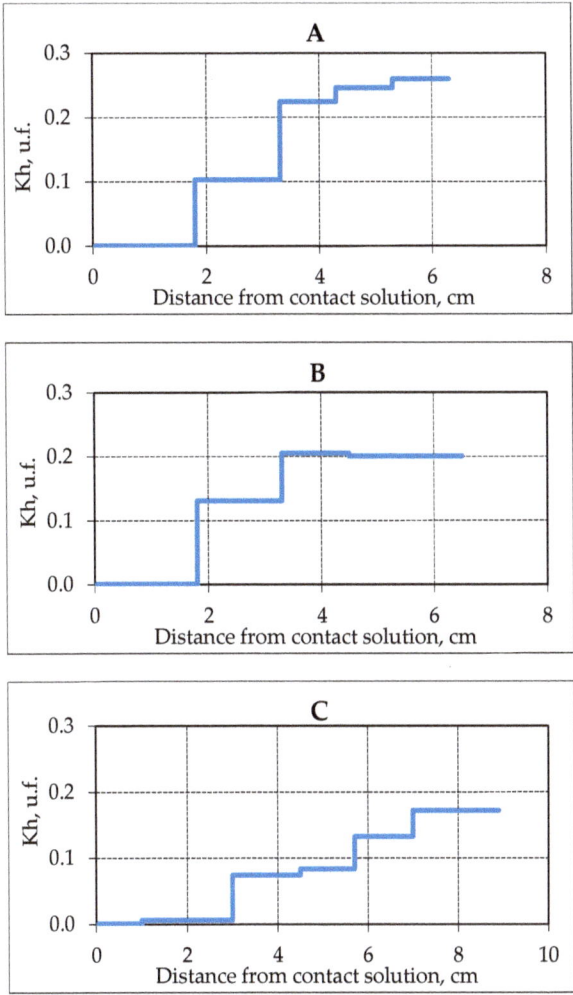

Figure 9. Changes in the hydration coefficient (K_h) in samples of frozen hydrate saturated sand (sand 2, W = 16%) after 3 h of contact with NaCl solution of various concentrations at a temperature of −3 °C. (**A–C**) solution concentration, respectively, 0.1, 0.2 and 0.4 N.

Data on the change in hydrate content of frozen hydrate-saturated sediments when interacting with a saline solution were used to estimate the kinetics of methane emission from frozen samples during decomposition of pore gas hydrates depending on temperature (Figure 10).

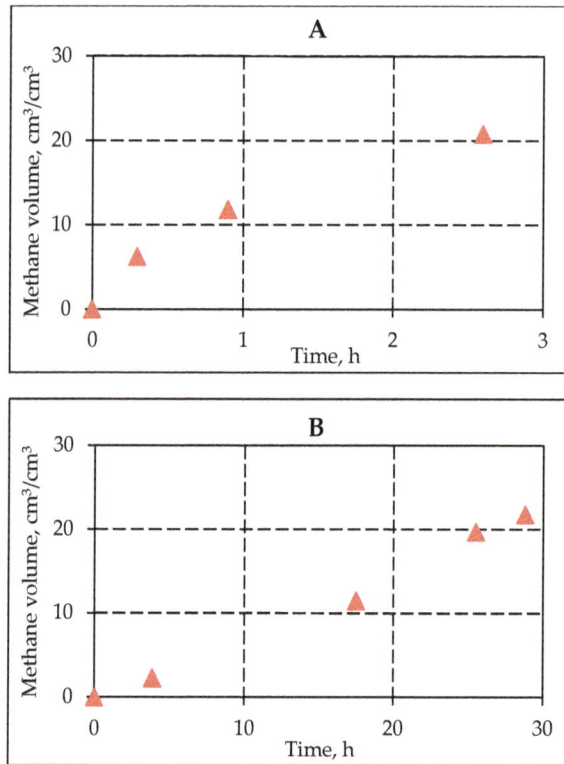

Figure 10. Methane emission in frozen hydrate saturated samples of sand (sand 1) during the gas hydrate dissociation in the pore space as a result of interaction with 0.2 N NaCl salt solution at temperatures −2.5 °C (**A**) and −4 °C (**B**).

The calculated data show that the intensity of methane release from frozen hydrate-containing samples depends on the salt ions migration rate. The more intense accumulation of salt ions at a higher negative temperature of −2.5 °C led to the fact that the relative emission of methane exceeded 10 cm^3/cm^3 just one hour after the start of interaction with the salt solution.

Experimental results demonstrated active salt migration in frozen hydrate saturated sediment while interacting with salt solutions, which leads to thawing of frozen sediment, destruction of intrapermafrost gas hydrate formations and methane emission.

4. Discussion

One of the possible causes of the decomposition of intrapermafrost gas hydrates on the Arctic shelf may be the migration of salt ions as a result of sea water interaction with underwater permafrost. At present, there are practically no special studies in the literature on the gas hydrate dissociation in frozen sediments porous media as a result of salt transfer. Although they may be important for understanding the nature of methane emissions during the degradation of permafrost on the Arctic shelf. However, since the 1980s it has been known that the interaction of frozen sediments with salt solutions leads to the active migration of salt ions [44–46]. In this case, several mechanisms for the salt ions transfer in frozen sediment were described, and also some regularities of their transfer depending on external conditions were considered. In general, it was shown that the transfer of salt ions in frozen sediment is primarily determined by the characteristics of the sediment itself (dispersion, chemical and mineral composition, ice content), as well as temperature conditions and the concentration of salt solutions, and is accompanied by the processes of moisture transfer and structure formation [45,47,48]. The experimental data on the interaction of frozen hydrate saturated sediments with salt solutions showed that the migration of salt ions in frozen hydrate-containing sediment can occur more intensively compared to frozen rocks non-containing hydrates, which is apparently due to the process of dissociation of the pore hydrate and the appearance of this liquid phase water. With an increase in the temperature of frozen hydrate bearing sediments and salt concentrations, the intensity of mass transfer processes increases. In this case, two fronts are observed: the hydrate decomposition front and the thawing front. The front of the hydrate dissociation in pore space is ahead of the thawing boundary, since the phase transition of pore ice into water requires higher accumulations of salt ions during their migration.

Schematic diagram presenting current understanding of the subsea permafrost – hydrate system existing in the ESAS is presented and discussed by [49]. Below we focus on the salt/hydrates effect which was not studied experimentally before. Our new experimental data can be used to illustrate interaction between cold bottom sea water and cold/thaw sediments/subsea permafrost, and hydrates. Downward salt migration can be considered a factor of relic hydrates dissociation (Figure 11).

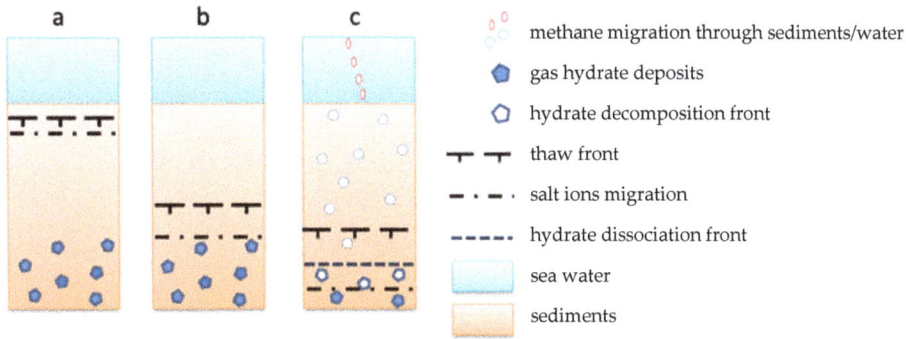

Figure 11. An estimated model of the interaction of cooled sea water with frozen sediments containing relict gas hydrates. (**a**) Initial state (interaction between cold seawater and frozen sediments); (**b**) Penetration of sea salt into permafrost (formation of salinity and thawing fronts); (**c**) Interaction of salt front with intra-permafrost gas hydrate accumulations (decomposition of gas hydrates, active methane emissions).

At the initial stage (Figure 11a), and after the transgression of the Arctic Sea, there is an active interaction of seawater with underlying frozen sediment, resulting in a change in the permafrost temperature and the formation of a thawing front [31]. Over time, the rate of thawing of the underwater permafrost decreases. At the same time, the penetration of seawater into the thawing zone is identified (Figure 11b). Since the frozen rocks are permeable to salt ions, the advance movement of salt ions from seawater relative to the thawing front is observed. During migration and accumulation of salt ions in frozen sediments, a decrease in the temperature of phase transitions and an increase in the content of the liquid phase in sediments occurs. As a result, deceleration rate of thawing front decreases. Salt ions migrating in frozen sediments upon reaching horizons containing relict hydrates will destabilize them and promote their active dissociation and release of methane (Figure 11c), thus forming the second phase transition phase—the front of gas hydrate dissociation. It occurs when a certain critical salt concentration is reached in the frozen hydrate-containing rock. This concentration, as shown by experimental studies, depends on the composition of hydrate-containing sediments and thermobaric conditions. During phase transitions (hydrate-ice-water) caused by salt migration the gas permeability of salted frozen sediments increases [50]. Methane formed during the dissociation of gas hydrates further migrates to the bottom surface and enters seawater, where it is partially dissolved, oxidized, and also diffused through the water column before it goes into the atmosphere (Figure 11c). The detailed explanation of the methane upward transport in the sediment-water column-atmosphere system can be found in [49].

This scheme makes it possible to explain the possible cause of the numerous gas manifestations recorded on the Arctic shelf in the areas of distribution of submarine permafrost as a result of the destabilization of the permafrost gas hydrate formations due to heat and mass transfer processes associated with the migration of salt ions from seawater.

5. Conclusions

New experimental results integrated with the observational data confirm a hypothesis about the primary role of submarine permafrost and hydrate destabilization in a massive methane release in the sediment-water-atmosphere system [11,12,31,49]. As a result of the degradation of relic submarine permafrost (containing hydrates), active dissociation of intrapermafrost gas hydrate can occur. As shown by experimental modeling, salt transfer plays an important role in this process.

Experimental studies revealed that salt migration during the interaction of frozen hydrate saturated sediments with salt solutions causes their migration though frozen horizons, which leads to decomposition of intrapermafrost gas hydrates and permafrost thawing. It was experimentally shown that in the process of salts transfer and accumulation, two fronts are formed: thawing of frozen sediments and gas hydrate decomposition. It is noted that the front of decomposition of gas hydrate is ahead of the front of thawing.

Experimental studies allowed us to trace the dynamics of the thawing front and the decomposition front of gas hydrates, caused by the migration of salts in frozen hydrate-containing sediments, and also to obtain some consistencies on the hydrate dissociation in pore space in frozen sediments depending on the ambient temperature and the salt solution concentration. According to the experimental modeling, methane emissions were estimated during the decomposition of pore gas hydrates in frozen sediments during salt transfer. The results of the experiments were used to build a model of the interaction of the hydrate saturated permafrost with sea water on the Arctic shelf.

Author Contributions: Conceptualization, E.C.; sampling, B.B., S.G.; experimental methodology, E.C., V.E., B.B. and S.G.; calculation methodology, E.C., V.E.; experiments, processing, and analysis, E.C., V.E. and B.B.; supervision, E.C., I.S.; writing manuscript and editing, E.C., V.E., B.B., N.S. and I.S.

Funding: The research was supported partly by the Russian Science Foundation (grants No. 16-17-00051, 18-77-10063 and 15-17-20032).

Conflicts of Interest: The authors declare no conflict of interest.

References

1. Yakushev, V.S.; Collett, T.S. Gas Hydrates in Arctic Regions: Risk to Drilling and Production. In Proceedings of the Second International Offshore and Polar Engineering Conference, San Francisco, CA, USA, 14–19 June 1992; Volume 1, pp. 669–673. [CrossRef]
2. Safronov, A.F.; Shits, E.Y.; Grigor'ev, M.N.; Semenov, M.E. Formation of Gas Hydrate Deposits in the Siberian Arctic Shelf. *Russ. Geol. Geophys.* **2010**, *51*, 83–87. [CrossRef]
3. Paull, C.K.; Ussler, W.; Dallimore, S.R.; Blasco, S.M.; Lorenson, T.D.; Melling, H.; Medioli, B.E.; Nixon, F.M.; McLaughlin, F.A. Origin of Pingo-like Features on the Beaufort Sea Shelf and Their Possible Relationship to Decomposing Methane Gas Hydrates. *Geophys. Res. Lett.* **2007**, *34*, L01603:1–L01603:5. [CrossRef]
4. Romanovskii, N.N.; Hubberten, H.W.; Gavrilov, A.V.; Eliseeva, A.A.; Tipenko, G.S.; Kholodov, A.L.; Romanovsky, V.E. Permafrost and Gas Hydrate Stability Zone Evolution on the Eastern Part of the Eurasia Arctic Sea Shelf in the Middle Pleistocene-Holocene (Published in Russian). *Earth's Cryosph.* **2003**, *7*, 51–64.
5. Andreassen, K.; Hubbard, A.; Winsborrow, M.; Patton, H.; Vadakkepuliyambatta, S.; Plaza-faverola, A.; Deryabin, A.; Mattingsdal, R.; Mienert, J. Gas Hydrate Regulate Methane Emissions from Arctic Petroleum Basins. In Proceedings of the 9th International Conference on Gas Hydrates (ICGH9), Denver, CO, USA, 25–30 June 2017; p. 2.

6.	Mau, S.; Romer, M.; Torres, M.E.; Bussmann, I.; Pape, T.; Damm, E.; Geprags, P.; Wintersteller, P.; Hsu, C.W.; Loher, M.; et al. Widespread Methane Seepage along the Continental Margin off Svalbard—From Bjornoya to Kongsfjorden. *Sci. Rep.* **2017**, *7*, 42997:1–42997:13. [CrossRef]

7.	Wood, W.T.; Gettrust, J.F.; Chapman, N.R.; Spence, G.D.; Hyndman, R.D. Decreased Stability of Methane Hydrates in Marine Sediments Owing to Phase-Boundary Roughness. *Nature* **2002**, *420*, 656–660. [CrossRef]

8.	Serov, P.; Vadakkepuliyambatta, S.; Mienert, J.; Patton, H.; Portnov, A.; Silyakova, A.; Panieri, G.; Carroll, M.L.; Carroll, J.; Andreassen, K.; et al. Postglacial Response of Arctic Ocean Gas Hydrates to Climatic Amelioration. *Proc. Natl. Acad. Sci. USA* **2017**, *114*, 6215–6220. [CrossRef] [PubMed]

9.	Dean, J.F.; Middelburg, J.J.; Röckmann, T.; Aerts, R.; Blauw, L.G.; Egger, M.; Jetten, M.S.M.; de Jong, A.E.E.; Meisel, O.H.; Rasigraf, O.; et al. Methane Feedbacks to the Global Climate System in a Warmer World. *Rev. Geophys.* **2018**, *56*, 207–250. [CrossRef]

10.	James, R.H.; Bousquet, P.; Bussmann, I.; Haeckel, M.; Kipfer, R.; Leifer, I.; Niemann, H.; Ostrovsky, I.; Piskozub, J.; Rehder, G.; et al. Effects of Climate Change on Methane Emissions from Seafloor Sediments in the Arctic Ocean: A Review. *Limnol. Oceanogr.* **2016**, S283–S299. [CrossRef]

11.	Shakhova, N.; Semiletov, I.; Leifer, I.; Sergienko, V.; Salyuk, A.; Kosmach, D.; Chernykh, D.; Stubbs, C.; Nicolsky, D.; Tumskoy, V.; et al. Ebullition and Storm-Induced Methane Release from the East Siberian Arctic Shelf. *Nat. Geosci.* **2014**, *7*, 64–70. [CrossRef]

12.	Shakhova, N.; Semiletov, I.; Salyuk, A.; Yusupov, V.; Kosmach, D.; Gustafsson, Ö. Extensive Methane Venting to the Atmosphere from Sediments of the East Siberian Arctic Shelf. *Science* **2010**, *327*, 1246–1250. [CrossRef]

13.	Makogon, Y.F. *Hydrates of Natural Gas (Published in Russian)*; Nedra: Moscow, Russia, 1974; p. 208. ISBN 978-364-214-233-8.

14.	Max, M. *Natural Gas Hydrate in Oceanic and Permafrost Environments*; Kluwer Academic Publishers: Washington, DC, USA, 2000; ISBN 978-1-4020-1362-1. [CrossRef]

15.	Shakhova, N.E.; Alekseev, V.A.; Semiletov, I.P. Erratum to: "Predicted Methane Emission on the East Siberian Shelf." *Dokl. Earth Sci.* **2013**, *452*, 1074. [CrossRef]

16.	Sergienko, V.I.; Lobkovskii, L.I.; Semiletov, I.P.; Dudarev, O.V.; Dmitrievskii, N.N.; Shakhova, N.E.; Romanovskii, N.N.; Kosmach, D.A.; Nikol'skii, D.N.; Nikiforov, S.L.; et al. The Degradation of Submarine Permafrost and the Destruction of Hydrates on the Shelf of East Arctic Seas as a Potential Cause of the "Methane Catastrophe": Some Results of Integrated Studies in 2011. *Dokl. Earth Sci.* **2012**, *446*, 1132–1137. [CrossRef]

17.	Solov'yev, V.A.; Ginsburg, G.D. Formation of Submarine Gas Hydrates. *Bull. Geol. Soc. Den.* **1994**, *41*, 86–94.

18.	Romanovskii, N.N.; Hubberten, H.-W.; Gavrilov, A.V.; Eliseeva, A.A.; Tipenko, G.S. Offshore Permafrost and Gas Hydrate Stability Zone on the Shelf of East Siberian Seas. *Geo-Mar. Lett.* **2005**, *25*, 167–182. [CrossRef]

19.	Chuvilin, E.; Davletshina, D. Formation and Accumulation of Pore Methane Hydrates in Permafrost: Experimental Modeling. *Geosciences* **2018**, *8*, 467. [CrossRef]

20.	Yakushev, V.S. *Natural Gas and Gas Hydrates in the Permafrost (Published in Russian)*; Gazprom VNIIGAZ: Moscow, Russia, 2009; p. 192.

21.	Solov'ev, V.A.; Ginsburg, G.D. *Submarine Gas Hydrates (Published in Russian)*; VNII Oceanologii: St. Petersburg, Russia, 1994; ISBN 5-7173-0290-8.

22.	Solov'ev, V. A Global Estimate of the Amount of Gas in Submarine Accumulations of Gas Hydrates. *Geol. Geophys.* **2002**, *43*, 648–661.

23.	Chuvilin, E.; Bukhanov, B.; Davletshina, D.; Grebenkin, S.; Istomin, V. Dissociation and Self-Preservation of Gas Hydrates in Permafrost. *Geosciences* **2018**, *8*, 431. [CrossRef]

24.	Yakushev, V.S. Gas Hydrates in Cryolithozone (Published in Russian). *Sov. Geol. Geophys.* **1989**, *1*, 100–105.

25.	Chuvilin, E.M.; Yakushev, V.S.; Perlova, E.V. Gas and Possible Gas Hydrates in the Permafrost of Bovanenkovo Gas Field, Yamal Peninsula, West Siberia. *Polarforschung* **2000**, *68*, 215–219.

26.	Hachikubo, A.; Takeya, S.; Chuvilin, E.; Istomin, V. Preservation Phenomena of Methane Hydrate in Pore Spaces. *Phys. Chem. Chem. Phys.* **2011**, *13*, 17449–17452. [CrossRef]

27.	Istomin, V.; Yakushev, V.; Makhonina, N.; Kwon, V.G.; Chuvilin, E.M. Self-Preservation Phenomenon of Gas Hydrate (Published in Russian). *Gas Ind.* **2006**, 36–46. Available online: https://istina.msu.ru/publications/article/2428015/ (accessed on 23 April 2019).

28.	Takeya, S.; Ebinuma, T.; Uchida, T.; Nagao, J.; Narita, H. Self-Preservation Effect and Dissociation Rates of CH4 Hydrate. *J. Cryst. Growth* **2002**, *237*, 379–382. [CrossRef]

29. Ershov, E.D.; Lebedenko, Y.P.; Chuvilin, E.M.; Istomin, V.A.; Yakushev, V.S. Features of the Existence of Gas Hydrates in the Cryolithozone (Published in Russian). *Rep. Acad. Sci. USSR* **1991**, *321*, 788–791.

30. Nicolsky, D.; Shakhova, N. Modeling Sub-Sea Permafrost in the East Siberian Arctic Shelf: The Dmitry Laptev Strait. *Environ. Res. Lett.* **2010**, *5*, 015006:1–015006:9. [CrossRef]

31. Shakhova, N.; Semiletov, I.; Gustafsson, O.; Sergienko, V.; Lobkovsky, L.; Dudarev, O.; Tumskoy, V.; Grigoriev, M.; Chernykh, D.; Koshurnikov, A.; et al. Current Rates and Mechanisms of Subsea Permafrost Degradation in the East Siberian Arctic Shelf. *Nat. Commun.* **2017**, *8*, 15872:1–15872:13. [CrossRef]

32. ACIA. *Future Climate Change: Modeling and Scenarios*; Cambridge University Press: Cambridge, UK, 2005; pp. 99–150.

33. Westbrook, G.K.; Thatcher, K.E.; Rohling, E.J.; Piotrowski, A.M.; Pälike, H.; Osborne, A.H.; Nisbet, E.G.; Minshull, T.A.; Lanoisellé, M.; James, R.H.; et al. Escape of Methane Gas from the Seabed along the West Spitsbergen Continental Margin. *Geophys. Res. Lett.* **2009**, *36*, L15608:1–L15608:5. [CrossRef]

34. Nicolsky, D.J.; Romanovsky, V.E.; Romanovskii, N.N.; Kholodov, A.L.; Shakhova, N.E.; Semiletov, I.P. Modeling Sub-Sea Permafrost in the East Siberian Arctic Shelf: The Laptev Sea Region. *J. Geophys. Res. Earth Surf.* **2012**, *117*, F03028:1–F03028:22. [CrossRef]

35. Delisle, G. Temporal Variability of Subsea Permafrost and Gas Hydrate Occurrences as Function of Climate Change in the Laptev Sea, Siberia. *Polarforschung* **2000**, *68*, 221–225.

36. Romanovskii, N.N.; Eliseeva, A.A.; Gavrilov, A.V.; Tipenko, G.S.; Hubberten, X. The Long-Term Dynamics of Frozen Strata and the Zone of Gas Hydrate Stability in the Rift Structures of the Arctic Shelf of Eastern Siberia (Report 2) (Published in Russian). *Earth's Cryosph.* **2006**, *10*, 29–38.

37. Chuvilin, E.M.; Tumskoy, V.E.; Tipenko, G.S.; Gavrilov, A.V.; Bukhanov, B.A.; Tkacheva, E.V.; Audibert-Hayet, A.; Cauquil, E. Relic Gas Hydrate and Possibility of Their Existence in Permafrost within the South-Tambey Gas Field. In Proceedings of the SPE Arctic and Extreme Environments Technical Conference and Exhibition, Moscow, Russia, 15–17 October 2013; pp. 166925:1–166925:9.

38. Chuvilin, E.; Bukhanov, B.; Grebenkin, S.; Tymskoy, V.; Shakhova, N.; Dudarev, O.; Semiletov, I. Thermal Conductivity of Bottom Sediments in the East Siberian Arctic Seas: A Case Study in the Buor-Khaya Bay. In Proceedings of the 7th Canadian Permafrost Conference, Queébec City, QC, Canada, 21–23 September 2015; pp. ABS557:1–ABS557:6.

39. Chuvilin, E.; Bukhanov, B. Thermal Conductivity of Frozen Sediments Containing Self-Preserved Pore Gas Hydrates at Atmospheric Pressure: An Experimental Study. *Geosciences* **2019**, *9*, 65. [CrossRef]

40. Chuvilin, E.M.; Guryeva, O.M. Experimental Study of Self-Preservation Effect of Gas Hydrates in Frozen Sediments. In Proceedings of the 9th International Conference on Permafrost, Fairbanks, AK, USA, 28 June–3 July 2008; Volume 28.

41. Shpolyanskaya, N.A.; Streletskaya, I.D.; Surkov, A. Cryolithogenesis within the Arctic Shelf (Modern and Ancient) (Published in Russian). *Earth's Cryosph.* **2006**, *10*, 49–60.

42. Chuvilin, E.; Bukhanov, B. Effect of Hydrate Formation Conditions on Thermal Conductivity of Gas-Saturated Sediments. *Energy Fuels* **2017**, *31*, 5246–5254. [CrossRef]

43. Chuvilin, E.M.; Kozlova, E.V.; Skolotneva, T.S. Experimental Simulation of Frozen Hydrate-Containing Sediments Formation. In Proceedings of the Fifth International Conference on Gas Hydrates, Trondheim, Norway, 13–16 June 2005; pp. 1561–1567.

44. Chuvilin, E.M. Migration of ions of chemical elements in freezing and frozen soils. *Polar Record.* **1999**, *35*, 59–66.

45. Chuvilin, E.M.; Ershov, E.D.; Smirnova, O.G. Ionic Migration in Frozen Soils and Ice. In Proceedings of the 7th International Permafrost Conference, Yellowknife, NT, Canada, 23–27 June 1998; pp. 167–171.

46. Lebedenko, Y.P. Cryogenic Migration of Ions and Bound Moisture in Ice-Saturated Frozen Rocks (Published in Russian). *Eng. Geol.* **1989**, *4*, 21–30.

47. Andersland, O.B.; Biggar, K.W. Site Investigations of Fuel Spill Migration into Permafrost. *J. Cold Reg. Eng.* **2002**, *13*, 165–166. [CrossRef]

48. Ershov, E.D.; She, Z.S.; Lebedenko, Y.; Chuvilin, E.M.; Kryuchkov, K.Y. Mass Transfer and Deformation Processes in Frozen Rocks Interacting with Aqueous Salt Solutions (Published in Russian). In Proceedings of the III Scientific and Technical Workshop "Engineering-Geological Study and Evaluation of Frozen, Freezing and Thawing Soils (IGK-92)", St. Petersburg, Russia, 1993; pp. 67–77. Available online: https://istina.msu.ru/publications/article/2627661/ (accessed on 23 April 2019).

49. Winiger, P.; Barrett, T.E.; Sheesley, R.J.; Huang, L.; Sharma, S.; Barrie, L.A.; Yttri, K.E.; Evangeliou, N.; Eckhardt, S.; Stohl, A.; et al. Source Apportionment of Circum-Arctic Atmospheric Black Carbon from Isotopes and Modeling. *Sci. Adv.* **2019**, *5*, eaau8052:1–eaau8052:10. [CrossRef]

50. Chuvilin, E.M.; Grebenkin, S.I.; Sacleux, M. Influence of Moisture Content on Permeability of Frozen and Unfrozen Soils. *Kriosf. Zemli* **2016**, *20*, 66–72.

geosciences

MDPI

Article

Role of Warming in Destabilization of Intrapermafrost Gas Hydrates in the Arctic Shelf: Experimental Modeling

Evgeny Chuvilin [1,*], Dinara Davletshina [1,2], Valentina Ekimova [1], Boris Bukhanov [1], Natalia Shakhova [3,4] and Igor Semiletov [3,5,6]

1   Skolkovo Institute of Science and Technology (Skoltech), 121205 Moscow, Russia;
    D.Davletshina@skoltech.ru (D.D.); Valentina.Ekimova@skoltech.ru (V.E.); B.Bukhanov@skoltech.ru (B.B.)
2   GSP-1, Lomonosov Moscow State University (MSU), 119991 Moscow, Russia
3   National Research Tomsk Polytechnic University, Tomsk Polytechnic University (TPU), 634050 Tomsk,
    Russia; shahova@tpu.ru (N.S.); ipsemiletov@alaska.edu (I.S.)
4   International Arctic Research Center, University Alaska Fairbanks, Fairbanks, AK 99775, USA
5   Pacific Oceanological Institute, Far Eastern Branch of Russian Academy of Sciences,
    690041 Vladivostok, Russia
6   Moscow Institute of Physics and Technology, 141701 Moscow Region, Russia
*   Correspondence: e.chuvilin@skoltech.ru

Received: 2 September 2019; Accepted: 18 September 2019; Published: 20 September 2019

check for updates

Abstract: Destabilization of intrapermafrost gas hydrates is one of the possible mechanisms responsible for methane emission in the Arctic shelf. Intrapermafrost gas hydrates may be coeval to permafrost: they originated during regression and subsequent cooling and freezing of sediments, which created favorable conditions for hydrate stability. Local pressure increase in freezing gas-saturated sediments maintained gas hydrate stability from depths of 200–250 m or shallower. The gas hydrates that formed within shallow permafrost have survived till present in the metastable (relict) state. The metastable gas hydrates located above the present stability zone may dissociate in the case of permafrost degradation as it becomes warmer and more saline. The effect of temperature increase on frozen sand and silt containing metastable pore methane hydrate is studied experimentally to reconstruct the conditions for intrapermafrost gas hydrate dissociation. The experiments show that the dissociation process in hydrate-bearing frozen sediments exposed to warming begins and ends before the onset of pore ice melting. The critical temperature sufficient for gas hydrate dissociation varies from −3.0 °C to −0.3 °C and depends on lithology (particle size) and salinity of the host frozen sediments. Taking into account an almost gradientless temperature distribution during degradation of subsea permafrost, even minor temperature increases can be expected to trigger large-scale dissociation of intrapermafrost hydrates. The ensuing active methane emission from the Arctic shelf sediments poses risks of geohazard and negative environmental impacts.

Keywords: Arctic shelf; permafrost; gas hydrate; temperature increase; hydrate dissociation; methane emission; environmental impact; geohazard

1. Introduction

A wealth of data on subsea permafrost in the Arctic shelf collected through Russian and international research projects [1–15] has revealed large-scale methane emission from bottom sediments into water and on into the atmosphere. The gases in the Arctic shelf are often attributed to increasing microbial methane generation, migration of gas through taliks and faults, as well as to decomposition of intrapermafrost and subpermafrost gas hydrates during progressive degradation of subsea permafrost [11,16–21]. The dissociation of hydrates related to subsea permafrost degradation has been largely discussed lately as the main mechanism maintaining the emanation of methane [4,14,18,22–26].

Gas hydrates (clathrates) are metastable ice- or snow-like solid compounds that form from water and low-molecular gas under certain pressures and temperatures [27,28]. One cubic meter of clathrate can store about 160 cubic meters of gas. Natural hydrates of gas (mainly methane) can be stable in marine sediments and in permafrost [29] and occur in the Arctic shelf below 250 m or 300 m [30,31]. Relict metastable hydrates can survive also at shallower depths in the Arctic subsea permafrost due to self-preservation [32]. Shallow intrapermafrost metastable gas hydrates are especially sensitive to environment changes leading to pressure decrease, temperature increase, and migration of salts [33].

The extent and structure of subsea permafrost in the Russian Arctic shelf remain poorly constrained because of drilling shortage. Some knowledge is available for the western Arctic sector from test boreholes and geophysical surveys [7,34], but the data from the eastern sector are limited to sporadic findings of shelf permafrost, most often by geophysical methods [35]. The Arctic shelf permafrost has been mapped recently with reference to modeling results, well log and core data, and geophysical surveys [36–41]. The modeling predicts that the Arctic shelf permafrost may be as thick as 700 m, which is favorable for the formation and preservation of intra- and subpermafrost gas hydrates [4,26,42,43].

The relation of methane emanation from bottom sediments with the presence of gas hydrates is known from different Arctic regions (Svalbard Archipelago, Norwegian Sea, Beaufort Sea, etc.). Gas flares up to 850 m high and 400 m thick were reported from areas of gas hydrate accumulation in the Angola basin west of Spitsbergen Island [44,45]. There is evidence of gas eruptions, anomalous phytoplankton abundances, bubbling in water, and meters-size pockmarks [14] in bottom sediments of the Norwegian Sea (Vestnes, Storegga, and Nyegga areas) [30,46]. Pingo-like cone-shaped edifices at gas vents on the bottom of the southern Beaufort Sea first described by [47] may result from gas hydrate dissociation caused by degradation of shelf permafrost rather than being associated with seeps from subbottom gas reservoirs as suggested by Paull and Sparrow [15,47]. More evidence of gas emanation in the Arctic shelf comes from Bennet Island and the Mackenzie delta where gas hydrates may occur both beneath and within permafrost [30,46,48–53].

Paleoclimate reconstructions and modeling of permafrost thickness dynamics indicate that gas hydrates in the subsea permafrost of the eastern Arctic sector formed simultaneously with the latter during Late Pliocene-Pleistocene regression about 20–25 kyr BP [37]. They could form at shallow depths within permafrost at favorable thermobaric conditions the emerged sediments became frozen, and then survived in the metastable state for a long time due to the self-preservation effect after the pressure and temperature had changed. The onshore permafrost became submerged during subsequent transgression which led to its warming and degradation both from below (by heat flux through the sea bottom) and from above (by interaction with sea water) [39]. Penetration of sea water and dissolved salts into frozen sediments [33] and increase in their temperature [4,38,41,54] apparently destabilized the intrapermafrost gas hydrates.

Warming of subsea permafrost triggers decomposition of intrapermafrost gas hydrates into gas (methane) and ice. The reaction consumes much of the heat and the sediments cool down, which prevents them from further thawing. Thus, the presence of intrapermafrost gas hydrates may account for the persistence of permafrost in the Arctic shelf and for the gradientless temperature distribution observed in some boreholes in the Laptev shelf [4]. The liberated methane passes into sea water, sinks to the bottom, and becomes a source of gas emission [26].

However, the thawing of frozen hydrate-bearing rocks and related dissociation of hydrates, which can maintain self-cooling of the sediments, remains poorly studied. In this respect, experimental modeling of the mechanism and patterns of pore gas hydrate dissociation induced by permafrost warming may shed light on destabilization of intrapermafrost gas hydrates and hazardous emission of methane in the Arctic shelf.

2. Methods

The effect of temperature on the dissociation of pore gas hydrates in frozen sediments was studied experimentally using a specially designed system (Figure 1) which can reproduce thermobaric conditions in a large range of temperatures and pressures. The system, with a working volume of ~420 cm^3, consists of a pressure cell that accommodates a metal container with samples; an analog-digital converter (ADC); and a personal computer (PC) for saving records of pressure and temperature changes [55]. The temperature was maintained, to an accuracy of 0.1 °C, by circulation of liquid from the HAAKE Phoenix C40P refrigerated bath along the "thermal coat" around the pressure cell. During our experiments, the temperature and pressure in the cell were accurate to 0.05 °C and 0.005 MPa, respectively.

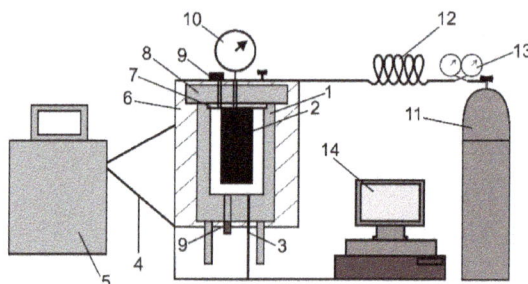

Figure 1. Sketch of experimental setup for modeling the dissociation of pore gas hydrates. **1** = pressure cell; **2** = container with soil samples; **3** = thermistor input sleeve; **4** = hose for circulating liquid; **5** = refrigerated bath; **6** = thermal coat; **7** = tephlon gasket; **8** = steel lid; **9** = pressure sensors; **10** = digital pressure gauge; **11** = gas bomb; **12** = gas tube; **13** = pressure regulator; **14** = PC with ADC.

To simulate the dissociation of pore gas hydrates, natural deformed soil samples of sand, silt, and clay silt (Table 1) were saturated with hydrate. The constituent minerals in the samples were identified by X-ray diffractometer ULTIMA-IV (Rigaku company, Tokyo, Japan), salinity was determined through water extracts from dry sediment samples, and the particle size distribution (Table 2) was determined following the procedure of State Standard [56].

Table 1. Mineralogy and salinity of soil samples.

Sample	Sampling Site	Mineralogy, %		Salinity, %
Sand	Arctic shelf (Buor Khaya Bay)	Quartz	55	0.4
		Albite	18	
		Microcline	9	
Silt	Arctic shelf (Buor Khaya Bay)	Quartz	42.5	0.1
		Albite	14.9	
		Illite	9.1	
		Chlorite	6.9	
		Microcline	5.9	
		Hydromica	3.9	
Clay silt	South Tambei GCF	Quartz	46.4	0.7
		Albite	25.3	
		Chlorite	10.4	
		Mirror stone	7.4	
		Orthoclase	6.5	
		Kaolinite	4.2	

Note: GCF = gas-condensate field. The listed mineral phases have percentages >1%.

Table 2. Particle size distribution in soil samples.

Sample	Particle Size Distribution, %			Lithology *
	1–0.05 mm	0.05–0.001 mm	<0.001 mm	
Sand	96.7	2.0	1.3	Fine sand
Silt	63.4	32.6	4.0	Silty sand
Clay silt	21.1	55.8	23.1	Clay silt (loam)

Note: * Lithology is according to classifications of E. Sergeev (sand) and V. Okhotin (silt and loam).

Samples were prepared from air-dry soil mixed with distilled water and crushed ice (in the case of silt) and left for 30 min at room temperature to achieve the wanted moisture content. The wet soil was compacted layer-by-layer in a cylindrical container (10 cm high and 4.6 cm in diameter) and placed into a pressure cell. The pressure cell with the samples was sealed tightly, vacuumed, frozen to −5–6 °C, and then filled with hydrate-forming gas (99.98% CH_4) at 4–6 MPa [55,57]. Pore hydrate accumulation was stimulated by cyclic freezing and thawing of the samples at above-equilibrium pressure.

Once hydrate formation in the samples decayed, the time-dependent kinetics of pore hydrate accumulation and phase transition parameters were analyzed by the pressure-volume-temperature (PVT) method [55].

The volume content of hydrate (Hv, %) was found as:

$$H_v = \frac{M_h \cdot \rho}{M_s \cdot \rho_h} \cdot 100\%$$

where Mh is the weight of pore gas hydrate (g); Ms is the weight of soil sample (g); is the sample density (g/cm³); h is the skeleton (hydrate) density of empty square lattice (without gas molecules by analogy with the pure ice structure); h for CH_4 was assumed to be 0.794 g/cm³ [55].

Hydrate saturation or percentage of pore space filled with hydrate (Sh, %) was inferred from the volume content of hydrate as:

$$S_h = \frac{H_v}{n}$$

where n is the sample porosity (u.f.), assuming a hydrate number of 5.9 for CH_4.

The fraction of water converted to hydrate, or hydrate coefficient (Kh, u.f.) was found as:

$$K_h = \frac{W_h}{W}$$

where Wh is the percentage of water converted to hydrate (% of dry sample weight) and W is the total amount of moisture (initial water content, %).

Then the frozen soil samples saturated with methane hydrate to a known level were exposed to non-equilibrium conditions at a constant negative temperature of −6 °C, by decreasing pressure in the cell to 0.6–1.6 MPa. These pressure and temperature conditions correspond to those in natural sediments at gas emanation sites. Pore gas hydrates underwent dissociation at a decaying rate because of the self-preservation effect [32]. The pressure in the cell was maintained at a constant low level by slow gas release as the pressure increased. Once the pore gas hydrates reached the metastable state, the self-preservation coefficient (Ksc) was calculated as a ratio of residual hydrate content in a frozen sample at below-equilibrium pressure to the initial hydrate content. At the next step, the temperature in the cell was increased for a few hours from the initial value −5 °C to +6 °C. The recorded time-dependent temperature and pressure changes were used to calculate the volume of releasing methane per 1 m³ of thawing hydrate-bearing soil.

3. Experimental Results

The soil samples, which were saturated with methane hydrate in laboratory and exposed to the self-preservation conditions by decreasing the pressure to below equilibrium (Table 3), showed variations during the experiment caused by dissociation of pore gas hydrates.

Table 3. Hydrate saturation in frozen hydrate-saturated samples.

Sample	Water Content, %	Density, g/cm³	Hydrate Saturation				
			Equilibrium		Self-Preservation		
			Sh initial, %	*Kh* initial, u.f.	*Sh* final, %	*Kh* final, u.f.	*Ksc*, %
Sand	14	1.80	47	0.59	4	0.05	8.5
Silt	15	1.82	25	0.29	6.9	0.05	26
Loam	18	1.20	11.5	0.29	7	0.17	61

The highest percentage of pore hydrate was measured in sand (*Sh* = 47%), while the fraction of water converted to hydrate (*Kh*) was 59%; the silt and clay silt samples showed lower respective values: 25% and 11.5% for *Sh* and <29% for *Kh*. The pressure decrease to below equilibrium led to partial dissociation and self-preservation of pore hydrate recorded in saturation decrease [32]. At the end of the dissociation process, the frozen samples contained 4–7% of residual hydrate. The self-preservation coefficient (*Ksc*) was the highest in clay silt (61%) and the lowest in sand (8.5%).

The frozen soil samples with metastable pore methane hydrate were heated to estimate the effect of warming on hydrate destabilization detectable in the time-dependent change of pressure, temperature, and hydrate coefficient (Figures 2–4). In addition, temperature corresponding to the onset of rapid pore hydrate dissociation (t_d) was determined for each experiment by the intersection of the tangents to the dramatic bend of the hydrate coefficient graph (blue line) and temperature graph (red line).

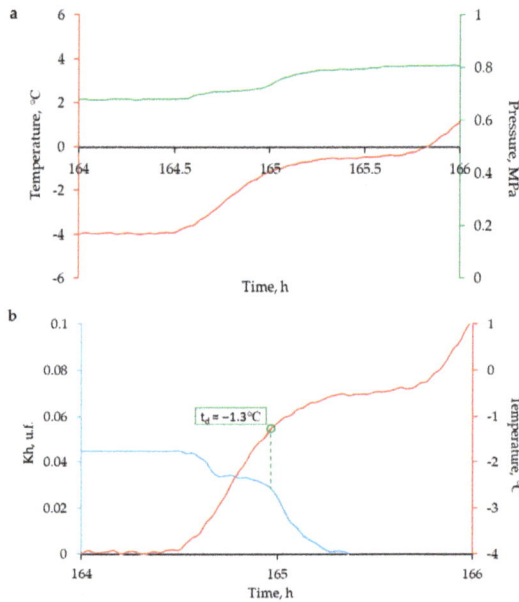

Figure 2. Time-dependent change of temperature and pressure (**a**), and hydrate coefficient (*Kh*) (**b**) in frozen hydrate-bearing sand exposed to warming. t_d is the temperature corresponding to the onset of rapid pore hydrate dissociation.

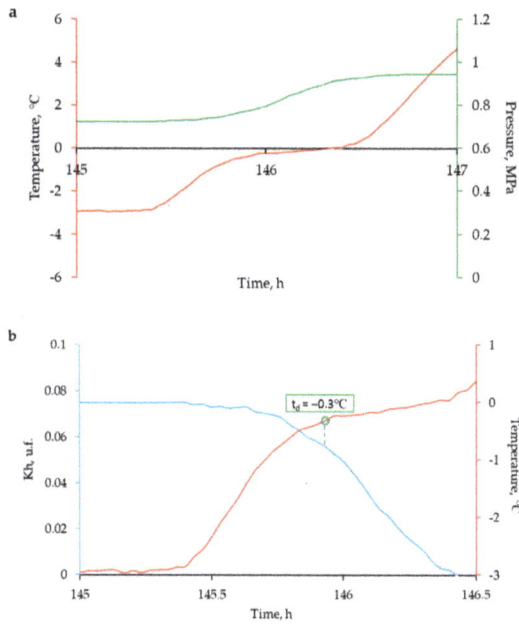

Figure 3. Time-dependent change of temperature and pressure (**a**), and hydrate coefficient (*Kh*) (**b**) in frozen hydrate-bearing silt exposed to warming. t_d is the temperature corresponding to the onset of rapid pore hydrate dissociation.

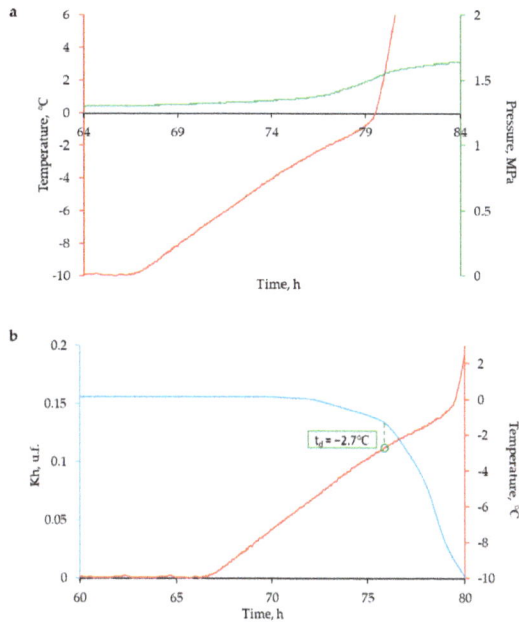

Figure 4. Time-dependent change of temperature and pressure (**a**), and hydrate coefficient (*Kh*) (**b**) in frozen hydrate-bearing clay silt exposed to warming. t_d is the temperature corresponding to the onset of rapid pore hydrate dissociation.

In the beginning of the temperature increase, the pressure in the cell became slightly (0.02–0.03 MPa) higher as a result of gas expansion, but then increased markedly upon dissociation of pore methane hydrate. The increase was 0.1 MPa for the sand sample, 0.15 MPa for silt, and 0.2 MPa for clay silt. Rapid dissociation of pore hydrate in the frozen samples began at different temperatures (Figure 2b, Figure 3b, Figure 4b): t_d was −1.3 °C in sand (Figure 2b), −0.3 °C in silt (Figure 3b), and −2.7 °C in clay silt (Figure 4b). These t_d values remained in the range of negative temperatures below the pore ice–water phase transition (melting point). The t_d temperature was the highest in non-saline silt (−0.3 °C) and shifted to lower values as salinity increased in the series 'silt–sand–clay silt' (Figure 5). Thus, the dissociation temperature of metastable pore hydrate is lower in saline and fine-grained soils than in coarser and less saline ones. Quite low t_d values (about −6.6 °C) were obtained for highly saline clay (Z up to 1.8%) sampled in the Yamal Peninsula near the Bovanenkovo gas and condensate field. Therefore, gas hydrates may have existed at colder conditions in the past but failed to survive as the permafrost temperature increased to about −3 °C [58].

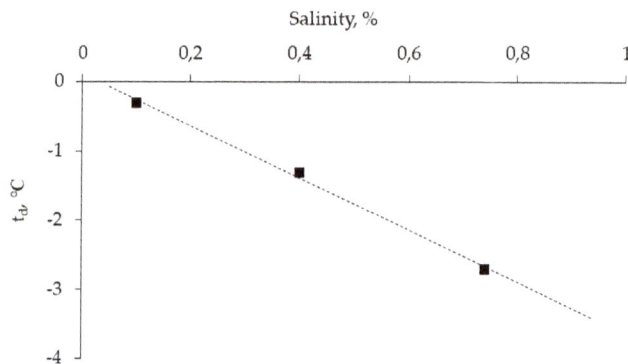

Figure 5. Temperature of rapid hydrate dissociation onset (t_d) in frozen hydrate-bearing samples as a function of salinity.

4. Discussion

The experimentally revealed behavior of pore gas hydrates in samples exposed to increasing temperatures has implications for the conditions of preservation and dissociation of gas hydrates in the Arctic shelf permafrost. This knowledge is of special interest for the eastern Arctic sector (Laptev shelf), where subsea permafrost is widespread and thick (600 m) and voluminous methane emission may result from dissociation of intrapermafrost gas hydrates. The composition and temperature of the permafrost in the area are poorly constrained for the lack of explicit geological evidence but the origin and evolution of shelf permafrost, as well as the local conditions for gas hydrate stability, were predicted by modeling [59–61].

The modeling results predict the following history of permafrost in the Arctic shelf. It formed during regression about 20–15 kyr BP as the emerged rocks became exposed to prolonged cooling [40], which produced a zone of gas hydrate stability both beneath and within the permafrost. The pore gas present in freezing sediments could partly convert to hydrate, including at depths shallower than 200 m [55]. The permafrost at that time had a temperature of −13 °C or −12 °C and its thickness reached 700 m or more [42,62,63]. The subsea permafrost gradually warmed up during the subsequent transgression (about 9 kyr BP) and has reached a temperature of −2 °C to −1.5 °C by present. Note that the temperature field has almost no gradient [40]. The temperature increase led to progressive degradation of the permafrost and destabilization of pore gas hydrates, including those in the metastable state. As shown by experiments, the destabilization begins at a certain temperature controlled by the lithology (particle size) and salinity of sediments.

The available published evidence and experimental results allow sketching a scenario of changes in warming subsea permafrost that contains relict metastable pore gas hydrates (Figure 6).

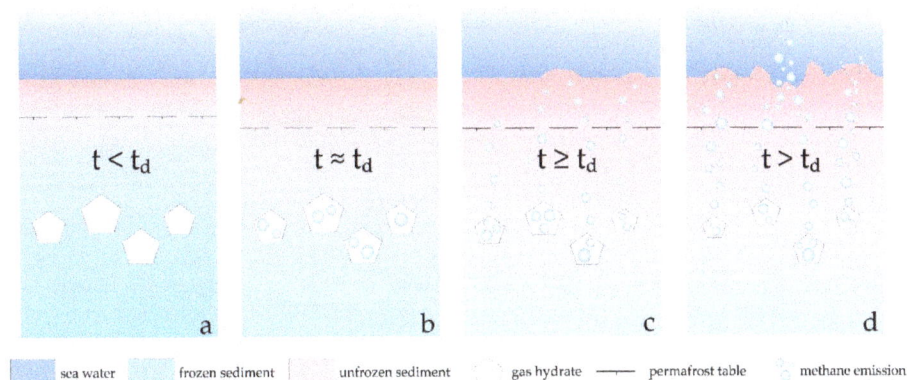

Figure 6. Evolution of hydrate-bearing subsea permafrost exposed to progressive temperature increase (a, b, c, d- details below in text).

In the beginning of transgression, the temperature of permafrost increases upon interaction with sea water. Metastable intrapermafrost gas hydrates hold as long as this temperature remains below the critical value t_d (Figure 6a) but begin to dissociate once the permafrost reaches the t_d level (Figure 6b). Gas hydrates dissociate and the liberated methane begins to rise toward the sea bottom and causes swelling of sediments while the permafrost temperature is equal to or slightly above the t_d level (Figure 6c). Finally, when the permafrost exceeds the critical temperature, rapid dissociation of gas hydrates can produce active methane emission and water bubbling detectable by acoustic emission survey [21], and pockmarks, sinkholes, and other efflux structures form on the sea bottom (Figure 6d).

The laboratory results place constraints on the amount of methane that can emit from permafrost containing metastable gas hydrates: 1 m^3 of frozen samples exposed to warming can release 1.5 m^3 to 5.5 m^3 of methane (Table 4).

Table 4. Amount of methane (CH$_4$) released upon warming of frozen hydrate-bearing samples.

Sample	Sh d., %	CH$_4$ Emission (m^3) from 1 m^3 of Thawing Hydrate-Bearing Rock
Sand	2.3	1.5
Silt	4.8	3.2
Clay silt	5.3	5.5

Since the critical temperature of the hydrate dissociation onset depends on lithology (pore size) and salinity of the host sediments, gas hydrates in heterogeneous subsea permafrost may exist in layers with different t_d temperatures. A part of accumulated gas hydrates can dissociate early during transgression at moderate permafrost warming (−3 °C to −5 °C), but some may hold at the present temperature level of relatively warm permafrost (−1 °C to −2 °C).

According to available field data [39], the Arctic subsea permafrost (e.g., that in the Laptev shelf) has low salinity, and the gas hydrates it stores may dissociate rapidly even upon minor warming. The critical temperature in non-saline sediments was predicted to be as low as t_d = −0.3 °C, while the Laptev shelf permafrost may become 0.5 °C warmer already in a few decades [26,62]. Thus, warming of subsea permafrost in the Arctic shelf to −0.5 °C to −1 °C may trigger large-scale gas hydrate dissociation in the near future. The possibility of violent methane release from bottom sediments poses risks of geohazard and negative environmental impacts.

5. Conclusions

Dissociation of gas hydrates in the subsea permafrost that formed during regression is one of possible causes of active methane emission in the Arctic shelf. The intrapermafrost gas hydrates became destabilized at a high sea stand, when hydrate-bearing permafrost underwent rapid degradation upon interaction with sea water which led to large-scale release of methane into the air.

The experimental study of the process in laboratory shows that active dissociation of intrapermafrost gas hydrates starts at some critical temperature depending on sediment lithology and salinity. The dissociation of metastable relict gas hydrates in frozen sediments exposed to warming begins and ends slightly before the onset of pore ice melting. The critical temperature sufficient for triggering pore hydrate dissociation in permafrost ranges from −3.0 °C to −0.3 °C, i.e., the active process may begin even at small increase of the negative temperature of frozen sediments. Taking into account an almost gradientless temperature field during the subsea permafrost degradation, large-scale dissociation of gas hydrates accompanied by active gas emission and methane bubbling detectable in water can be expected in the near future.

Thus, the theoretical calculations and the physical modeling predict that minor warming of subsea permafrost poses risks of hazardous dissociation of metastable gas hydrates in the Arctic shelf.

Author Contributions: Conceptualization, experimental methodology, supervision, E.C.; carry out experiments, E.C., B.B., and D.D.; processing and analysis, E.C., D.D., and V.E.; writing manuscript and editing, E.C., D.D., V.E., B.B., N.S., and I.S.

Funding: The research was supported by the Russian Science Foundation (grants No. 16−17−00051, 18−77−10063, and 15-17-20032), the Russian Foundation for Basic Research (grant No. 17−05−00995), the Russian government (#14, Z50.31.0012/03.19.2014), and Tomsk Polytechnic University (Competitiveness Enhancement Program grant, Project Number TPU CEP_SESE-299\2019).

Acknowledgments: The authors acknowledge engineer S.Grebenkin (Skoltech) for technical support of experiments.

Conflicts of Interest: The authors declare no conflict of interest.

References

1. Grigoriev, N.F. *Permafrost in the Seaside Zone of Yakutia (Published in Russian)*; Nauka: Moscow, Russia, 1966; p. 180.
2. Wood, W.T.; Gettrust, J.F.; Chapman, N.R.; Spence, G.D.; Hyndman, R.D. Decreased Stability of Methane Hydrates in Marine Sediments Owing to Phase-Boundary Roughness. *Nature* **2002**, *420*, 656–660. [CrossRef] [PubMed]
3. Shakhova, N.E.; Nicolsky, D.Y.; Semiletov, I.P. Current State of Subsea Permafrost on the East Siberian Shelf: Tests of Modeling Results Based on Field Observations. *Dokl. Earth Sci.* **2010**, *429*, 1518–1521. [CrossRef]
4. Shakhova, N.; Semiletov, I.; Gustafsson, O.; Sergienko, V.; Lobkovsky, L.; Dudarev, O.; Tumskoy, V.; Grigoriev, M.; Chernykh, D.; Koshurnikov, A.; et al. Current Rates and Mechanisms of Subsea Permafrost Degradation in the East Siberian Arctic Shelf. *Nat. Commun.* **2017**, *8*, 15872:1–15872:13. [CrossRef] [PubMed]
5. Yusupov, V.I.; Salyuk, A.N.; Karnaukh, V.N.; Semiletov, I.P.; Shakhova, N.E. Detection of Areas of Methane Bubble Discharge on the Laptev Sea Shelf in the Eastern Arctic (Published in Russian). *Rep. Acad. Sci.* **2010**, *430*, 820–823.
6. Anisimov, O.A.; Borzenkova, I.I.; Lavrov, S.A.; Strelchenko, Y.G. Modern Dynamics of Underwater Permafrost and Methane Emission on the Shelf of the Seas of the Eastern Arctic (Published in Russian). *Ice Snow* **2012**, *2*, 97–105.
7. Loktev, A.; Bondarev, V.; Kulikov, S.; Rokos, S. Russian Arctic Offshore Permafrost. In *Offshore Site Investigation and Geotechnics: Integrated Technologies—Present and Future*; Society of Underwater Technology: London, UK, 2012; pp. 1–8.

8. Sergienko, V.I.; Lobkovskii, L.I.; Semiletov, I.P.; Dudarev, O.V.; Dmitrievskii, N.N.; Shakhova, N.E.; Romanovskii, N.N.; Kosmach, D.A.; Nikol'skii, D.N.; Nikiforov, S.L.; et al. The Degradation of Submarine Permafrost and the Destruction of Hydrates on the Shelf of East Arctic Seas as a Potential Cause of the Methane Catastrophe: Some Results of Integrated Studies in 2011. *Dokl. Earth Sci.* **2012**, *446*, 1132–1137. [CrossRef]

9. Anisimov, O.A.; Kokorev, V.A. Comparative Analysis of Land, Marine, and Satellite Observations of Methane in the Lower Atmosphere in the Russian Arctic under Conditions of Climate Change. *Izv. Atmos. Ocean. Phys.* **2015**, *51*, 979–991. [CrossRef]

10. Koloskov, E.N.; Firsov, Y.G. The Use of Modern Hydrographic Technologies for the Study of Topography and Bottom Gas Occurrence in the Northern Seas of Russia (Published in Russian). *Bull. State Univ. Marit. River Fleet Admiral SB Makarov* **2015**, *3*, 54–62.

11. James, R.H.; Bousquet, P.; Bussmann, I.; Haeckel, M.; Kipfer, R.; Leifer, I.; Niemann, H.; Ostrovsky, I.; Piskozub, J.; Rehder, G.; et al. Effects of Climate Change on Methane Emissions from Seafloor Sediments in the Arctic Ocean: A Review. *Limnol. Oceanogr.* **2016**, *61*, S283–S299. [CrossRef]

12. Leifer, I.; Chernykh, D.; Shakhova, N.; Semiletov, I. Sonar Gas Flux Estimation by Bubble Insonification: Application to Methane Bubble Flux from Seep Areas in the Outer Laptev Sea. *Cryosphere* **2017**, *11*, 1333–1350. [CrossRef]

13. Thornton, B.F.; Geibel, M.C.; Crill, P.M.; Humborg, C.; Mörth, C.M. Methane Fluxes from the Sea to the Atmosphere across the Siberian Shelf Seas. *Geophys. Res. Lett.* **2016**, *43*, 5869–5877. [CrossRef]

14. Andreassen, K.; Hubbard, A.; Winsborrow, M.; Patton, H.; Vadakkepuliyambatta, S.; Plaza-Faverola, A.; Gudlaugsson, E.; Serov, P.; Deryabin, A.; Mattingsdal, R.; et al. Massive Blow-out Craters Formed by Hydrate-Controlled Methane Expulsion from the Arctic Seafloor. *Science* **2017**, *356*, 948–953. [CrossRef] [PubMed]

15. Sparrow, K.J.; Kessler, J.D.; Southon, J.R.; Garcia-Tigreros, F.; Schreiner, K.M.; Ruppel, C.D.; Miller, J.B.; Lehman, S.J.; Xu, X. Limited Contribution of Ancient Methane to Surface Waters of the U.S. Beaufort Sea Shelf. *Sci. Adv.* **2018**, *4*, eaao4842:1–eaao4842:7. [CrossRef] [PubMed]

16. Serov, P.; Portnov, A.; Mienert, J.; Semenov, P.; Ilatovskaya, P. Methane Release from Pingo-like Features across the South Kara Sea Shelf, an Area of Thawing Offshore Permafrost. *J. Geophys. Res. F Earth Surf.* **2015**, *120*, 1515–1529. [CrossRef]

17. Vonk, J.E.; Tank, S.E.; Bowden, W.B.; Laurion, I.; Vincent, W.F.; Alekseychik, P.; Amyot, M.; Billet, M.F.; Canario, J.; Cory, R.M.; et al. Reviews and Syntheses: Effects of Permafrost Thaw on Arctic Aquatic Ecosystems. *Biogeoscience* **2015**, *12*, 10719–10815. [CrossRef]

18. Lobkovskiy, L.I.; Nikiforov, S.L.; Dmitrevskiy, N.N.; Libina, N.V.; Semiletov, I.P.; Ananiev, R.A.; Meluzov, A.A.; Roslyakov, A.G. Gas Extraction and Degradation of the Submarine Permafrost Rocks on the Laptev Sea Shelf. *Oceanology* **2015**, *55*, 283–290. [CrossRef]

19. Ostanin, I.; Anka, Z.; di Primio, R. Role of Faults in Hydrocarbon Leakage in the Hammerfest Basin, SW Barents Sea: Insights from Seismic Data and Numerical Modelling. *Geosciences* **2017**, *7*, 28. [CrossRef]

20. Baranov, B.V.; Lobkovsky, L.I.; Dozorova, K.A.; Tsukanov, N.V. The Fault System Controlling Methane Seeps on the Shelf of the Laptev Sea. *Dokl. Earth Sci.* **2019**, *486*, 571–574. [CrossRef]

21. Shakhova, N.; Semiletov, I.; Salyuk, A.; Yusupov, V.; Kosmach, D.; Gustafsson, Ö. Extensive Methane Venting to the Atmosphere from Sediments of the East Siberian Arctic Shelf. *Science* **2010**, *327*, 1246–1250. [CrossRef]

22. Romanovskii, N.N.; Hubberten, H.W.; Gavrilov, A.V.; Eliseeva, A.A.; Tipenko, G.S. Offshore Permafrost and Gas Hydrate Stability Zone on the Shelf of East Siberian Seas. *Geo-Marine Lett.* **2005**, *25*, 167–182. [CrossRef]

23. Burwicz, E.; Rüpke, L.; Wallmann, K. Estimation of the Global Amount of Submarine Gas Hydrates Formed via Microbial Methane Formation Based on Numerical Reaction-Transport Modeling and a Novel Parameterization of Holocene Sedimentation. *Geochim. Cosmochim. Acta* **2011**, *75*, 4562–4576. [CrossRef]

24. Perlstein, G.Z.; Sergeev, D.O.; Tipenko, G.S.; Tumskoy, V.E.; Khimenkov, A.N.; Vlasov, A.N.; Merzlyakov, V.P.; Stanilovskaya, Y.V. Hydrocarbon Gases and Cryolithozone of the Arctic Shelf (Published in Russian). *Arct. Ecol. Econ.* **2015**, *2*, 35–44.

25. Ruppel, C.D.; Kessler, J.D. The Interaction of Climate Change and Methane Hydrates. *Rev. Geophys.* **2017**, *55*, 126–168. [CrossRef]

26. Shakhova, N.; Semiletov, I.; Chuvilin, E. Understanding the Permafrost–Hydrate System and Associated Methane Releases in the East Siberian Arctic Shelf. *Geosciences* **2019**, *9*, 251. [CrossRef]

27. Sloan, E.D. *Clathrate Hydrates of Natural Gases, Second Edition, Revised and Expanded*; CRC Press: New York, NY, USA, 1998; p. 705. ISBN 9780824799373.

28. Koh, C.A.; Sloan, E.D.; Sum, A.K.; Wu, D.T. Fundamentals and Applications of Gas Hydrates. *Annu. Rev. Chem. Biomol. Eng.* **2011**, *2*, 237–257. [CrossRef]

29. Max, M. *Natural Gas Hydrate In Oceanic and Permafrost Environments*; Kluwer Academic Publishers: Washington, WA, USA, 2000; p. 419. ISBN 978-1-4020-1362-1. [CrossRef]

30. Solov'ev, V.A.; Ginsburg, G.D. *Submarine Gas Hydrates (Published in Russian)*; VNII Oceanologii: St Petersburg, Russia, 1994; p. 194. ISBN 5-7173-0290-8.

31. Max, M.D.; Johnson, A.H.; Dillon, W.P. *Natural Gas Hydrate—Arctic Ocean Deepwater Resource Potential*; Springer: Dordrecht, The Netherlands, 2013; p. 113. ISBN 9783319025070. [CrossRef]

32. Chuvilin, E.; Bukhanov, B.; Davletshina, D.; Grebenkin, S.; Istomin, V. Dissociation and Self-Preservation of Gas Hydrates in Permafrost. *Geosciences* **2018**, *8*, 431. [CrossRef]

33. Chuvilin, E.; Ekimova, V.; Bukhanov, B.; Grebenkin, S.; Shakhova, N.; Semiletov, I. Role of Salt Migration in Destabilization of Intra Permafrost Hydrates in the Arctic Shelf: Experimental Modeling. *Geosciences* **2019**, *9*, 188. [CrossRef]

34. Loktev, A.; Tokarev, M.; Chuvilin, E. Problems and Technologies of Offshore Permafrost Investigation. *Proced. Eng.* **2017**, *189*, 459–465. [CrossRef]

35. Chuvilin, E.; Bukhanov, B.; Tymskoy, V.; Shakhova, N.; Dudarev, O.; Semiletov, I. Thermal Conductivity of Bottom Sediments in the Area of Buor-Haya Bay (Laptev Sea Shelf) (Published in Russian). *Earth's Cryosph.* **2013**, *XVII*, 24–36.

36. Romanovskii, N.N.; Gavrilov, A.V.; Tumskoy, V.E.; Kholodov, A.L.; Siegert, C.; Hubberten, H.W.; Sher, A.V. Environmental Evolution in the Laptev Sea Region during Late Pleistocene and Holocene. *Polarforschung* **2000**, *68*, 237–245.

37. Romanovskii, N.N.; Hubberten, H.W. Results of Permafrost Modelling of the Lowlands and Shelf of the Laptev Sea Region, Russia. *Permafr. Periglac. Process.* **2001**, *12*, 191–202. [CrossRef]

38. Nicolsky, D.; Shakhova, N. Modeling Sub-Sea Permafrost in the East Siberian Arctic Shelf: The Dmitry Laptev Strait. *Environ. Res. Lett.* **2010**, *5*, 015006. [CrossRef]

39. Nicolsky, D.J.; Romanovsky, V.E.; Romanovskii, N.N.; Kholodov, A.L.; Shakhova, N.E.; Semiletov, I.P. Modeling Sub-Sea Permafrost in the East Siberian Arctic Shelf: The Laptev Sea Region. *J. Geophys. Res. Earth Surf.* **2012**, *117*, F03028. [CrossRef]

40. Romanovskii, N.N.; Tumskoi, V.E. Retrospective Approach to the Estimation of the Contemporary Extension and Structure of the Shelf Cryolithozone in East Arctic (Published in Russian). *Earth's Cryosph.* **2011**, *15*, 3–14.

41. Overduin, P.P.; Schneider von Deimling, T.; Miesner, F.; Grigoriev, M.N.; Ruppel, C.; Vasiliev, A.; Lantuit, H.; Juhls, B.; Westermann, S. Submarine Permafrost Map in the Arctic Modeled Using 1-D Transient Heat Flux (SuPerMAP). *J. Geophys. Res. Ocean.* **2019**. [CrossRef]

42. Romanovskii, N.N.; Eliseeva, A.A.; Gavrilov, A.V.; Tipenko, G.S.; Hubberten, X. The Long-Term Dynamics of Frozen Strata and the Zone of Gas Hydrate Stability in the Rift Structures of the Arctic Shelf of Eastern Siberia (Report 2) (Published in Russian). *Earth's Cryosph.* **2006**, *10*, 29–38.

43. Delisle, G. Temporal Variability of Subsea Permafrost and Gas Hydrate Occurrences as Function of Climate Change in the Laptev Sea, Siberia. *Polarforschung* **2000**, *68*, 221–225.

44. Mazarovich, A.O. Real and Potential Geological Hazards on the Ocean Floor, Slopes, and Shelf. *Her. Russ. Acad. Sci.* **2012**, *82*, 320–325. [CrossRef]

45. Sahling, H.; Römer, M.; Pape, T.; Bergès, B.; dos Santos Fereirra, C.; Boelmann, J.; Geprägs, P.; Tomczyk, M.; Nowald, N.; Dimmler, W.; et al. Gas Emissions at the Continental Margin West of Svalbard: Mapping, Sampling, and Quantification. *Biogeosciences* **2014**, *11*, 6029–6046. [CrossRef]

46. Bondur, V.G.; Kuznetsova, T.V.; Vorobyev, V.E.; Zamshin, V. Detection of Gas Shows on the Russian Shelf Based on Satellite Imagery Data (Published in Russian). *Georesour. Geoenergy Geopolit.* **2014**, *1*, 1–23.

47. Paull, C.K.; Ussler, W.; Dallimore, S.R.; Blasco, S.M.; Lorenson, T.D.; Melling, H.; Medioli, B.E.; Nixon, F.M.; McLaughlin, F.A. Origin of Pingo-like Features on the Beaufort Sea Shelf and Their Possible Relationship to Decomposing Methane Gas Hydrates. *Geophys. Res. Lett.* **2007**, *34*, L01603:1–L01603:5. [CrossRef]

48. Biastoch, A.; Treude, T.; Rpke, L.H.; Riebesell, U.; Roth, C.; Burwicz, E.B.; Park, W.; Latif, M.; Böning, C.W.; Madec, G.; et al. Rising Arctic Ocean Temperatures Cause Gas Hydrate Destabilization and Ocean Acidification. *Geophys. Res. Lett.* **2011**, *38*, 8. [CrossRef]

49. Mau, S.; Romer, M.; Torres, M.E.; Bussmann, I.; Pape, T.; Damm, E.; Geprags, P.; Wintersteller, P.; Hsu, C.W.; Loher, M.; et al. Widespread Methane Seepage along the Continental Margin off Svalbard—From Bjornoya to Kongsfjorden. *Sci. Rep.* **2017**, *7*, 42997:1–42997:13. [CrossRef]
50. Westbrook, G.K.; Thatcher, K.E.; Rohling, E.J.; Piotrowski, A.M.; Pälike, H.; Osborne, A.H.; Nisbet, E.G.; Minshull, T.A.; Lanoisellé, M.; James, R.H.; et al. Escape of Methane Gas from the Seabed along the West Spitsbergen Continental Margin. *Geophys. Res. Lett.* **2009**, *36*, L15608. [CrossRef]
51. Majorowicz, J.A.; Hannigan, P.K. Natural Gas Hydrates in the Offshore Beaufort-Mackenzie Basin—Study of a Feasible Energy Source II. *Nat. Resour. Res.* **2000**, *9*, 201–214. [CrossRef]
52. Bogoyavlensky, V.; Kishankov, A.; Yanchevskaya, A.; Bogoyavlensky, I. Forecast of Gas Hydrates Distribution Zones in the Arctic Ocean and Adjacent Offshore Areas. *Geosciences* **2018**, *8*, 453. [CrossRef]
53. Moridis, G.J.; Collett, T.S.; Dallimore, S.R.; Satoh, T.; Hancock, S.; Weatherill, B. Numerical Studies of Gas Production from Several CH4 Hydrate Zones at the Mallik Site, Mackenzie Delta, Canada. *J. Pet. Sci. Eng.* **2004**, *43*, 219–238. [CrossRef]
54. Overduin, P.; Liebner, S.; Knoblauch, C.; Kneier, F.; Günther, F.; Schirrmeister, L.; Wetterich, S. Subsea Permafrost Degradation and Inferred Methane Release in Shallow Coastal Water of the Central Laptev Sea. In Proceeding of PERGAMON Final Symposium, Kiel, Germany, 4–7 November 2013.
55. Chuvilin, E.; Davletshina, D. Formation and Accumulation of Pore Methane Hydrates in Permafrost: Experimental Modeling. *Geosciences* **2018**, *8*, 467. [CrossRef]
56. GOST 12536-2014 Soils. *Laboratory Determination of the Grain (Grain-Size) and Microaggregate Composition (Publsihed in Russian)*; Standartinform: Moscow, Russia, 2014.
57. Chuvilin, E.; Bukhanov, B. Effect of Hydrate Formation Conditions on Thermal Conductivity of Gas-Saturated Sediments. *Energy Fuels* **2017**, *31*, 5246–5254. [CrossRef]
58. Yakushev, V.S.; Semenov, A.P.; Bogoyavlensky, V.I.; Medvedev, V.I.; Bogoyavlensky, I.V. Experimental Modeling of Methane Release from Intrapermafrost Relic Gas Hydrates When Sediment Temperature Change. *Cold Reg. Sci. Technol.* **2018**, *149*, 46–50. [CrossRef]
59. Denisov, S.N.; Arzhanov, M.M.; Eliseev, A.V.; Mokhov, I. Assessing the Response of the Subanalytic Methane. *Rep. Acad. Sci.* **2011**, *441*, 685.
60. Romanovskii, N.N.; Hubberten, H.W.; Gavrilov, A.V.; Eliseeva, A.A.; Tipenko, G.S.; Kholodov, A.L.; Romanovsky, V.E. Permafrost and Gas Hydrate Stability Zone Evolution on the Eastern Part of the Eurasia Arctic Sea Shelf in the Middle Pleistocene-Holocene (Published in Russian). *Earth's Cryosph.* **2003**, *7*, 51–64.
61. Tipenko, G.S.; Romanovsky, N.N.; Kholodov, A.L. Modeling the Dynamics of the Submarine Cryolithozone and the Zone of Stability of Gas Hydrates: A Mathematical Solution, Numerical Implementation and the Results of Test Calculations (Published in Russian). *Earth's Cryosph.* **1999**, *3*, 71–78.
62. Malakhova, V.V. Mathematical Modeling of the Long-Term Dynamics of the Submarine Permafrost of the Arctic Shelf (Published in Russian). *Interexpo Geo-Siberia* **2014**, *4*, 1:1–1:5.
63. Malakhova, V.V.; Golubeva, E.; Eliseev, A.V.; Platov, G. Arctic Ocean Estimation of Possible Climate Change Impact on Methane Hydrate in the Arctic Ocean. *IOP Conf. Ser. Earth Environ. Sci.* **2018**, *211*. [CrossRef]

geosciences

MDPI

Article

A Quick-Look Method for Initial Evaluation of Gas Hydrate Stability below Subaqueous Permafrost

Umberta Tinivella [1,*], **Michela Giustiniani** [1] **and Héctor Marín-Moreno** [1,2]

1   Geophysical Department, OGS (Istituto Nazionale di Oceanografia e di Geofisica Sperimentale),
    Borgo Grotta Gigante 42/C, 34010 Sgonico, Trieste, Italy
2   NOC (National Oceanography Centre), University of Southampton Waterfront Campus, European Way,
    Southampton SO14 3ZH, UK
*   Correspondence: utinivella@inogs.it; Tel.: +39-040-2140-219; Fax: +39-040-327521

Received: 26 June 2019; Accepted: 23 July 2019; Published: 26 July 2019

check for
updates

Abstract: Many studies demonstrated the coexistence of subaqueous permafrost and gas hydrate. Subaqueous permafrost could be a factor affecting the formation/dissociation of gas hydrate. Here, we propose a simple empirical approach that allows estimating the steady-state conditions for gas hydrate stability in the presence of subaqueous permafrost. This approach was derived for pressure, temperature, and salinity conditions typical of subaqueous permafrost in marine (brine) and lacustrine (freshwater) environments.

Keywords: modeling; gas hydrate; subaqueous permafrost

1. Introduction

Gas hydrate is a naturally occurring "ice-like" material of water molecules containing gas that forms at high pressure and low temperature, and it is present worldwide in permafrost regions (e.g., References [1,2]) and in marine sediments of outer continental margins [3–5], as well as beneath ice sheets [6,7]. Methane hydrates make up to 80% of the total inventory of gas hydrates [8]. This locked methane could be a potential future energy resource, and field experiments suggest that it may be produced with existing conventional oil and gas production technology [9–13]. At present, ocean warming-induced hydrate dissociation may be occurring in permafrost regions and in shallow marine sediments in polar continental margins (e.g., References [5,14,15]), as well as in ice sheets that stored methane in hydrate form during the last glaciation [6,7]. A significant release of methane from dissociated gas hydrate could create a positive feedback loop of warming [16,17], as suggested for past hyperthermal events such as the Paleocene Eocene Thermal Maximum (e.g., Reference [18]).

Permafrost is defined as the ground that remains at or below 0 °C for more than two years; its extent reaches about 20% of the land of the Earth [1]. Over half of Canada and Russia, most of Alaska, and northeast China are underlain by permafrost. Climate is one of the main drivers for permafrost distribution [1]; note that heat flow within permafrost is mainly due to conduction, because most of the pore fluid is in a solid state even if the presence of unfrozen fluid affects mass redistribution with temperature and pressure changes. A way to indirectly understand if the permafrost constitutes fully frozen bearing sediments, i.e., not unfrozen fluid, is to determine if the temperature remains almost constant with depth. The thickness of permafrost is mainly controlled by the geothermal heat flow and by the lithology [1]. As reported by several authors, gas hydrate can exist below permafrost, but also within it (Reference [19] and references therein), due to the temperature and pressure conditions favorable to its stability sustained by the presence of permafrost (e.g., References [20,21]).

If the surface temperature increases due to climate change, gas hydrate could dissociate, releasing large quantities of methane into the atmosphere [8]. In addition, during warm periods, the sea level increases and the permafrost slowly degrades beneath warm and salty water. The subaqueous permafrost (SAP) in marine and lacustrine environments differs from onshore permafrost because it is generally relict, warmer, and degrading (e.g., Reference [1]); however, several tens of thousands of years are necessary to melt it. Presently, SAP exists both in marine (e.g., Reference [22] and references therein) and lacustrine [23–26] environments, as confirmed by drilling (e.g., Reference [22]), and there is increasing interest in better understanding its response to changes in climate as indicated by several studies [22,27–32]. The most studied SAP is located in the Arctic shelf, and several authors (e.g., References [1,3–5,27,29]) suggested that the SAP thickness could reach locally 1000 m with coexistence of fully frozen and unfrozen fluid-bearing sediments. Note that the integration of sophisticated data acquisition techniques is required to detect the presence of SAP, but the extreme environmental conditions hinder the acquisition of data and, thus, the mapping of SAP thickness and distribution [33]. Cleary, this topic remains a challenge for the scientific and industrial communities, and modeling offers a useful alternative to overcome the lack of data [34–37].

Hydrate located in SAP is considered only a small fraction of the global hydrate inventory, but it is suggested to be a highly susceptible reservoir to the effects of global warming [38–40]. Clearly, degradation of SAP and the consequent destabilization of gas hydrate could increase the flux of methane to the ocean and, perhaps, to the atmosphere. Some authors [4,31] suggested that SAP-associated gas hydrate deposits are present at water depths of ~120–130 m. Generally, permafrost is composed of fully frozen bearing sediments for water depths of less than 60 m and both frozen and unfrozen fluid-bearing sediments for greater water depths [22,41]. Sub-aerial emergence of portions of the Arctic continental shelf over repeated Pleistocene glaciations exposed the shelf to temperature conditions, which favored the formation of permafrost and gas hydrate. After the Last Glacial Maximum (~19 ka), coastal inundation from sea-level rise [42,43] thawed the SAP across the Arctic (e.g., References [30,38]). Thus, some authors [44,45] suggested that "relict" permafrost and gas hydrate may exist on the continental shelf of the Arctic Ocean, even if a limited number of direct measurements of permafrost occurrence on the shelf exist. At the West Yamal shelf, high-resolution seismic data indicate a continuous SAP extending to water depths up to ~20 m offshore, and a presence of both ice- and unfrozen fluid-bearing sediments extending further offshore to ~115 m water depth [31]. Similarly, in the western Laptev Sea, evidence from a coastal and offshore drilling program confirms the existence of frozen sediments on the shelf [46] with a discontinuous SAP controlled by the dynamics of coastal inundation. In this site, the presence of unfrozen and saline permafrost suggests that permafrost may not be as cold or thick as predicted by thermal modeling. As noticed by Romanovskii et al. [29], the reduction of the thickness of the fully frozen bearing sediments is more pronounced than the reduction of the total thickness of SAP during transgressions. Temperature changes in climate and transgression–regression cycles may affect the thicknesses of the SAP and gas hydrate stability zone (GHSZ) differently. In the East Siberian Seas, at present, permafrost degradation may be occurring in the outer part of the shelf, whereas GHSZ may be stable or even thickening, indicating that the dynamic of permafrost thickness and the variation of the GHSZ are similar but not identical [29]. Several authors [5,31,32] suggested that shallow permafrost sediments in some areas of the Arctic shelf are charged with methane gas, and sustained warming may increase the gas-venting rate in the future.

SAP can be present in the lacustrine environment even if it is highly sensitive to climate change [46]. Warmer climate accelerates the complete permafrost thaw and enhances seasonal flow within the active layer. Hydrate could also exist in lakes with permafrost deposits, and its presence is strongly controlled by lithology, porosity, and lake size and shape, as suggested by Majorowicz et al. [27]. They modeled the behavior of talik permafrost and gas hydrate below shallow lakes and compared the results with similar models of the Beaufort Mackenzie Basin. In particular, they suggested that, below a lake of any size, where the underlying lithology is sand, the change on thermal conditions only cannot produce a spread talik or dissociate significant gas hydrate accumulations, but just a talik of about a few tens of meters. Regarding the effect of the porosity, their results suggest that permafrost degradation is facilitated for porosities <40%, and, for higher porosities, gas hydrate can be stable even where deep taliks formed [27]. Data from the Qalluuraq Lake indicate a very high concentration of methane in the seepage gas, which could be related to hydrate dissociation [47,48]. Here, we propose a simple empirical approach that allows assessing, under steady-state conditions, if hydrate below SAP could be stable for different thermodynamic conditions typical of SAP in shallow waters both in marine and lacustrine environments. To our knowledge, this is the first empirical method that allows a quick look and easy initial estimation of the conditions sufficient to have the stability of hydrate below SAP in absence of direct geological or geophysical data.

2. Materials and Methods

We evaluated the sufficient conditions to have hydrate stability below the bottom of the SAP under steady-state conditions (Figure 1). We numerically estimated the intersection between the hydrate stability curve and the temperature profile versus pressure below sea level in order to obtain the pressure at the GHSZ base. We used the following input parameters: (i) water depths ranging from 50 to 150 m, (ii) SAP thickness from 0 to 500 m, (iii) saturation of ice in the SAP from 80 to 100%, (iv) SAP temperature of −1, −1.5, or −2 °C, (v) geothermal gradient (GG) from 20 to 40 °C/km, and (vi) water salinity of 0 (freshwater) or 3.5 wt.% (brine). These parameter ranges are based on the literature where SAP was identified [22,29–32]. Portnov et al. [44] reported the presence of SAP in shallow waters at about 20 m in the South Kara Sea shelf, but SAP is discontinuous for water depths greater than about 60 m [29]. In the absence of SAP, hydrate is stable in the Arctic Ocean for water depths greater than 250 m [22,49–51]. Romanovskii et al. [29] reported a maximum SAP thickness of about 700 m, even if other authors considered a maximum permafrost thickness up to 500 m [31]. We considered a range of geothermal gradients, which span the variability in thermal structure reported globally in SAP sediments (e.g., References [29,32,52]). We imposed an annual mean SAP temperature from −1 °C to −2 °C (e.g., References [22,53]), which needs to be at or below zero to allow the formation of ice in the lacustrine and marine permafrost zones, respectively [22]. The pore water was assumed to be freshwater and brine to model the lacustrine and marine environments, respectively.

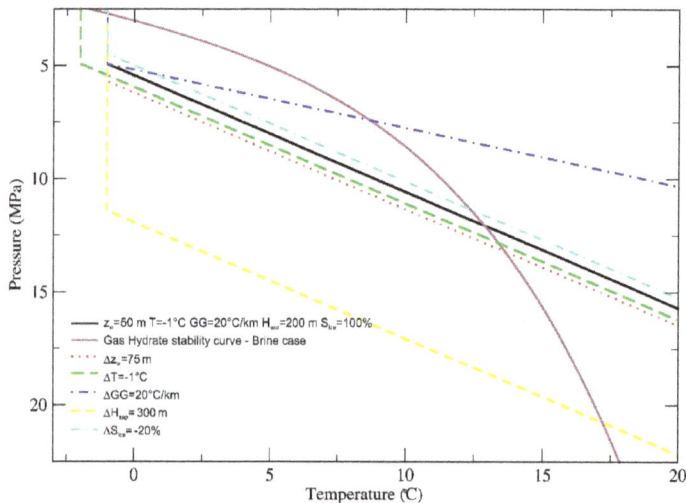

Figure 1. Thickness of the gas hydrate stability zone (GHSZ) for different combinations of controlling parameters in the marine environment (brine with 3.5 wt.% salinity). The black solid line represents the reference case. The other curves represent a variation of one parameter with respect to the reference case. Red dotted line: $\Delta z_w = 75$ m. Green dashed line: $\Delta T = -1$ °C. Yellow dashed line: $\Delta H_{SAP} = 300$ m. Light-blue dashed-dotted line: $\Delta S_{ice} = -20\%$. Δ means variation with respect to the reference case. Explanation of the parameters is reported in the text.

The saturation of ice-bearing permafrost, which indicates the thickness of ice-bearing permafrost with respect to the total thickness of SAP, was introduced to model the discontinuity of the SAP [5]. We assumed a minimum ice-bearing SAP saturation of 80% because this is the minimum amount of ice necessary to reach a stable permafrost system [31]. For example, if the SAP is assumed equal to 100 m and composed of 80 m of fully frozen bearing sediments and 20 m of unfrozen fluid-bearing sediments, it means that there is 80% ice saturation. Note that we assumed the thickness of the active layer (ground zone that freezes and thaws each year) to be much thinner than the thickness of perennial permafrost and, thus, the active layer was not modeled. Regarding gas composition, most parts of the Arctic permafrost are composed of pure methane (e.g., References [4,29,51,54]), although other gases such as CO_2 could also be present at the base of the permafrost driven by vertical fluid flow from deep sources.

We evaluated the hydrate stability by using Moridis et al.'s [55] stability boundary for pure methane hydrate. This is a conservative assumption, because the base of methane HSZ is shallower compared to that of hydrate formed by a mixture of different type of gases. Moridis's [55] stability boundary is defined for pure water; therefore, we applied Dickens and Quinby-Hunt's [56] relationship to account for a water salinity of 3.5% weight total (wt.%) of sodium chloride. For the conversion, we assume a pure water fusion temperature of 273.2 K, a pure water fusion enthalpy of 6008 J·mol^{-1}, an enthalpy of hydrate dissociation of 54,200 J·mol^{-1}, six water molecules in the hydrate formula ($CH_4 \cdot 6H_2O$), and Blangden's law [57] to calculate the fusion temperature of water in an electrolyte solution of 3.5 wt.% salinity. For Blangden's law, we assumed a water cryoscopic constant of 1853 K·g·mol^{-1} and a sodium chloride van't Hoff factor of 2. In permafrost regions, the correlation between pressure and depth is affected by poor data on pressure regime (e.g., Reference [2]). However, some authors underlined that hydrostatic pressures should not be used in permafrost environments because most pores are filled with ice, likely generating a pore pressures above the hydrostatic (e.g., Reference [58]). In this case, the depth of the base of the GHSZ would be deeper than in the hydrostatic case [59]. Based on these considerations, we assumed the following pressure formulation modified after Liu et al. [2]:

$$P = P_{SAP} + \rho_w \, g \, (H - H_{SAP}),$$

$$P_{SAP} = P_a + P_w + \rho_s \, g \, H_{SAP} \, S_{ice} + \rho_w \, g \, H_{SAP} \, (1 - S_{ice}), \tag{1}$$

$$P_w = \rho_w \, g \, z_w,$$

where P is the pore pressure below SAP at the depth H below seabed, P_{SAP} is the pore pressure at the base of the SAP with a thickness H_{SAP}, P_a is the atmospheric pressure, P_w is the hydrostatic pressure of the water column above sediments (z_w), ρ_w is the water density (1046 kg/m^3, e.g., References [50,59]), ρ_s is the bulk sediment density of the fully frozen SAP assumed as 2200 kg/m^3, g is the gravitational acceleration constant (9.81 m/s^2), and S_{ice} is the saturation of ice-bearing SAP.

To model the temperature versus depth profile, we propose the following formula:

$$T = T_{SAP} + \Delta T_{SAP} + GG \, (H - H_{SAP}), \tag{2}$$

$$\Delta T_{SAP} = GG \, H_{SAP} \, (1 - S_{ice}),$$

where T is the temperature at the depth H below seabed, T_{SAP} is the SAP temperature, and GG is the geothermal gradient. To model the heat flow in the SAP due to the presence of unfrozen water that allows fluid circulation, we included the term ΔT_{SAP} that is the temperature increase in the SAP due to the coexistence of fully frozen and unfrozen fluid-bearing sediments, as given by the parameter S_{ice}. Equation (2) was verified by using well data (4D12 and 4D13) in the East Siberian Arctic shelf (e.g., Reference [60]); the temperature increase from top to bottom of the SAP, evaluated using Equation (2), is in agreement with the temperature measurements.

The hydrate stability curve was compared with the temperature/pressure curve to estimate the depth of the base of the GHSZ for the marine (brine) and lacustrine (freshwater) environments. Figure 1 shows the influence of the controlling parameters (geothermal gradient, SAP temperature, water depth, and saturation of ice-bearing SAP) in the thickness of the GHSZ for SAP in a marine environment. Figure 2 shows the pressure at the base of the GHSZ versus the pressure at the base of SAP for different combinations of the controlling parameters for both marine and lacustrine environments. To easily evaluate the stability of the gas hydrate below the SAP and the depth of its base, we fit the curves in Figure 2 using the following relationship:

$$P_{GHSZ} = a1 + a2\, P_{SAP} + a3\, P_{SAP}^2, \tag{3}$$

where P_{SAP} is the pressure at the base of SAP (see Equation (1)), and P_{GHSZ} is the pressure at the base of the GHSZ, which is given by the intersection of the hydrate stability curve with the temperature/pressure curve of sediments. The parameters a1, a2, and a3 are reported in Tables 1 and 2 for the cases of freshwater and brine, respectively. The expression reported in Equation (3) was the function that better reproduced the theoretical curves (i.e., minimum standard deviation) and simplified the estimation of the hydrate stability thickness below SAP. The fitting was performed by using our codes and the open-source software XMGRACE.

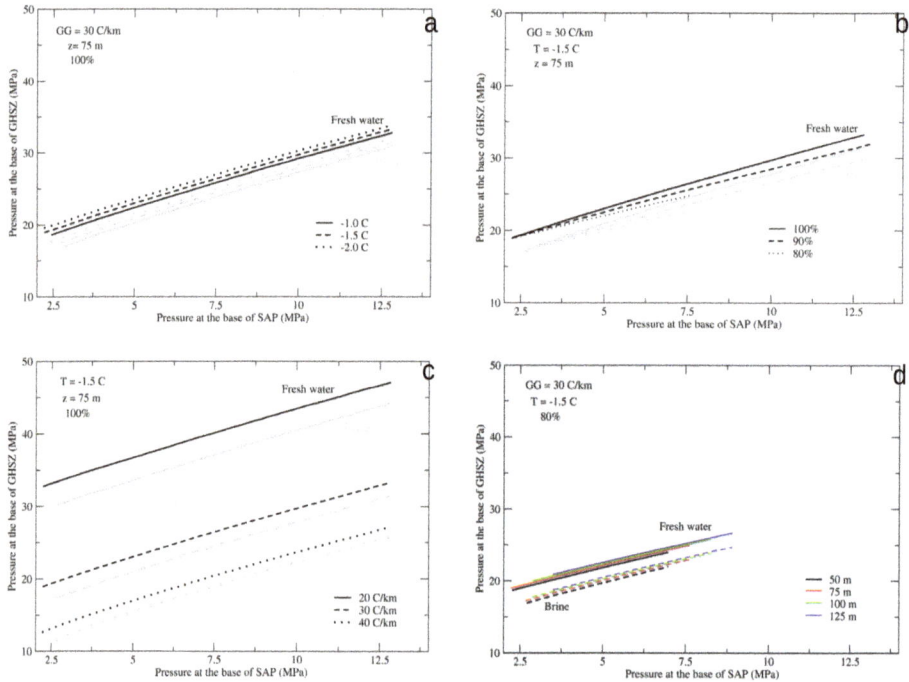

Figure 2. P_{GHSZ} versus P_{SAP} for freshwater and brine cases. In each panel, one parameter is changed, while the others are fixed.

Table 1. Fitting parameters a1, a2, and a3 for modeled bottom water temperatures (T), geothermal gradient (GG), and water depths (WD) for the freshwater case. The standard deviation (SD) is also reported, as well as the range of pressure in which the hydrate stability is satisfied, indicated as Min/Max P_{SAP} and Min/Max P_{GHSZ}.

T (°C)	GG (C/km)	WD (m)	S_{ice} (%)	a1 (MPa)	a2	a3 (MPa^{-1})	SD (MPa)	Min P_{SAP} (MPa)	Max P_{SAP} (MPa)	Min P_{GHSZ} (MPa)	Max P_{GHSZ} (MPa)
−1.0	20	50	100	28.7	1.47	−0.008	0.008	2.5	12.2	32.2	45.5
−1.0	20	50	90	28.9	1.30	−0.006	0.009	2.5	12.4	32.1	44.0
−1.0	20	50	80	29.0	1.17	−0.008	0.004	2.5	7.0	31.9	36.8
−1.0	20	75	100	28.7	1.47	−0.007	0.012	2.4	12.8	32.2	46.3
−1.0	20	75	90	29.0	1.30	−0.006	0.008	2.5	13.0	32.2	44.9
−1.0	20	75	80	29.3	1.13	−0.005	0.003	2.5	7.6	32.1	37.7
−1.0	20	100	100	28.7	1.46	−0.007	0.014	2.9	13.4	32.8	47.1
−1.0	20	100	90	29.2	1.29	−0.006	0.013	2.9	13.6	32.8	45.7
−1.0	20	100	80	29.6	1.12	−0.004	0.003	2.9	8.2	32.8	38.6
−1.0	20	125	100	28.7	1.45	−0.006	0.011	3.5	14.1	33.7	47.9
−1.0	20	125	90	29.3	1.29	−0.006	0.011	3.5	14.3	33.7	46.5
−1.0	20	125	80	29.8	1.12	−0.004	0.003	3.5	8.9	33.7	39.4
−1.0	30	50	100	14.9	1.56	−0.014	0.014	2.5	12.2	18.7	31.9
−1.0	30	50	90	15.0	1.44	−0.017	0.002	2.5	8.6	18.5	26.3
−1.0	30	50	80	15.1	1.33	−0.021	0.001	2.5	5.0	18.3	21.2
−1.0	30	75	100	15.0	1.55	−0.013	0.022	2.4	12.8	18.6	32.8
−1.0	30	75	90	15.2	1.43	−0.015	0.005	2.5	9.3	18.6	27.1
−1.0	30	75	80	15.4	1.30	−0.018	0.001	2.5	5.6	18.5	22.1
−1.0	30	100	100	15.1	1.53	−0.012	0.017	2.9	13.4	19.3	33.6
−1.0	30	100	90	15.3	1.41	−0.014	0.004	2.9	9.9	19.3	28.0
−1.0	30	100	80	15.7	1.26	−0.014	0.001	2.9	6.2	19.2	23.0
−1.0	30	125	100	15.1	1.52	−0.011	0.013	3.5	14.1	20.3	34.3
−1.0	30	125	90	15.5	1.38	−0.011	0.002	3.5	10.5	20.2	28.8
−1.0	30	125	80	16.0	1.25	−0.014	0.001	3.5	6.9	20.2	23.9
−1.0	40	50	100	8.7	1.65	−0.019	0.045	2.5	12.2	12.6	26.0
−1.0	40	50	90	8.5	1.65	−0.034	0.001	2.5	6.9	12.4	18.3
−1.0	40	50	80	8.6	1.57	−0.044	0.000	2.5	4.1	12.2	14.2
−1.0	40	75	100	8.8	1.63	−0.018	0.065	2.4	12.8	12.6	26.7
−1.0	40	75	90	8.8	1.60	−0.029	0.003	2.5	7.5	12.5	19.2
−1.0	40	75	80	9.0	1.49	−0.035	0.000	2.5	4.7	12.4	15.2
−1.0	40	100	100	8.9	1.60	−0.016	0.047	2.9	13.4	13.3	27.5
−1.0	40	100	90	9.0	1.55	−0.024	0.002	2.9	8.1	13.2	20.0
−1.0	40	100	80	9.2	1.49	−0.036	0.000	2.9	5.3	13.2	16.1
−1.0	40	125	100	9.1	1.55	−0.014	0.026	3.5	14.1	14.3	28.3
−1.0	40	125	90	9.3	1.49	−0.020	0.001	3.5	8.8	14.3	20.9
−1.0	40	125	80	9.7	1.37	−0.024	0.000	3.5	6.0	14.2	17.1
−1.5	20	50	100	29.6	1.45	−0.007	0.011	2.2	12.2	32.9	46.3
−1.5	20	50	90	29.9	1.29	−0.006	0.007	2.3	12.4	32.7	44.9
−1.5	20	50	80	30.1	1.12	−0.005	0.004	2.3	9.7	32.6	40.5
−1.5	20	75	100	29.6	1.46	−0.007	0.014	2.2	12.8	32.8	47.1
−1.5	20	75	90	30.0	1.28	−0.006	0.013	2.2	13.0	32.8	45.7
−1.5	20	75	80	30.3	1.13	−0.005	0.006	2.2	10.3	32.8	41.4
−1.5	20	100	100	29.7	1.44	−0.006	0.011	2.9	13.4	33.7	47.9
−1.5	20	100	90	30.1	1.28	−0.006	0.011	2.9	13.6	33.7	46.5
−1.5	20	100	80	30.5	1.12	−0.005	0.006	2.9	10.9	33.7	42.2
−1.5	20	125	100	29.7	1.44	−0.006	0.009	3.5	14.1	34.6	48.7
−1.5	20	125	90	30.2	1.27	−0.005	0.009	3.5	14.3	34.6	47.4
−1.5	20	125	80	30.7	1.12	−0.005	0.008	3.5	11.6	34.6	43.0
−1.5	30	50	100	15.6	1.54	−0.013	0.012	2.2	12.2	19.0	32.5
−1.5	30	50	90	15.9	1.37	−0.011	0.012	2.3	12.4	18.8	31.1
−1.5	30	50	80	15.9	1.27	−0.015	0.001	2.3	7.0	18.7	24.0
−1.5	30	75	100	15.7	1.53	−0.012	0.024	2.2	12.8	19.0	33.3
−1.5	30	75	90	16.0	1.35	−0.010	0.018	2.2	13.0	18.9	31.9
−1.5	30	75	80	16.2	1.25	−0.013	0.001	2.2	7.6	18.9	24.9
−1.5	30	100	100	15.7	1.51	−0.011	0.011	2.9	13.4	19.9	34.1

Table 1. *Cont.*

T (°C)	GG (C/km)	WD (m)	S_{ice} (%)	a1 (MPa)	a2	a3 (MPa^{-1})	SD (MPa)	Min P_{SAP} (MPa)	Max P_{SAP} (MPa)	Min P_{GHSZ} (MPa)	Max P_{GHSZ} (MPa)
−1.5	30	100	90	16.2	1.34	−0.009	0.009	2.9	13.6	19.9	32.7
−1.5	30	100	80	16.5	1.21	−0.011	0.002	2.9	8.2	19.9	25.7
−1.5	30	125	100	15.8	1.49	−0.010	0.015	3.5	14.1	20.9	34.9
−1.5	30	125	90	16.3	1.32	−0.008	0.010	3.5	14.3	20.9	33.5
−1.5	30	125	80	16.8	1.19	−0.009	0.001	3.5	8.9	20.8	26.6
−1.5	40	50	100	9.3	1.62	−0.018	0.052	2.2	12.2	12.7	26.4
−1.5	40	50	90	9.3	1.52	−0.022	0.007	2.3	9.5	12.6	21.8
−1.5	40	50	80	9.3	1.45	−0.032	0.000	2.3	5.4	12.5	16.2
−1.5	40	75	100	9.3	1.61	−0.018	0.060	2.2	12.8	12.7	27.2
−1.5	40	75	90	9.5	1.49	−0.020	0.012	2.2	10.1	12.7	22.6
−1.5	40	75	80	9.6	1.42	-0.029	0.001	2.2	6.0	12.7	17.2
-1.5	40	100	100	9.5	1.57	-0.015	0.044	2.9	13.4	13.8	27.9
−1.5	40	100	90	9.8	1.45	−0.017	0.009	2.9	10.8	13.7	23.4
−1.5	40	100	80	10.0	1.37	−0.024	0.000	2.9	6.7	13.7	18.1
−1.5	40	125	100	9.6	1.54	−0.013	0.023	3.5	14.1	14.8	28.7
−1.5	40	125	90	10.0	1.40	−0.014	0.007	3.5	11.4	14.7	24.3
−1.5	40	125	80	10.4	1.28	−0.016	0.001	3.5	7.3	14.7	18.9
−2.0	20	50	100	30.6	1.44	−0.007	0.013	2.0	12.2	33.4	47.1
−2.0	20	50	90	30.8	1.27	−0.005	0.011	2.0	12.4	33.4	45.7
−2.0	20	50	80	31.0	1.12	−0.005	0.009	2.0	12.6	33.3	44.3
−2.0	20	75	100	30.6	1.44	−0.006	0.011	2.2	12.8	33.7	47.9
−2.0	20	75	90	30.9	1.27	−0.006	0.011	2.2	13.0	33.7	46.5
−2.0	20	75	80	31.2	1.11	−0.004	0.008	2.2	13.2	33.7	45.1
−2.0	20	100	100	30.6	1.44	−0.006	0.009	2.9	13.4	34.6	48.7
−2.0	20	100	90	31.0	1.26	−0.005	0.009	2.9	13.6	34.6	47.4
−2.0	20	100	80	31.4	1.11	−0.004	0.011	2.9	13.8	34.6	46.0
−2.0	20	125	100	30.6	1.42	−0.006	0.012	3.5	14.1	35.6	49.5
−2.0	20	125	90	31.2	1.26	−0.005	0.013	3.5	14.3	35.5	48.2
−2.0	20	125	80	31.7	1.10	−0.004	0.009	3.5	14.5	35.5	46.8
−2.0	30	50	100	16.3	1.52	−0.012	0.016	2.0	12.2	19.3	33.0
−2.0	30	50	90	16.5	1.35	−0.010	0.013	2.0	12.4	19.2	31.7
−2.0	30	50	80	16.7	1.20	−0.010	0.002	2.3	8.8	19.4	26.5
−2.0	30	75	100	16.4	1.51	−0.011	0.013	2.2	12.8	19.6	33.8
−2.0	30	75	90	16.7	1.34	−0.010	0.016	2.2	13.0	19.6	32.5
−2.0	30	75	80	16.9	1.20	−0.010	0.003	2.2	9.4	19.5	27.3
−2.0	30	100	100	16.4	1.50	−0.011	0.013	2.9	13.4	20.6	34.6
−2.0	30	100	90	16.8	1.33	−0.009	0.008	2.9	13.6	20.5	33.3
−2.0	30	100	80	17.2	1.17	−0.008	0.003	2.9	10.0	20.5	28.1
−2.0	30	125	100	16.5	1.47	−0.009	0.008	3.5	14.1	21.5	35.4
−2.0	30	125	90	17.0	1.31	−0.008	0.009	3.5	14.3	21.5	34.1
−2.0	30	125	80	17.5	1.16	−0.008	0.003	3.5	10.7	21.4	29.0
−2.0	40	50	100	9.9	1.60	−0.018	0.053	2.0	12.2	12.9	26.7
−2.0	40	50	90	10.1	1.43	−0.015	0.038	2.0	12.4	12.8	25.4
−2.0	40	50	80	10.0	1.36	−0.023	0.001	2.3	7.0	13.0	18.4
−2.0	40	75	100	9.9	1.58	−0.016	0.047	2.2	12.8	13.3	27.5
−2.0	40	75	90	10.2	1.41	−0.014	0.034	2.2	13.0	13.2	26.2
−2.0	40	75	80	10.3	1.33	−0.021	0.002	2.2	7.6	13.2	19.3
−2.0	40	100	100	10.1	1.54	−0.014	0.026	2.9	13.4	14.3	28.3
−2.0	40	100	90	10.5	1.37	−0.012	0.021	2.9	13.6	14.3	27.0
−2.0	40	100	80	10.7	1.28	−0.017	0.001	2.9	8.2	14.2	20.1
−2.0	40	125	100	10.2	1.51	−0.012	0.025	3.5	14.1	15.3	29.1
−2.0	40	125	90	10.7	1.34	−0.010	0.014	3.5	14.3	15.2	27.8
−2.0	40	125	80	11.1	1.23	−0.013	0.002	3.5	8.9	15.2	21.0

Table 2. Fitting parameters a1, a2, and a3 for modeled bottom water temperatures (T), geothermal gradient (GG), and water depths (WD) for the pore water with a 3.5 wt.% salinity (brine). The standard deviation (SD) is also reported, as well as the range of pressure in which the hydrate stability is satisfied, indicated as Min/Max P_{SAP} and Min/Max P_{GHSZ}.

T (°C)	GG (°C/km)	WD (m)	S_{ice} (%)	a1 (MPa)	a2	a3 (MPa^{-1})	SD (MPa)	Min P_{SAP} (MPa)	Max P_{SAP} (MPa)	Min P_{GHSZ} (MPa)	Max P_{GHSZ} (MPa)
−1.0	20	50	100	25.3	1.53	−0.009	0.012	2.9	12.2	29.6	42.5
−1.0	20	50	90	25.5	1.36	−0.008	0.009	2.9	12.4	29.4	41.1
−1.0	20	50	80	25.6	1.23	−0.010	0.002	2.9	7.0	29.2	33.7
−1.0	20	75	100	25.3	1.53	−0.010	0.007	2.9	12.8	29.6	43.3
−1.0	20	75	90	25.6	1.36	−0.008	0.010	2.9	13.0	29.5	41.9
−1.0	20	75	80	25.9	1.21	−0.009	0.002	2.9	7.6	29.3	34.6
−1.0	20	100	100	25.3	1.53	−0.009	0.014	2.9	13.4	29.6	44.2
−1.0	20	100	90	25.8	1.34	−0.007	0.012	2.9	13.6	29.6	42.8
−1.0	20	100	80	26.2	1.19	−0.007	0.003	2.9	8.2	29.5	35.5
−1.0	20	125	100	25.4	1.51	−0.008	0.014	3.5	14.1	30.5	45.0
−1.0	20	125	90	25.9	1.34	−0.007	0.009	3.5	14.3	30.5	43.6
−1.0	20	125	80	26.4	1.18	−0.006	0.003	3.5	8.9	30.5	36.4
−1.0	30	50	100	12.5	1.65	−0.017	0.021	2.9	12.2	17.1	30.1
−1.0	30	50	90	12.5	1.55	−0.021	0.003	2.9	8.6	16.8	24.3
−1.0	30	50	80	12.5	1.47	−0.030	0.000	2.9	5.0	16.6	19.0
−1.0	30	75	100	12.5	1.63	−0.016	0.032	2.9	12.8	17.0	30.9
−1.0	30	75	90	12.7	1.52	−0.019	0.002	2.9	9.3	16.9	25.2
−1.0	30	75	80	12.8	1.45	−0.029	0.000	2.9	5.6	16.8	20.0
−1.0	30	100	100	12.6	1.63	−0.015	0.031	2.9	13.4	17.0	31.7
−1.0	30	100	90	12.9	1.50	−0.017	0.004	2.9	9.9	17.0	26.0
−1.0	30	100	80	13.1	1.39	−0.022	0.001	2.9	6.2	17.0	21.0
−1.0	30	125	100	12.7	1.59	−0.013	0.024	3.5	14.1	18.1	32.5
−1.0	30	125	90	13.1	1.47	−0.015	0.003	3.5	10.5	18.0	26.9
−1.0	30	125	80	13.5	1.34	−0.018	0.001	3.5	6.9	18.0	21.9
−1.0	40	50	100	6.7	1.76	−0.024	0.073	2.9	12.2	11.5	24.6
−1.0	40	50	90	6.2	1.87	−0.048	0.003	2.9	6.9	11.2	16.8
−1.0	40	50	80	5.7	2.00	−0.085	0.000	2.9	4.1	10.9	12.5
−1.0	40	75	100	6.8	1.74	−0.023	0.092	2.9	12.8	11.5	25.4
−1.0	40	75	90	6.5	1.81	−0.042	0.005	2.9	7.5	11.3	17.7
−1.0	40	75	80	6.3	1.86	−0.069	0.000	2.9	4.7	11.1	13.5
−1.0	40	100	100	6.9	1.71	−0.021	0.121	2.9	13.4	11.4	26.2
−1.0	40	100	90	6.7	1.74	−0.036	0.006	2.9	8.1	11.4	18.6
−1.0	40	100	80	6.9	1.70	−0.049	0.000	2.9	5.3	11.4	14.5
−1.0	40	125	100	7.1	1.66	−0.018	0.062	3.5	14.1	12.6	27.0
−1.0	40	125	90	7.2	1.64	−0.028	0.003	3.5	8.8	12.5	19.4
−1.0	40	125	80	7.6	1.52	−0.032	0.000	3.5	6.0	12.5	15.5
−1.5	20	50	100	26.3	1.52	−0.009	0.007	2.7	12.2	30.3	43.3
−1.5	20	50	90	26.5	1.34	−0.008	0.009	2.7	12.4	30.1	41.9
−1.5	20	50	80	26.7	1.19	−0.007	0.005	2.7	9.7	29.9	37.5
−1.5	20	75	100	26.3	1.51	−0.009	0.012	2.7	12.8	30.2	44.2
−1.5	20	75	90	26.6	1.33	−0.007	0.011	2.7	13.0	30.1	42.8
−1.5	20	75	80	26.9	1.19	−0.008	0.004	2.7	10.3	30.0	38.3
−1.5	20	100	100	26.3	1.50	−0.008	0.014	2.9	13.4	30.5	45.0
−1.5	20	100	90	26.8	1.33	−0.007	0.009	2.9	13.6	30.5	43.6
−1.5	20	100	80	27.2	1.17	−0.006	0.006	2.9	10.9	30.5	39.2
−1.5	20	125	100	26.4	1.49	−0.008	0.010	3.5	14.1	31.5	45.8
−1.5	20	125	90	26.9	1.32	−0.006	0.015	3.5	14.3	31.4	44.5
−1.5	20	125	80	27.4	1.16	−0.006	0.005	3.5	11.6	31.4	40.0
−1.5	30	50	100	13.3	1.61	−0.016	0.020	2.7	12.2	17.4	30.6
−1.5	30	50	90	13.4	1.44	−0.014	0.016	2.7	12.4	17.2	29.2
−1.5	30	50	80	13.4	1.35	−0.019	0.001	2.7	7.0	17.0	22.0
−1.5	30	75	100	13.3	1.61	−0.016	0.022	2.7	12.8	17.4	31.4
−1.5	30	75	90	13.6	1.43	−0.013	0.014	2.7	13.0	17.3	30.1
−1.5	30	75	80	13.7	1.35	−0.019	0.001	2.7	7.6	17.2	22.9

Table 2. *Cont.*

T (°C)	GG (°C/km)	WD (m)	S_ice (%)	a1 (MPa)	a2	a3 (MPa⁻¹)	SD (MPa)	Min P_SAP (MPa)	Max P_SAP (MPa)	Min P_GHSZ (MPa)	Max P_GHSZ (MPa)
−1.5	30	100	100	13.3	1.59	−0.014	0.029	2.9	13.4	17.7	32.2
−1.5	30	100	90	13.8	1.41	−0.012	0.019	2.9	13.6	17.7	30.9
−1.5	30	100	80	14.0	1.32	−0.017	0.002	2.9	8.2	17.6	23.7
−1.5	30	125	100	13.4	1.56	−0.012	0.019	3.5	14.1	18.7	33.1
−1.5	30	125	90	14.0	1.39	−0.011	0.014	3.5	14.3	18.7	31.7
−1.5	30	125	80	14.3	1.28	−0.013	0.002	3.5	8.9	18.7	24.6
−1.5	40	50	100	7.3	1.72	−0.023	0.069	2.7	12.2	11.7	25.0
−1.5	40	50	90	7.2	1.65	−0.029	0.015	2.7	9.5	11.4	20.4
−1.5	40	50	80	7.0	1.68	−0.047	0.001	2.7	5.4	11.2	14.7
−1.5	40	75	100	7.4	1.70	−0.021	0.089	2.7	12.8	11.7	25.8
−1.5	40	75	90	7.5	1.62	−0.026	0.021	2.7	10.1	11.5	21.2
−1.5	40	75	80	7.4	1.61	−0.041	0.001	2.7	6.0	11.4	15.6
−1.5	40	100	100	7.5	1.67	−0.019	0.089	2.9	13.4	12.0	26.6
−1.5	40	100	90	7.7	1.57	−0.023	0.017	2.9	10.8	12.0	22.1
−1.5	40	100	80	7.9	1.52	−0.033	0.001	2.9	6.7	11.9	16.6
−1.5	40	125	100	7.7	1.62	−0.016	0.047	3.5	14.1	13.1	27.4
−1.5	40	125	90	8.1	1.50	−0.018	0.009	3.5	11.4	13.1	22.9
−1.5	40	125	80	8.3	1.45	−0.027	0.000	3.5	7.3	13.0	17.4
−2.0	20	50	100	27.3	1.49	−0.008	0.010	2.5	12.2	30.9	44.2
−2.0	20	50	90	27.5	1.32	−0.007	0.009	2.5	12.4	30.7	42.8
−2.0	20	50	80	27.7	1.16	−0.006	0.007	2.5	12.6	30.6	41.4
−2.0	20	75	100	27.3	1.48	−0.008	0.012	2.4	12.8	30.9	45.0
−2.0	20	75	90	27.6	1.32	−0.007	0.009	2.5	13.0	30.8	43.6
−2.0	20	75	80	27.9	1.15	−0.006	0.009	2.5	13.2	30.7	42.2
−2.0	20	100	100	27.3	1.48	−0.008	0.010	2.9	13.4	31.5	45.8
−2.0	20	100	90	27.7	1.31	−0.006	0.015	2.9	13.6	31.4	44.5
−2.0	20	100	80	28.2	1.15	−0.005	0.008	2.9	13.8	31.4	43.0
−2.0	20	125	100	27.4	1.47	−0.007	0.014	3.5	14.1	32.4	46.7
−2.0	20	125	90	27.9	1.30	−0.006	0.008	3.5	14.3	32.4	45.3
−2.0	20	125	80	28.4	1.13	−0.005	0.011	3.5	14.5	32.4	43.9
−2.0	30	50	100	14.0	1.60	−0.015	0.019	2.5	12.2	17.7	31.1
−2.0	30	50	90	14.2	1.42	−0.013	0.014	2.5	12.4	17.6	29.8
−2.0	30	50	80	14.2	1.31	−0.016	0.002	2.5	8.8	17.4	24.5
−2.0	30	75	100	14.0	1.58	−0.014	0.024	2.4	12.8	17.7	32.0
−2.0	30	75	90	14.3	1.41	−0.012	0.016	2.5	13.0	17.6	30.6
−2.0	30	75	80	14.5	1.28	−0.014	0.003	2.5	9.4	17.6	25.3
−2.0	30	100	100	14.1	1.56	−0.013	0.020	2.9	13.4	18.4	32.8
−2.0	30	100	90	14.5	1.39	−0.011	0.018	2.9	13.6	18.3	31.4
−2.0	30	100	80	14.8	1.26	−0.012	0.003	2.9	10.0	18.3	26.2
−2.0	30	125	100	14.2	1.54	−0.011	0.018	3.5	14.1	19.4	33.6
−2.0	30	125	90	14.7	1.36	−0.010	0.012	3.5	14.3	19.4	32.2
−2.0	30	125	80	15.1	1.23	−0.011	0.003	3.5	10.7	19.3	27.0
−2.0	40	50	100	8.0	1.69	−0.022	0.067	2.5	12.2	11.9	25.4
−2.0	40	50	90	8.1	1.52	−0.019	0.052	2.5	12.4	11.7	24.1
−2.0	40	50	80	7.9	1.51	−0.031	0.001	2.7	7.0	11.8	16.9
−2.0	40	75	100	8.0	1.67	−0.020	0.091	2.4	12.8	11.8	26.2
−2.0	40	75	90	8.3	1.50	−0.018	0.067	2.5	13.0	11.8	24.9
−2.0	40	75	80	8.2	1.48	−0.029	0.003	2.5	7.6	11.7	17.8
−2.0	40	100	100	8.1	1.63	−0.018	0.062	2.9	13.4	12.6	27.0
−2.0	40	100	90	8.6	1.46	−0.015	0.045	2.9	13.6	12.5	25.6
−2.0	40	100	80	8.7	1.41	−0.024	0.001	2.9	8.2	12.5	18.7
−2.0	40	125	100	8.3	1.59	−0.015	0.040	3.5	14.1	13.6	27.7
−2.0	40	125	90	8.9	1.41	−0.013	0.026	3.5	14.3	13.6	26.4
−2.0	40	125	80	9.1	1.34	−0.018	0.002	3.5	8.9	13.6	19.5

3. Results

The curves in Figure 2 summarize the results of our empirical approach. As can be observed in Tables 1 and 2, the parameter a3 is generally low, indicating a linear relationship between the pressure at the base of SAP and the pressure at the base of the GHSZ. Comparing the results obtained in the freshwater and brine cases, we deduced that a1 was higher for the freshwater case. This trend is clearly observed in Figure 2, in which the intercepts of the freshwater cases were higher. On the other hand, the parameter a2 had an opposite trend, as observed by the highest slopes of the curves related to brine (Figure 2).

Regarding the correlation of a1 and a2 with the other variables, they showed an opposite trend. We noted that (i) a1/a2 increased/decreased if SAP temperature decreased; (ii) a1/a2 decreased/increased if GG increased; (iii) a1/a2 increased/decreased if water depth increased; (iv) a1/a2 increased/decreased if the saturation of ice-bearing SAP decreased.

The ranges of pressures at the base of SAP (P_{SAP}) and GHSZ (P_{GHSZ}) are reported for freshwater and brine in Tables 1 and 2, respectively. Figure 1 shows that the range of P_{SAP} in the freshwater case was smaller than that in the brine case; on the contrary, the range of P_{GHSZ} in the brine case was smaller than that in the freshwater case. Regarding the pressure dependence with GG, T_{SAP}, z_w, and S_{ice}, the minimum P_{SAP} at which the hydrate was stable was independent of GG, while it increased if T_{SAP} or z_w increased and if S_{ice} decreased. The maximum P_{SAP} was independent of both GG and T_{SAP}, while it increased if water depth increased. The minimum and maximum P_{GHSZ} increased if (i) GG or T_{SAP} decreased, or (ii) z_w or S_{ice} increased.

4. Conclusions

In this paper, we proposed an empirical approach that considers the dominant physical parameters controlling the stability of hydrate under steady-state conditions in SAP environments. It is a simple method that can be easily and reliably applied to assess if the sufficient conditions to have hydrate stability below SAP are satisfied. Because of the growing interest in SAP environments, this approach is particularly useful in SAP areas with environmental conditions that hinder the acquisition of data, to allow an initial and quick estimation of the thickness of the gas hydrate stability zone.

Author Contributions: Conceptualization, U.T., M.G., and H.M.-M.; simulations, U.T.; analysis, U.T. and M.G.; writing—original draft, U.T. and M.G.; writing—review and editing, U.T., M.G., and H.M.-M.

Funding: This research was partially supported by (I) the Italian Ministry of Education, Universities and Research (Decreto MIUR No. 631 dd. 8 August 2016) under the extraordinary contribution for Italian participation in activities related to the international infrastructure PRACE—The Partnership for Advanced Computing in Europe (www.prace-ri.eu) and (2) TALENTS FVG Programme Activity 1-Incoming Mobility Scheme-European Social Fund, Operational Programme 2007–2013, Objective 2 Regional Competitiveness and Employment, Axis 5 Transnational cooperation (decision n. 7629 of 26/11/2013).

Acknowledgments: We thank the three anonymous reviewers for their insightful comments and suggestions that helped us to improve the manuscript.

Conflicts of Interest: The authors declare no conflicts of interest.

References

1. Osterkamp, T.E.; Burn, C.R. Permafrost. In *Encyclopedia of Atmospheric Sciences*; Holton, J.R., Ed.; Academic Press: Oxford, UK, 2003; pp. 1717–1729. [CrossRef]
2. Liu, S.; Jiang, Z.; Liu, H.; Pang, S.; Xia, Z.; Jin, Z.; Wang, J.; Wei, X. The natral-gas exploration prospects of the Nayixiong formation in the Kaixinling-Wuli permafrost, Qinghai-Tibet plateau. *Mar. Pet. Geol.* **2016**, *72*, 179–192. [CrossRef]
3. Osterkamp, T.E. Sub-sea Permafrost. In *Encyclopedia of Ocean Sciences*, 1st ed.; Academic Press: New York, NY, USA, 2001; pp. 2902–2912.
4. Collett, T.S.; Lee, M.W.; Agena, W.F.; Miller, J.J.; Lewis, K.A.; Zyrianova, M.V.; Boswell, R.; Inks, T.L. Permafrost-associated natural gas hydrate occurrences on the Alaska North Slope. *Mar. Petr. Geol.* **2011**, *28*, 279–294. [CrossRef]
5. Frederick, J.M.; Buffett, B.A. Taliks in relict submarine permafrost and gas hydrate deposits: Pathways for methane escape under present and future conditions. *J. Geophys. Res. Earth Surf.* **2014**, *119*, 106–122. [CrossRef]

6. Portnov, A.; Vadakkepuliyambatta, S.; Mienert, J.; Hubbard, A. Ice-sheet-driven methane storage and release in the Arctic. *Nat. Commun.* **2016**, *7*, 101314. [CrossRef]
7. Crémière, A.; Lepland, A.; Chand, S.; Sahy, D.; Condon, D.J.; Noble, S.R.; Martma, T.; Thorsnes, T.; Sauer, S.; Brunstad, H. Timescales of methane seepage on the Norwegian margin following collapse of the Scandinavian Ice Sheet. *Nat. Commun.* **2016**, *7*, 11509. [CrossRef]
8. Kvenvolden, K.A. Gas hydrates—Geological perspective and global change. *Rev. Geophys.* **1993**, *31*, 173–187. [CrossRef]
9. Anderson, L.G.; Björk, G.; Jutterström, S.; Pipko, I.; Shakhova, N.; Semiletov, I.; Wåhlström, I. East Siberian Sea, an Arctic region of very high biogeochemical activity. *Biogeosciences* **2011**, *8*, 1745–1754. [CrossRef]
10. Dallimore, S.R.; Wright, J.F.; Nixon, F.M.; Kurihara, M.; Yamamoto, K.; Fujii, T.; Fujii, K.; Numasawa, M.; Yasuda, M.; Imasato, Y. Geologic and porous media factors affecting the 2007 production response characteristics of the JOGMEC/NRCAN/AURORA Mallik gas hydrate production research well. In Proceedings of the 6th International Conference on Gas Hydrates, Vancouver, BC, Canada, 6–10 July 2008; p. 10.
11. Dallimore, S.R.; Wright, J.F.; Yamamoto, K. Appendix D: Update on Mallik. In *Energy from Gas Hydrates: Assessing the Opportunities and Challenges for Canada*; Council of Canadian Academies: Ottawa, ON, Canada, 2008; pp. 196–200.
12. Moridis, G.J.; Collett, T.S.; Pooladi-Darwish, M.; Hancock, S.; Santamarina, C.; Boswell, R.; Kneafsey, T.; Rutqvist, J.; Kowalsky, M.J.; Reagan, M.T.; et al. Challenges, Uncertainties and Issues Facing Gas Production From Hydrate Deposits in Geologic Systems. *SPE Res. Eval. Eng* **2011**, *14*, 76–112. [CrossRef]
13. Yamamoto, K.; Kanno, T.; Wang, X.-X.; Tamaki, M.; Fujii, T.; Chee, S.-S.; Wang, X.-W.; Pimenov, V.; Shako, V. Thermal responses of a gas hydrate-bearing sediment to a depressurization operation. *R. Soc. Chem.* **2017**, *7*, 5554–5577. [CrossRef]
14. Marín-Moreno, H.; Minshull, T.A.; Westbrook, G.K.; Sinha, B. Estimates of future warming-induced methane emissions from hydrate offshore west Svalbard for a range of climate models. *Geochem. Geophys. Geosyst.* **2015**, *16*, 1307–1323. [CrossRef]
15. Marin-Moreno, H.; Giustiniani, M.; Tinivella, U. The Potential Response of the Hydrate Reservoir in the South Shetland Margin, Antarctic Peninsula, to Ocean Warming over the 21st Century. *Polar Res.* **2015**, *34*, 27443. [CrossRef]
16. Nisbet, E. Some northern sources of atmospheric methane: Production, history, and future implications. *Can. J. Earth Sci.* **1989**, *26*, 1603–1611. [CrossRef]
17. Nisbet, E.G. The end of the ice age. *Can. J. Earth Sci.* **1990**, *27*, 148–157. [CrossRef]
18. Dickens, G.R. Down the Rabbit Hole: Toward appropriate discussion of methane release from gas hydrate systems during the Paleocene-Eocene thermal maximum and other past hyperthermal events. *Clim. Past* **2011**, *7*, 831–846. [CrossRef]
19. Chuvilin, E.; Davletshina, D. Formation and Accumulation of Pore Methane Hydrates in Permafrost: Experimental Modeling. *Geosciences* **2018**, *8*, 467. [CrossRef]
20. Makogon, Y.F. A Gas Hydrate Formation in the Gas Saturated Layers under Low Temperature. *Gas Ind.* **1965**, *5*, 14–15.
21. Chuvilin, E.; Bukhanov, B.; Davletshina, D.; Grebenkin, S.; Istomin, V. Dissociation and Self-Preservation of Gas Hydrates in Permafrost. *Geosciences* **2018**, *8*, 431. [CrossRef]
22. Bogoyavlensky, V.; Kishankov, A.; Yanchevskaya, A.; Bogoyavlensky, I. Forecast of Gas Hydrates Distribution Zones in the Arctic Ocean and Adjacent Offshore Areas. *Geosciences* **2018**, *8*, 453. [CrossRef]
23. Istomin, V.A.; Yakushev, V.S. *Gas Hydrates in Nature*; Nedra: Moscow, Russia, 1992; p. 235. ISBN 5-247-02442-7.
24. Rivkin, F.M.; Levantovskaya, N.P. Dynamics of sub-river channel taliks and the formation of gas hydrates. *Kriosf. Zemli.* **2002**, *6*, 36–42.
25. Kraev, G.; Schulze, E.-D.; Yurova, A.; Kholodov, A.; Chuvilin, E.; Rivkina, E. Cryogenic Displacement and Accumulation of Biogenic Methane in Frozen Soils. *Atmosphere* **2017**, *8*, 105. [CrossRef]
26. Istomin, V.A.; Chuvilin, E.M.; Sergeeva, D.V.; Buhkanov, B.A.; Stanilovskaya, Y.V.; Green, E.; Badetz, C. Thermodynamic calculation of freezing temperature of gas-saturated pore water in talik zones. In Proceedings of the 5th European Conference Permafrost, Chamonix, France, 23 June–1 July 2018; pp. 480–481.
27. Majorowicz, J.; Osadetz, K.; Safanda, J. Models of Talik, Permafrost and Gas Hydrate Histories—Beaufort Mackenzie Basin, Canada. *Energies* **2015**, *8*, 6738–6764. [CrossRef]

28. Chuvilin, E.; Bukhanov, B. Thermal conductivity of frozen sediments containing self-preserved pore gas hydrates at atmospheric pressure: An experimental study. *Geosciences* **2019**, *9*. [CrossRef]
29. Romanovskii, N.N.; Hubberten, H.-W.; Gavrilov, A.V.; Eliseeva, A.A.; Tipenko, G.S. Offshore permafrost and gas hydrate stability zone on the shelf of East Siberian Seas. *Geo Mar. Lett.* **2005**, *25*, 167–182. [CrossRef]
30. Malakhova, V.V.; Eliseev, A.V. Influence of rift zones and thermokarst lakes on the formation of subaqueous permafrost and the stability zone of methane hydrates of the Laptev sea shelf in the Pleistocene. *Ice Snow* **2018**, *58*, 231–242. [CrossRef]
31. Portnov, A.; Mienert, J.; Serov, P. Modeling the evolution of climate sensitive Arctic subsea permafrost in regions of extensive gas expulsion at the West Yamal shelf. *Biogeosciences* **2014**, *119*, 2082–2094. [CrossRef]
32. Malakhova, V.V. Estimation of the subsea permafrost thickness in the Arctic Shelf. In Proceedings of the 24th International Symposium on Atmospheric and Ocean Optics: Atmospheric Physics, Tomsk, Russia, 2–5 July 2018; p. 108337T. [CrossRef]
33. Loktev, A.S.; Tokarev, M.Y.; Chuvilin, E.M. Problems and technologies of offshore permafrost investigation. *Procedia Eng.* **2017**, *189*, 459–465. [CrossRef]
34. Romanovskii, N.N.; Hubberten, H.-W.; Gavrilov, A.V.; Tumskoy, V.E.; Tipenko, G.S.; Grigoriev, M.N.; Siegert, C. Thermokarst and Land-Ocean Interactions, Laptev Sea, Region Russia. *Permafr. Periglac. Process.* **2000**, *11*, 157–162. [CrossRef]
35. Nikolsky, D.; Shakhova, N. Modeling sub-sea permafrost in the East Siberian Arctic Shelf: The Dmitry Laptev Strait. *Environ. Res. Lett.* **2010**, *5*, 015006. [CrossRef]
36. Nikolsky, D.; Romanovsky, V.E.; Romanovskii, N.N.; Kholodov, A.L.; Shakhova, N.E.; Semiletov, I.P. Modeling sub-sea permafrost in the East Siberian Arctic Shelf: The Laptev Sea region. *J. Geophys. Res.* **2012**, *117*, F03028. [CrossRef]
37. Overduin, P.P.; Schneider von Deimling, T.; Miesner, F.; Grigoriev, M.N.; Ruppel, C.; Vasiliev, A.; Lantuit, H.; Juhls, B.; Westermann, S. Submarine Permafrost Map in the Arctic Modeled Using 1-D Transient Heat Flux (SuPerMAP). *J. Geophys. Res. Ocean.* **2019**, *124*. [CrossRef]
38. Ruppel, C. Methane Hydrates and Contemporary Climate Change. *Nat. Educ. Knowl.* **2011**, *3*, 29.
39. Ruppel, C. Permafrost-Associated Gas Hydrate: Is it Really ~1% of the Global System? *J. Chem. Eng. Data* **2014**. [CrossRef]
40. Streletskaya, I.D.; Vasiliev, A.A.; Oblogov, G.E.; Streletskiy, D.A. Methane content in ground ice and sediments of the Kara Sea coast. *Geosciences* **2018**, *8*, 434. [CrossRef]
41. Romanovskii, N.; Hubberten, H.-W.; Gavrilov, A.; Tumskoy, V.; Kholodov, A.L. Permafrost of the east Siberian Arctic shelf and coastal lowlands. *Quat. Sci. Rev.* **2004**, *23*, 1359–1369. [CrossRef]
42. Peltier, W.R. Ice age paleotopography. *Science* **2004**, *265*, 195–201. [CrossRef] [PubMed]
43. Kendall, R.A.; Mitrovica, J.X.; Milne, G.A. On post-glacial sea level—II. Numerical formulation and comparative results on spherically symmetric models. *Geophys. J. Int.* **2005**, *161*, 679–706. [CrossRef]
44. Portnov, A.; Smith, A.J.; Mienert, J.; Cherkashov, G.; Rekant, P.; Semenov, P.; Serov, P.; Vanshtein, B. Offshore permafrost decay and massive seabed methane escape in water depths >20 m at the South Kara Sea shelf. *Geo. Res. Lett.* **2013**, *40*, 1–6. [CrossRef]
45. Chuvilin, E.M.; Yakushev, V.S.; Perlova, E.V. Gas hydrates in the permafrost of Bovanenkovo gas field, Yamal Peninsula, West Siberia. *Polarforschung* **2000**, *68*, 215–219.
46. Wellman, T.P.; Voss, C.I.; Walvoord, M.A. Impacts of climate, lake size, and supra- and sub-permafrost groundwater flow on lake-talik evolution, Yukon Flats, Alaska (USA). *Hydrogeol. J.* **2013**, *21*, 281–298. [CrossRef]
47. Ruo, H.; Wooller, M.J.; Pohlman, J.W.; Quensen, J.; Tiedje, J.M.; Beth Leigh, M. Shifts in Identity and Activity of Methanotrophs in Arctic Lake Sediments in Response to Temperature Changes. *Appl. Environ. Microbiol.* **2012**, *78*, 4715–4723.
48. Wooller, M.J.; Gaglioti, B.; Fulton, T.L.; Lopez, A.; Shapiro, B. Post-glacial dispersal patterns of Northern pike inferred from an 8,800 year old pike (Esox cf. lucius) skull from interior Alaska. *Quat. Sci. Rev.* **2015**, *120*, 118–125. [CrossRef]
49. Giustiniani, M.; Tinivella, U.; Jakobsson, M.; Rebesco, M. Arctic Ocean Gas Hydrate Stability in a Changing Climate. *J. Geol. Res.* **2013**. [CrossRef]
50. Marín-Moreno, H.; Giustiniani, M.; Tinivella, U.; Piñero, E. The challenges of quantifying the carbon stored in Arctic marine gas hydrate. *Mar. Pet. Geol.* **2016**, *71*, 76–82. [CrossRef]

51. Makogon, Y.F. Production from natural gas hydrate deposits. *Gazov. Promishlennost* **1984**, *10*, 24–26.
52. Bugge, T.; Elvebakk, G.; Fanavoll, S.; Mangerud, G.; Smelror, M.; Weiss, H.M.; Gjelberg, J.; Kristensen, S.E.; Nilsen, K. Shallow stratigraphic drilling applied in hydrocarbon exploration of the Nordkapp Basin, Barents Sea. *Mar. Pet. Geol.* **2002**, *19*, 13–37. [CrossRef]
53. Rachold, V.; Yu, D.; Bolshiyanov, M.N.; Grigoriev, H.-W.; Hubberten, R.; Junker, V.V.; Kunitsky, F.; Merker, P.P.; Overduin, P.; Schneider, W. Near-shore Arctic Subsea Permafrost in Transition. *EOS: Trans. Am. Geophys. Union* **2007**, *88*, 149–156. [CrossRef]
54. Shakhova, N.; Semiletov, I. Methane release and coastal environment in the East Siberian Arctic shelf. *J. Mar. Syst.* **2007**, *66*, 227–243. [CrossRef]
55. Moridis, G.J.; Kowalsky, M.B.; Pruess, K. TOUGH+HYDRATE v1.2 User's Manual: A Code for The Simulation of System Behavior in Hydrate-Bearing Geological Media. 2012. Available online: https://tough.lbl.gov/assets/docs/TplusH_Manual_v1.pdf (accessed on 26 July 2019).
56. Dickens, G.R.; Quinby-Hunt, M.S. Methane hydrate stability in pore water: A simple theoretical approach for geophysical applications. *JGR* **1997**, *102*, 773–783. [CrossRef]
57. Ladd, M. *Introduction to Physical Chemistry*; Cambridge University Press: Cambridge, UK, 1998.
58. Betlem, P.; Senger, K.; Hodson, A. 3D thermobaric modelling of the gas hydrate stability zone onshore central Spitsbergen, Arctic Norway. *Mar. Pet. Geol.* **2019**, *100*, 246–262. [CrossRef]
59. Tinivella, U.; Giustiniani, M. Variations in BSR depth due to gas hydrate stability versus pore pressure. *Glob. Planet. Chang.* **2013**, *100*, 119–128. [CrossRef]
60. Shakhova, N.; Semiletov, I.; Gustafsson, O.; Sergienko, V.; Lobkovsky, L.; Dudarev, O.; Tumskoy, V.; Grigoriev, M.; Mazurov, A.; Salyuk, A.; et al. Current rates and mechanisms of subsea permafrost degradation in the East Siberian Arctic Shelf. *Nat. Commun.* **2017**, *8*, 15872. [CrossRef] [PubMed]

geosciences

MDPI

Review

Gas Seeps at the Edge of the Gas Hydrate Stability Zone on Brazil's Continental Margin

Marcelo Ketzer [1,*], **Daniel Praeg** [2,3,4], **Maria A.G. Pivel** [5], **Adolpho H. Augustin** [2], **Luiz F. Rodrigues** [2], **Adriano R. Viana** [6] **and José A. Cupertino** [2]

1   Department of Biology and Environmental Science, Linnaeus University, 391-82 Kalmar, Sweden
2   Institute of Petroleum and Natural Resources, PUCRS—Pontifícia Universidade Católica do Rio Grande do Sul, Porto Alegre 90619-900, Brazil; daniel.praeg@pucrs.br (D.P.); adolpho.augustin@pucrs.br (A.H.A.); frederico.rodrigues@pucrs.br (L.F.R.); jose.cupertino@pucrs.br (J.A.C.)
3   LAGEMAR, Department of Geology, Universidade Federal Fluminense (UFF), Niteroi 24210-346, Brazil
4   Géoazur, UMR7329 CNRS, rue Albert Einstein 250, 06560 Valbonne, France
5   Centro de Estudos de Geologia Costeira e Oceânica, Instituto de Geociências, Universidade Federal do Rio Grande do Sul, Porto Alegre 91501-970, Brazil; maria.pivel@ufrgs.br
6   PETROBRAS—E&P Exploration, Rio de Janeiro 20031-170, Brazil; aviana@petrobras.com.br
*   Correspondence: marcelo.ketzer@lnu.se

Received: 6 March 2019; Accepted: 8 April 2019; Published: 28 April 2019

check for updates

Abstract: Gas hydrate provinces occur in two sedimentary basins along Brazil's continental margin: (1) The Rio Grande Cone in the southeast, and (2) the Amazon deep-sea fan in the equatorial region. The occurrence of gas hydrates in these depocenters was first detected geophysically and has recently been proven by seafloor sampling of gas vents, detected as water column acoustic anomalies rising from seafloor depressions (pockmarks) and/or mounds, many associated with seafloor faults formed by the gravitational collapse of both depocenters. The gas vents include typical features of cold seep systems, including shallow sulphate reduction depths (<4 m), authigenic carbonate pavements, and chemosynthetic ecosystems. In both areas, gas sampled in hydrate and in sediments is dominantly formed by biogenic methane. Calculation of the methane hydrate stability zone for water temperatures in the two areas shows that gas vents occur along its feather edge (water depths between 510 and 760 m in the Rio Grande Cone and between 500 and 670 m in the Amazon deep-sea fan), but also in deeper waters within the stability zone. Gas venting along the feather edge of the stability zone could reflect gas hydrate dissociation and release to the oceans, as inferred on other continental margins, or upward fluid flow through the stability zone facilitated by tectonic structures recording the gravitational collapse of both depocenters. The potential quantity of venting gas on the Brazilian margin under different scenarios of natural or anthropogenic change requires further investigation. The studied areas provide natural laboratories where these critical processes can be analyzed and quantified.

Keywords: gas hydrates; gas seeps; ocean acidification

1. Introduction

Contemporary climate change is significantly impacting the marine environment. Among the most drastic impacts are declining oxygen levels [1], acidification [2], and loss of biodiversity [3]. Current estimates suggest that the ocean surface and intermediate waters (<700 m) could be 1–4 °C warmer at the end of this century [4]. Such an increase in temperature would promote the destabilization of gas hydrate on continental margins worldwide, even if warming is accompanied by pressure increase owing to sea-level rise [5]. Gas hydrate is still being discovered beneath the ocean margins and contain a large reservoir of organic carbon (mainly methane) on the Earth's surface ($0.5–12.7 \times 10^{21}$ g [6,7]), and the release of even a fraction of it to the oceans and atmosphere could potentially lead to a positive feedback in greenhouse gas emissions [8,9]. Estimates of the amount of carbon that could be released from sediments to oceans and seas owing to hydrate dissociation are poorly constrained, in part as we do not fully understand the mechanisms by which gas may move through the gas hydrate stability zone (GHSZ) and/or escape from its 'feather edge' on the upper continental slope. The feather edge of the GHSZ (i.e., the region on the upper continental slopes where it thins to vanishing at the seafloor [10]) contains about 3.5% of the global hydrate inventory and is particularly susceptible to hydrate dissociation in response to ocean warming or sea level change [10,11]. Global models of the gas hydrate system response to scenarios of climate-driven change indicate that methane release to the seafloor from hydrate dissociation is greatest along the feather edge and could exceed methane fluxes from other sources by the year 2100 (i.e., $>30–50$ Tg·CH_4·year^{-1} [5]). Other simulations indicate that gas hydrate is sensitive to rapid temperature increases [5,12–15], and a 5 °C increase in bottom water temperatures along continental slopes could add ca. 2000 Gt·CH_4 to sediments worldwide [16], part of which may flow toward the seafloor. It is possible that the dissociation of hydrate will also be linked to the additional release of methane by opening pathways to free gas ascending from underneath the hydrate stability zone [17].

Not all methane released by hydrate dissociation will reach the seafloor and atmosphere because it will be consumed by anaerobic oxidation via sulphate reduction in sediments [18,19], and dissolution and aerobic oxidation in the water column [20–22]. The oxidation of large quantities of methane in the water column, however, may contribute to a decrease in the ocean's pH [23,24]. In addition to its possible contribution to greenhouse gas emissions and ocean acidification, massive destabilization of gas hydrate may be associated to the triggering of submarine landslides and related tsunamis [25].

Dissociation of marine gas hydrate at the feather edge of the gas hydrate stability zone has been reported in several locations around the world [26–28]. This phenomenon received particular interest at high latitude regions owing to the high amplitude of warming in polar oceans [29–31]. The aim of the present article is to review evidence of venting from two gas hydrate provinces along Brazil's continental margin: (1) The Rio Grande Cone and (2) the upper Amazon deep-sea fan. The latter is, to our knowledge, the only case in the world where gas venting near the feather edge of the GHSZ was reported in equatorial regions. We review recent discoveries within these provinces and compare them to the calculated feather edge of the GHSZ. The results provide insights into the role of climate-driven changes in ocean conditions versus processes controlling fluid flow within continental margins, and suggest further investigations on the Brazilian margin in relation to the global carbon cycle and ongoing climate change.

2. Gas Hydrate and Gas Venting Structures on Brazil's Continental Margin

The presence of gas hydrate on Brazil's continental margin was first reported in the 1980s from bottom simulating reflectors (BSRs) observed in two deep-water depocenters: (1) The Rio Grande cone in the Pelotas Basin, western South Atlantic margin [32–34], and (2) the Amazon deep-sea fan in the Foz do Amazonas Basin, Equatorial Atlantic margin [35–37] (Figure 1). Other less well defined BSRs in Brazil were observed in the Campos and Santos basins [33,38]. The Rio Grande Cone forms a protuberance in the continental slope and contains a deposit of up to 12 km of sediments (Barremian to recent) in an area of approximately 250,000 km². Sediments were sourced from the Rio de la Plata river [39] and/or from large contourite systems occurring to the south [40] and to the north [41]. The Amazon deep-sea fan also forms a protuberance in the continental slope and contains of a deposit of up to 10 km of sediments (Late Miocene to recent) sourced mainly from the Amazon River [42], and distributed in an area of 330,000 km² [43].

Sediment loading in the Rio Grande Cone and in the Amazon deep-sea fan resulted in large-scale gravitational collapse, expressed as paired belts of extensional and compressional structures rooted on deep detachment surfaces. In contrast to the mainly stratified internal character of the Rio Grande cone, the Amazon fan is characterized by giant mass transport deposits recording sediment failure from the upper slope [35,44]. High rates of sedimentation (including organic matter), particularly in the upper Amazon fan, contribute to low geothermal gradients (15–19 °C/km, based on BSR depth and bottom-hole temperatures; [37]), which, together with low bottom water temperatures, result in a potential thick gas hydrate stability zone. The quantity of methane trapped in gas hydrate in the two areas has been estimated to be 22 trillion m^3 (ca. 780 tcf) and 12 trillion m^3 (ca. 430 tcf), respectively [45].

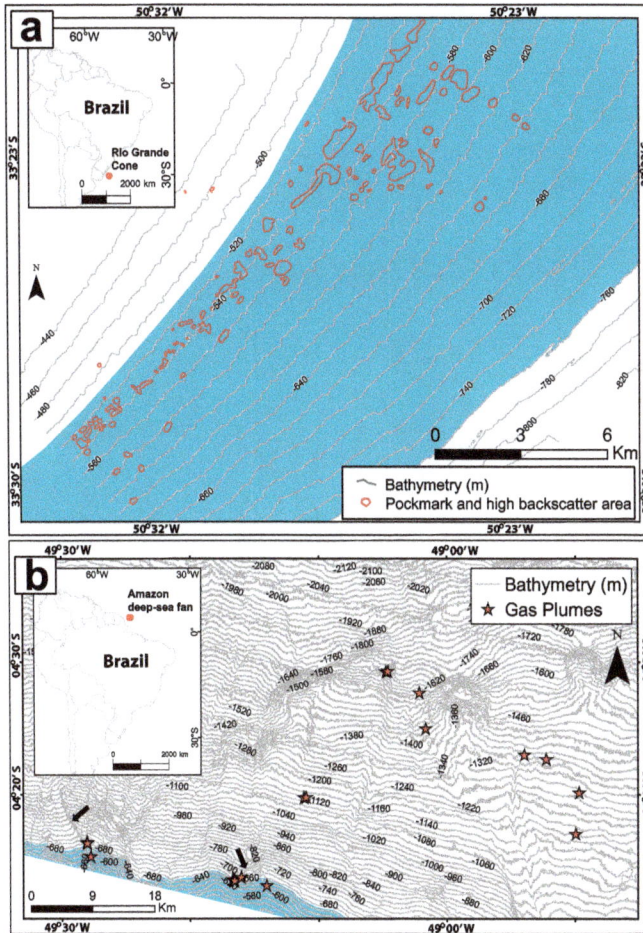

Figure 1. Maps showing bathymetric contours derived from multibeam data across the two areas in Brazil where there is evidence of gas seepage near the feather edge of the methane hydrate stability zone: (**a**) The Rio Grande Cone, Pelotas Basin, western South Atlantic (modified from [46]) and (**b**) the Amazon deep-sea fan, Foz do Amazonas Basin, Equatorial Atlantic Ocean (modified from [47]). The areas in light blue mark the depth range of the feather edge of the gas hydrate stability zone calculated for the two regions using pure methane in seawater [48], and water temperature measurements obtained for the regions from the World Ocean Database (WOD18 [49]; see Figure 2). Red areas in (**a**) are pockmarks and high backscatter areas aligned with both subjacent NE-SW faults and the feather edge of the GHSZ (for details see [46]). Red stars in (**b**) are water column gas plumes, and black arrows indicate trace of faults on the seafloor near gas seeps at the edge of the methane hydrate stability zone (for details see [47]).

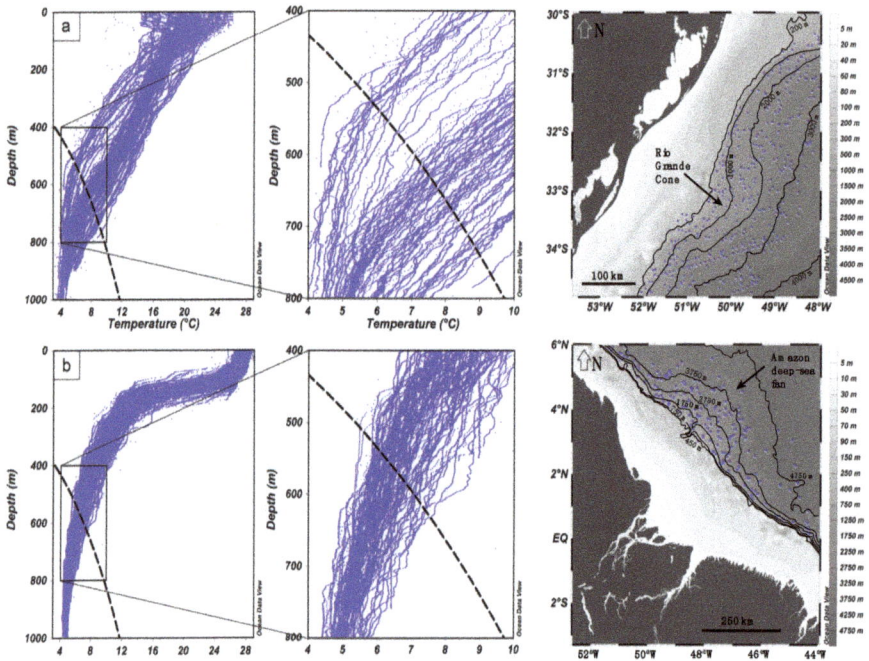

Figure 2. Water temperature profiles for (**a**) the Rio Grande Cone and (**b**) Amazon deep-sea fan from the World Ocean Database (WOD18 [49]). The dashed line indicates the phase boundary for pure methane in seawater [47]. Blue dots in the bathymetric maps to the right indicate the location of the temperature profiles.

Direct evidence of natural gas hydrate and gas seeps on the seafloor were found in the Rio Grande Cone [46] and in the Amazon deep-sea fan [47]—in both areas in association with gas venting from the GHSZ. Living chemosynthesis-based communities in pockmarks in the Rio Grande Cone [50,51], in addition to acoustic disturbance caused by the presence of free gas at shallow depths (<10 m) below seafloor in sub-bottom profiles [52], are indicative of active methane seeps. Direct evidence of gas seepage was identified in the Amazon deep-sea fan by the presence of acoustic anomalies in the water column using multi-beam echo sounder backscatter data, which is also supported by the presence of remnants of chemosynthetic organisms at the seafloor [47]. In both areas, gas vents occur both within the GHSZ and along its edge, and are inferred to be influenced by normal and thrust faults related to the gravitational tectonics induced by sediment loading [46,47]. Gas hydrate was recovered and sampled in both areas in piston cores at water depths from 550–1400 m below sea level (mbsl) in the Rio Grande Cone [46], and at 1000–1800 mbsl in the Amazon deep-sea fan [47]. In both cases, dominantly CH_4 composition with $\delta^{13}C < -66.7‰$ V-PDB (Vienna—Pee Dee Belemnite standard) for the Rio Grande Cone and $\delta^{13}C < -77.3‰$ V-PDB for the Amazon deep-sea fan indicate a biogenic origin for the methane trapped in hydrate.

3. The Edge of the Stability Zone and Seafloor Gas Vents

The theoretical depth range of the feather edge of the methane hydrate stability zone (MHSZ) can be calculated using an equilibrium equation for pure methane hydrate in seawater [48]:

$$T = 11.726 + 20.5 \times \log_{10}z - 2.2 \times (\log_{10}z)^2 \tag{1}$$

where T is the temperature of the phase boundary (°C) and z is the depth (km).

The upper depth limit of the MHSZ can be estimated using the above equation and historical (1958 to 2018) water column temperature data from the World Ocean Database (WOD18 [49]), which show high variability in measured temperatures above 900 m water depth in both areas of interest (5.4 to 9.3 °C in the Rio Grande Cone and 5.4 to 7.6 °C in the Amazon fan; Figure 2). The feather edge of the MHSZ is thus constrained within a range of depths, between 510 and 760 mbsl on the Rio Grande Cone and between 500 and 670 mbsl on the Amazon deep-sea fan. Comparison of these depth ranges with recently published observations based on water column, seafloor, and sub-bottom acoustic imagery and samples from piston cores [45,46,51] indicates that gas seeps occur both within and near the edge of the GHSZ in the two studied areas (Figure 1).

On the Rio Grande cone, multibeam bathymetric and backscatter imagery reveal pockmark fields in two main locations: On the mid-slope in water depths of ca. 1300 m (ca. 320 km² with a pockmark density of ca. 1/km², or 51% of the vent sites), and on the upper slope in depths of 520–660 m (ca. 38 km², with a pockmark density of 8/km²; or 49% of the vent sites; Figure 1). In both areas, pockmarks are high backscatter features of variable relief, with their long axes parallel to subjacent extensional faults [46]. The upper pockmark field lies within the estimated depth range of the MHSZ feather edge (510–760 m) and comprises a slope-parallel zone 20 km long by 3 km wide, widening to 6 km in the NW, including a central zone where pockmarks cover most of the seafloor (Figure 1). Sub-bottom profiles across pockmarks show acoustic blanking, consistent with free gas rising to the seafloor through chimney-like features [46,52]. Methane concentration and sulphate profiles in pore waters obtained from piston cores samples in a cross-section perpendicular to the pockmark field corroborate with the presence of shallow gas, notably focused at the 545 m isobath [52], where there is a major concentration of pockmarks in the field [46]. The sulphate methane interface is considerably shallower within the pockmark field (3–4 mbsf) compared to a background area downslope within the GHSZ at ca. 1300 mbsl (ca. >10 mbsf [52]). The data suggest, therefore, the existence of active methane seeps within the pockmark field. The presence of authigenic carbonate concretions (centimeters in diameter) in piston cores [46], for which the radiometric ages are yet to be determined, indicates that the seeps may have been active for thousands of years (assuming a growth rate of ca. 0.4–0.8 cm/kyr [53]). Foraminiferal stable carbon isotope and sediment mineralogy found in ancient pockmarks in the northern portion of the Pelotas Basin (200 km north of the Rio Grande Cone) indicate seafloor methane release during the last glacial period (40–20 cal ka BP, Before Present, [54]).

On the upper Amazon deep-sea fan, multibeam water column and seafloor data acquired across water depths of 650–2600 m reveal the existence of at least 53 gas plumes rising up to 900 m into the water column from seafloor venting features that include both pockmarks and mounds [47]. Most of the gas vents (60%) are located within the MHSZ along lineaments corresponding to faults that may have acted as pathways for fluid migration [47], whereas some (40%) are located in water depths of 650–715 m within the feather edge of the MHSZ (500–670 m) along about 50 km of its length (Figure 1b). The latter features include 23 water column gas plumes that rise up to 350 m into the water column from seafloor mounds 10–20 m high [47]. The gas bubble plumes were not sampled, but 24 dissolved and free gas and three gas hydrate samples in piston cores at plume sites revealed a dominantly methane composition (with the absence of heavier hydrocarbons), and a strong depletion in ^{13}C (δ^{13}C from −102.2 to −73.7‰, V-PDB), indicating a biogenic origin [55].

4. Discussion

Brazil's continental margin contains at least two major gas hydrate provinces, located in depocenters in which rapid deposition drives tectonism, fluid migration, and methanogenesis, providing ideal conditions for near-surface gas hydrate accumulation [46,47]. In both provinces, gas hydrates have been sampled from seafloor venting structures that indicate gas seepage is taking place within the GHSZ and along its feather edge. Gas venting to the oceans within the GHSZ has been reported from many other deep-water settings [27,28] and linked to the formation of chimney-like features within the GHSZ [56]. Mechanisms to account for such features all involve the upward migration of warm gas-rich fluids, which modify the base of the GHSZ and/or sediment properties within it [57–59]. Such mechanisms have yet to be tested against evidence from the deep-water gas vents on the Brazilian margin. Nonetheless, the fact that many of the vents observed within the two areas investigated to date are associated with seafloor faults related to gravity tectonics (e.g., 51% in the Rio Grande Cone and 60% in the Amazon deep-sea fan) [46,47] provides evidence that processes internal to continental margin depocenters may be important in the creation of fluid migration pathways into and through the GHSZ.

Evidence of structurally-influenced gas venting through the GHSZ on the Brazilian margin raises questions regarding the origin of the gas vents observed along its feather edge. Gas seeps observed in similar settings around the world have been suggested to record the dissociation of gas hydrates in response to ocean warming [26,27]. This is also possible on the Brazilian margin, as recorded variations in water temperatures, which record cyclic or progressive changes in the water masses that impinge on the upper slope over decadal timescales, imply that the edge of the GHSZ has migrated across large depth ranges (Figures 1 and 2). The Rio Grande cone and Amazon fan include seeps within the areas corresponding to these depth ranges, which could record gas hydrate dissociation. However, seeps in both areas contain evidence of deeper structural influences, notably on the Rio Grande cone where a slope-parallel field of elongate pockmarks is aligned with both subjacent faults and the edge of the GHSZ. The location of seeps relative to the edge of the GHSZ might be related to a complex interplay among structures focusing gas flow to the seafloor and the geomechanics of sediments during hydrate dissociation [60], the geometry of the edge of the GHSZ [61], gas migration along the base of the GHSZ to escape at its edge [62], and the dynamics of bottom water temperature changes—including multidecadal warming and shoaling of Antarctic Intermediate Water observed in the region [63]. The dynamics of the Antarctic Intermediate Water in particular may be an important component for the stability of gas hydrate deposits in the South Atlantic Ocean margins (and the Rio Grande Cone), where warming between 0.01 and 0.02 $°C\cdot year^{-1}$ has been observed since the 1970s [63]. Using the equilibrium equation for pure methane hydrate in seawater [48], it is possible to estimate that such a warming rate could dislocate the feather edge of the GHSZ by several hundreds of meters downslope (tens of meters deeper) in a few decades. Similar warming rates (0.007 $°C\cdot year^{-1}$) of the North Pacific Intermediate Water resulting in dislocation of the edge of the GHSZ was observed in the Cascadian margin, North Pacific Ocean [14]. More detailed temporal studies relating the downslope dislocation of the feather edge of the GHSZ to bottom water temperature measurements and associated seafloor seeping structures offshore Brazil must be undertaken to determine whether gas hydrate dissociation is linked to long-term (millennia) trends, following the last glacial maximum, and/or short-term (decades) trends related to anthropogenic warming [12,17,26].

Numerical modeling indicates that ocean warming over the last decades may have released significant quantities of methane from global gas hydrate systems [15,56,64]. These models have several limitations and assumptions, but provide valuable insights about the potential release of CH_4 from hydrate dissociation [5]. On the Cascadia margin, warming along 273 km of the GHSZ feather edge between 1970 and 2013 is estimated to have released 4.35 Tg of methane to the water column, and may release another 45–80 Tg by 2100 [14]. Modeling of gas hydrate dissociation off North Carolina, in the western Atlantic, suggests that 2.5 Gt of methane was produced over an area of 10,000 km^2 by bottom water warming [65]. There are no published models of gas hydrate dissociation offshore South America, thus the possible magnitude of methane release along Brazil's continental margin is unknown. However, if the quantities of carbon released to the oceans are of the same order of magnitude as those modeled for other regions of the world experiencing the same warming rates of bottom waters, and if the observed seepage features can be extrapolated for tens to hundreds of kilometers beyond the limits of the study area, such as in the Cascadia margin [14], the gas venting known to be ongoing in at least two gas hydrate provinces offshore Brazil may be of concern and deserves further investigation.

Over longer timescales, there is a possible connection between gas hydrate dissociation at the edge of the GHSZ and the triggering of landslides within the studied areas. The Amazon fan contains a record of recurrent giant landslides sourced on its upper slopes, which have been suggested to record massive dissociation of gas hydrate linked to glacial/interglacial changes in sea level [66], although they have alternatively been linked to collapse tectonics on the upper fan [44]. Mass wasting deposits of a smaller scale are observed near the feather edge of the GHSZ on the Rio Grande cone [52]. A megaslide complex is recognized to the south of the Rio Grande cone, with headwall scarps near the shelf-break and upper slope, although there is no clear association with gas hydrate systems in the area and gravity tectonics has been proposed as the main triggering mechanism [67]. Along most of the continental slope, gas hydrate stability has decreased since the last glacial maximum (LGM) due to the warming of South Atlantic bottom waters by at least 3.5 °C at a depth of 657 m on the upper Rio Grande cone [68]. However, investigations of a pockmark field at shallower depths (475 m) at 200 km to the north of the Pelotas Basin provide evidence of seafloor methane release at the LGM, linked to the impingement of warmer bottom waters on the uppermost slope owing to sea level lowering [54]. The effects of depth-dependent changes in ocean temperatures during the last sea level cycle on gas hydrate stability remain poorly understood offshore Brazil, as on other continental margins, and it is of interest to undertake modeling to examine the possible connection between hydrate dissociation and landslides in the Rio Grande Cone and Amazon deep-sea fan, and explore their geohazard potential [69].

5. Conclusions

Brazil's vast continental slope area remains to be fully investigated, but includes two proven gas hydrate provinces: the Amazon deep-sea fan in the far north and the Rio Grande cone in the far south. In both areas, gas venting is observed to take place within the gas hydrate stability zone and near its feather edge, as calculated from bottom water temperatures. Further studies are necessary to determine the extent to which gas venting from the GHSZ is driven by subsurface fluid flow linked to the internal dynamics of the gravitationally collapsing depocenters, or by gas hydrate dissociation at the feather edge of the stability zone linked to changes in ocean conditions over glacial and/or anthropogenic timescales. It is also important to model the quantities of methane that may be transferred from sediments to the oceans in both areas under different scenarios. Considering the possible existence of gas hydrate provinces in other basins along the Brazilian margin, the area provides a natural laboratory for further investigations of gas hydrate dynamics.

Author Contributions: Conceptualization, M.K.; Investigation, M.K., M.A.G.P., A.H.A., L.F.R., A.R.V., D.P., and J.A.C.; Data curation, M.K., M.A.G.P., and A.H.A.; Writing—original draft preparation, M.K.; Writing—review and editing, M.K., M.A.G.P., A.H.A., L.F.R., A.R.V., D.P., and J.A.C.

Funding: D.P. was supported in part by funding from the European Union's Horizon 2020 research and innovation program under Marie Skłodowska-Curie grant agreement No. 656821, and in part from the Brazilian Coordination of Superior Level Staff Improvement CAPES-IODP (2018–2019). This research received no other external funding.

Acknowledgments: The authors thank the editors for the invitation to participate in the Special Issue of the journal *Geosciences* "Gas Hydrate: Environmental and Climate Impacts".

Conflicts of Interest: The authors declare no conflict of interest.

References

1. Breitburg, D.; Levin, L.A.; Oschlies, A.; Grégoire, M.; Chavez, F.P.; Conley, D.J.; Garçon, V.; Gilbert, D.; Gutiérrez, D.; Isensee, K.; et al. Declining oxygen in the global ocean and coastal waters. *Science* **2018**, *359*, eaam7240. [CrossRef]

2. Dutkiewicz, S.; Morris, J.J.; Follows, M.J.; Scott, J.; Levitan, O.; Dyhrman, S.T.; Berman-Frank, I. Impact of ocean acidification on the structure of future phytoplankton communities. *Nat. Clim. Chang.* **2015**, *5*, 1002–1006. [CrossRef]

3. Bruno, J.F.; Bates, A.E.; Cacciapaglia, C.; Pike, E.P.; Amstrup, S.C.; van Hooidonk, R.; Henson, S.A.; Aronson, R.B. Climate change threatens the world's marine protected areas. *Nat. Clim. Chang.* **2018**, *8*, 499–503. [CrossRef]

4.   Pachauri, R.K.; Allen, M.R.; Barros, V.R.; Broome, J.; Cramer, W.; Christ, R.; Church, J.A.; Clarke, L.; Dahe, Q.; Dasgupta, P.; et al. *Climate Change 2014: Synthesis Report. Contribution of Working Groups I, II and III to the Fifth Assessment Report of the IPCC*; Intergovernmental Panel on Climate Change: Geneva, Switzerland, 2014; p. 112.

5.   Hunter, S.J.; Goldobin, D.S.; Haywood, A.M.; Ridgwell, A.; Rees, J.G. Sensitivity of the global submarine hydrate inventory to scenarios of future climate change. *Earth Planet. Sci. Lett.* **2013**, *367*, 105–115. [CrossRef]

6.   Pinero, E.; Marquardt, M.; Hensen, C.; Haeckel, M.; Wallmann, K. Estimation of the global inventory of methane hydrates in marine sediments using transfer functions. *Biogeosciences* **2013**, *10*, 959–975. [CrossRef]

7.   Dickens, G.R. Down the Rabbit Hole: Toward appropriate discussion of methane release from gas hydrate systems during the Paleocene-Eocene thermal maximum and other past hyperthermal events. *Clim. Past* **2011**, *7*, 831–846. [CrossRef]

8.   Archer, D. Methane hydrate stability and anthropogenic climate change. *Biogeosciences* **2007**, *4*, 521–544. [CrossRef]

9.   Giustiniani, M.; Tinivella, U.; Jakobsson, M.; Rebesco, M. Arctic Ocean Gas Hydrate Stability in a Changing Climate. *J. Geol. Res.* **2013**, *2013*, 783969. [CrossRef]

10.   Ruppel, C.D. Methane Hydrates and Contemporary Climate Change. *Nat. Educ. Knowl.* **2011**, *3*, 29.

11.   Tinivella, U.; Giustiniani, M.; Accettella, D. BSR versus Climate Change and Slides. *J. Geol. Res.* **2011**, *2011*, 390547. [CrossRef]

12.   Thatcher, K.E.; Westbrook, G.K.; Sarkar, S.; Minshull, T.A. Methane release from warming-induced hydrate dissociation in the West Svalbard continental margin: Timing, rates, and geological controls. *J. Geophys. Res. Solid Earth* **2013**, *118*, 22–38. [CrossRef]

13.   Ferré, B.; Mienert, J.; Feseker, T. Ocean temperature variability for the past 60 years on the Norwegian-Svalbard margin influences gas hydrate stability on human time scales. *J. Geophys. Res.* **2012**, *117*, C10017. [CrossRef]

14.   Hautala, S.; Solomon, E.; Johnson, H.P.; Harris, R.N.; Miller, U.K. Dissociation of Cascadia margin gas hydrates in response to contemporary ocean warming. *Geophys. Res. Lett.* **2014**, *41*, 8486–8494. [CrossRef]

15.   Stranne, C.; O'Regan, M.; Dickens, G.R.; Crill, P.; Miller, C.; Preto, P.; Jakobsson, M. Dynamic simulations of potential methane release from East Siberian continental slope sediments. *Geochem. Geophys. Geosyst.* **2016**, *17*, 872–886. [CrossRef]

16.   Hornbach, M.J.; Saffer, D.M.; Holbrook, W.S. Critically pressured free-gas reservoirs below gas-hydrate provinces. *Nature* **2004**, *427*, 142–144. [CrossRef]

17.   Berndt, C.; Feseker, T.; Treude, T.; Kraster, S.; Liebetrau, V.; Niemann, H.; Bertics, V.J.; Dumke, I.; Dünnbier, K.; Ferré, B.; et al. Temporal constraints on hydrate-controlled methane seepage off Svalbard. *Science* **2014**, *343*, 284–287. [CrossRef]

18.   Reeburgh, W.S. Oceanic methane biogeochemistry. *Chem. Rev.* **2007**, *107*, 486–513. [CrossRef]

19.   Regnier, P.; Dale, A.W.; Arndt, S.; La Rowe, D.E.; Mogollón, J.; Van Cappellen, P. Quantitative analysis of anaerobic oxidation of methane (AOM) in marine sediments: A modeling perspective. *Earth Sci. Rev.* **2011**, *106*, 105–130. [CrossRef]

20.   McGinnis, D.F.; Greinert, J.; Artemov, Y.; Beaubien, S.E.; Wuest, A. Fate of rising methane bubbles in stratified waters: How much methane reaches the atmosphere? *J. Geophys. Res.* **2006**, *111*, C09007. [CrossRef]

21.   Mau, S.; Valentine, D.L.; Clark, J.F.; Reed, J.; Camilli, R.; Washburn, L. Dissolved methane distributions and air-sea flux in the plume of a massive seep field, Coal Oil Point, California. *Geophys. Res. Lett.* **2007**, *34*, L22603. [CrossRef]

22.   Leonte, M.; Kessler, J.D.; Kellermann, M.Y.; Arrington, E.C.; Valentine, D.L.; Sylva, S.P. Rapid rates of aerobic methane oxidation at the feather edge of gas hydrate stability in the waters of Hudson Canyon, US Atlantic Margin. *Geochim. Cosmochim. Acta* **2017**, *204*, 375–387. [CrossRef]

23.   Biastoch, A.; Treude, T.; Rüpke, L.H.; Riebesell, U.; Roth, C.; Burwicz, E.B.; Park, W.; Latif, M.; Böning, C.W.; Madec, G.; et al. Rising Arctic Ocean temperatures cause gas hydrate destabilization and ocean acidification. *Geophys. Res. Lett.* **2011**, *38*, L08602. [CrossRef]

24.   Boudreau, B.P.; Luo, Y.; Meysman, F.J.R.; Middelburg, J.J.; Dickens, G.R. Gas hydrate dissociation prolongs acidification of the Anthropocene oceans. *Geophys. Res. Lett.* **2015**, *42*, 9337A–9344A. [CrossRef]

25.   Maslin, M.; Owen, M.; Betts, R.; Day, S.; Jones, T.D.; Ridgwell, A. Gas hydrates: Past and future geohazard? *Philos. Trans. R. Soc. A* **2010**, *368*, 2369–2393. [CrossRef]

26. Westbrook, G.K.; Thatcher, K.E.; Rohling, E.J.; Piotrowski, A.M.; Pälike, H.; Osborne, A.H.; Nisbet, E.G.; Minshull, T.A.; Lanoiselle, M.; James, R.H.; et al. Escape of methane gas from the seabed along the West Spitsbergen continental margin. *Geophys. Res. Lett.* **2009**, *36*, 1–5. [CrossRef]

27. Skarke, A.; Ruppel, C.; Kodis, M.; Brothers, D.; Lobecker, E. Widespread methane leakage from the sea floor on the northern US Atlantic margin. *Nat. Geosci.* **2014**, *7*, 657–661. [CrossRef]

28. Johnson, H.P.; Miller, U.K.; Salmi, M.S.; Solomon, E.A. Analysis of bubble plume distributions to evaluate methane hydrate decomposition on the continental slope. *Geochem. Geophys. Geosyst.* **2015**, *16*, 3825–3839. [CrossRef]

29. Screen, J.A.; Simmonds, I. The central role of diminishing sea ice in recent Arctic temperature amplification. *Nature* **2010**, *464*, 1334–1337. [CrossRef]

30. Serreze, M.C.; Barry, R.G. Processes and impacts of Arctic amplification: A research synthesis. *Glob. Planet. Chang.* **2011**, *77*, 85–96. [CrossRef]

31. Spielhagen, R.F.; Werner, K.; Sørensen, S.A.; Zamelczyk, K.; Kandiano, E.; Budeus, G.; Husum, K.; Marchitto, T.M.; Hald, M. Enhanced modern heat transfer to the Arctic by warm Atlantic water. *Science* **2011**, *331*, 450–453. [CrossRef] [PubMed]

32. Fontana, L.R. Evidencias geofísicas da Presença de Hidratos de Gas na Bacia de Pelotas. In Proceedings of the 1st Congress of the Brazilian Geophysical Society, Rio de Janeiro, Brazil, 20 November 1989.

33. Fontana, R.L.; Mussumeci, A. Hydrates offshore Brazil. In *Annals of the New York Academy of Sciences, International Conference on Natural Gas Hydrates*; New York Academy of Sciences: New York, NY, USA, 1994; Volume 715, pp. 106–113.

34. Oliveira, S.; Vilhena, O.; da Costa, E. Time-frequency spectral signature of Pelotas Basin deep water gas hydrates system. *Mar. Geophys. Res.* **2010**, *31*, 89–97. [CrossRef]

35. Manley, P.L.; Flood, R.D. Cyclic sediment deposition within Amazon deep-sea fan. *Am. Assoc. Petrol. Geol. Bull.* **1988**, *72*, 912–925.

36. Tanaka, M.D.; Silva, C.G.; Clennell, M.B. Gas hydrates on the Amazon Submarine Fan, Foz do Amazonas Basil, Brazil. In Proceedings of the AAPG Search and Discovery Article #90013, AAPG Annual Meeting, Salt Lake City, UT, USA, 11–14 May 2003.

37. Berryman, J.; Kearns, H.; Rodriguez, K. Foz do Amazonas Basin—A case for oil generation from geothermal gradient modelling. *First Break* **2015**, *33*, 91–95.

38. Matsuda, N.S.; Freire, A.F.M. The Bottom Simulating Reflector (BSR) along the Brazilian Atlantic Coast: A New Perspective for Gas Hydrates Exploration in the Southern Hemisphere. In Proceedings of the AAPG International Conference and Exhibition, Cape Town, South Africa, 26–29 October 2008.

39. Martins, L.R.; Melo, U.; França, A.M.C.; Santana, C.I.; Martins, I.R. Distribuicao Faciologica da Margem Continental Sul Riograndense. In Proceedings of the XXVI Congresso Brasileiro de Geologia, Belem, Brazil, 26 October 1972; Volume 2, pp. 115–132.

40. Hernándes-Molina, F.J.; Soto, M.; Piola, A.R.; Tomasini, J.; Preu, B.; Thompson, P.; Badalini, G.; Creaser, A.; Violante, R.A.; Morales, E.; et al. A contourite depositional system along the Uruguayan continental margin: Sedimentary, oceanographic and paleoceanographic implications. *Mar. Geol.* **2016**, *378*, 333–349. [CrossRef]

41. Viana, A.R. Seismic expression of shallow- to deep-water contourites along the south-eastern Brazilian margin. *Mar. Geophys. Res.* **2002**, *22*, 509–521. [CrossRef]

42. Figueiredo, J.; Hoorn, C.; van der Ven, P.; Soares, E. Late Miocene onset of the Amazon River and the Amazon deep-sea fan: Evidence from the Foz do Amazonas Basin. *Geology* **2009**, *37*, 619–622. [CrossRef]

43. Damuth, J.E.; Kumar, N. Amazon cone, morphology, sediments, age, and growth pattern. *Geol. Soc. Am. Bull.* **1975**, *86*, 863–878. [CrossRef]

44. Reis, A.T.; Araújo, E.; Silva, C.G.; Cruz, A.M.; Gorini, C.; Droz, L.; Migeon, S.; Perovano, R.; King, I.; Bache, F. Effects of a regional décollement level for gravity tectonics on late Neogene to recent large-scale slope instabilities in the Foz do Amazonas Basin, Brazil. *Mar. Petrol. Geol.* **2016**, *75*, 29–52. [CrossRef]

45. Sad, A.R.E.; Silveira, D.P.; Machado, D.A.P.; Silva, S.R.P.; Maciel, R.R. Marine gas hydrates evidence along the Brazilian coast. In Proceedings of the AAPG International Conference and Exhibition, Rio de Janeiro, Brazil, 8–11 November 1998.

46. Miller, D.J.; Ketzer, J.M.; Viana, A.R.; Kowsmann, R.O.; Freire, A.F.; Oreiro, S.G.; Augustin, A.H.; Lourega, R.V.; Rodrigues, L.F.; Heemann, R.; et al. Natural gas hydrates in the Rio Grande Cone (Brazil): A new province in the western South Atlantic. *Mar. Petrol. Geol.* **2015**, *67*, 187–196. [CrossRef]

47. Ketzer, J.M.; Augustin, A.; Rodrigues, L.F.; Oliveira, R.; Praeg, D.; Pivel, M.A.G.; Reis, A.T.; Silva, C.G.; Leonel, B. Gas seeps and gas hydrates in the Amazon deep-sea fan. *Geo-Mar. Lett.* **2018**, *38*, 429–438. [CrossRef]

48. Dickens, G.R.; Quinby-Hunt, M.S. Methane hydrate stability in sea- water. *Geophys. Res. Lett.* **1994**, *21*, 2115–2118. [CrossRef]

49. Boyer, T.P.; Baranova, O.K.; Coleman, C.; Garcia, H.E.; Grodsky, A.; Locarnini, R.A.; Mishonov, A.V.; O'Brien, T.D.; Paver, C.R.; Reagan, J.R.; et al. *World Ocean Database 2018*; Mishonov, A., Ed.; NOAA: Silver Spring, MD, USA, 2018; 209p.

50. Giongo, A.; Haag, T.; Lopes Simao, T.L.; Medina-Silva, R.; Utz, L.R.P.; Bogo, M.R.; Bonatto, S.; Zamberlan, P.; Augustin, A.H.; Lourega, R.V.; et al. Discovery of a chemosynthesis-based community in the western South Atlantic Ocean. *Deep Sea Res. Part Oceanograph. Res. Pap.* **2016**, *112*, 45–56. [CrossRef]

51. Medina-Silva, R.; Oliviera, R.R.; Trindade, F.J.; Trindade, F.J.; Borges, L.G.A.; Simao, T.L.L.; Augustin, A.H.; Valdez, F.P.; Constant, M.; Simundi, C.; et al. Microbiota associated with tubes of Escarpia sp. from cold seeps in the southwestern Atlantic Ocean constitutes a community distinct from that of surrounding marine sediment and water. *Antonie Leeuwenhoek Int. J. Gen. Mol. Microbiol.* **2017**, *111*, 533–550. [CrossRef]

52. Rodrigues, L.F.; Ketzer, J.M.; Lourega, R.V.; Augustin, A.H.; Sbrissa, G.; Miller, D.J.; Heemann, R.; Viana, A.R.; Freire, A.F.M.; Morad, S. The influence of methane fluxes on the sulfate/methane interface in sediments from the Rio Grande Cone Gas Hydrate Province, southern Brazil. *Braz. J. Geol.* **2017**, *47*, 369–381. [CrossRef]

53. Bayon, G.; Henderson, G.M.; Bohn, M. U–Th stratigraphy of a cold seep carbonate crust. *Chem. Geol.* **2009**, *260*, 47–56. [CrossRef]

54. Portilho-Ramos, R.C.; Cruz, A.P.S.; Barbosa, C.F.; Rathburn, A.E.; Mulitza, S.; Venancio, I.M.; Schwenk, T.; Rühlemann, C.; Vidal, L.; Chiessi, C.M.; et al. Methane release from the southern Brazilian margin during the last glacial. *Sci. Rep.* **2018**, *8*, 5948. [CrossRef]

55. Rodrigues, L.F.; Ketzer, J.M.; Oliveira, R.R.; Santos, V.H.J.M.; Augustin, A.H.; Cupertino, J.A.; Viana, A.R.; Leonel, B.; Dorle, W. Molecular and isotopic composition of hydrate-bound, dissolved and free gases in the Amazon deep-sea fan and slope sediments, Brazil. *Geosciences* **2019**, *9*, 73. [CrossRef]

56. Haacke, R.R.; Hyndman, R.D.; Park, K.P.; Yoo, D.G.; Stoian, I.; Schmidt, U. Migration and venting of deep gases into the ocean through hydrate choked chimneys offshore Korea. *Geology* **2009**, *37*, 531–534. [CrossRef]

57. Liu, X.; Flemings, P.B. Dynamic multiphase flow model of hydrate formation in marine sediments. *J. Geophys. Res. Solid Earth* **2007**, *112*, B03101. [CrossRef]

58. Hornbach, M.J.; Ruppel, C.; Van Dover, C.L. Three dimensional structure of fluid conduits sustaining an active deep marine cold seep. *Geophys. Res. Lett.* **2007**, *34*, L05601. [CrossRef]

59. Riedel, M.; Tréhu, A.M.; Spence, G.D. Characterizing the thermal regime of cold vents at the northern Cascadia margin from bottom simulating reflector distributions, heat probe measurements and borehole temperature data. *Mar. Geophys. Res.* **2010**, *31*, 1–16. [CrossRef]

60. Stranne, C.; O'Regan, M.; Jakobsson, M. Modeling fracture propagation and seafloor gas release during seafloor warming-induced hydrate dissociation. *Geophys. Res. Lett.* **2017**, *44*, 8510–8519. [CrossRef]

61. Gorman, A.R.; Senger, K. Defining the updip extent of the gas hydrate stability zone on continental margins with low geothermal gradients. *J. Geophys. Res.* **2010**, *115*, B07105. [CrossRef]

62. Horozal, S.; Bahk, J.J.; Urgeles, R.; Kim, G.Y.; Cukur, D.; Kim, S.P.; Lee, G.H.; Lee, S.H.; Ryu, B.J.; Kim, J.H. Mapping gas hydrate and fluid flow indicators and modeling gas hydrate stability zone (GHSZ) in the Ulleung Basin, East (Japan) Sea: Potential linkage between the occurrence of mass failures and gas hydrate dissociation. *Mar. Pet. Geol.* **2017**, *80*, 171–191. [CrossRef]

63. Schmidtko, S.; Johnson, G. Multidecadal Warming and Shoaling of Antarctic Intermediate Water. *J. Clim.* **2012**, *25*, 207–221. [CrossRef]

64. Stranne, C.; O'Regan, M.; Jakobsson, M. Overestimating climate warming-induced methane gas escape from the seafloor by neglecting multiphase flow dynamics. *Geophys. Res. Lett.* **2016**, *43*, 8703–8712. [CrossRef]

65. Phrampus, B.J.; Hornback, M.J. Recent changes to the Gulf Stream causing widespread gas hydrate destabilization. *Nature* **2012**, *490*, 527–529. [CrossRef]

66. Maslin, M.; Vilela, C.; Mikkelsen, N.; Grootes, P. Causes of catastrophic sediment failures of the Amazon fan. *Quat. Sci. Rev.* **2005**, *24*, 2180–2193. [CrossRef]

67. Reis, A.T.; Silva, C.G.; Gorini, M.A.; Leao, R.; Pinto, N.; Perovano, R.; Santos, M.V.M.; Guerra, J.V.; Jeck, I.K.; Tavares, A.A. The Chui Megaslide Complex: Regional-Scale Submarine Landslides on the Southern Brazilian Margin. In *Submarine Mass Movements and their Consequences, Advances in Natural and Technological Hazards Research 41*; Lamarche, G., Mountjoy, J., Bull, S., Hubble, T., Krastel, S., Lane, E., Micallef, A., Moscardelli, L., Mueller, C., Pecher, I., et al., Eds.; Springer: Cham, Switzerland, 2016.

68. Chiessi, C.M.; Mulitza, S.; Paul, A.; Pätzold, J.; Groeneveld, J.; Wefer, G. South Atlantic interocean exchange as the trigger for the Bølling warm event. *Geology* **2008**, *36*, 919–922. [CrossRef]

69. Milkov, A.V.; Sassen, R.; Novikova, I.; Mikhailov, E. Gas hydrates at minimum stability water depths in the Gulf of Mexico: Significance to Geohazard Assessment. *Gulf Coast Assoc. Geol. Soc. Trans. L* **2000**, *50*, 217–224.

geosciences

MDPI

Article

Molecular and Isotopic Composition of Hydrate-Bound, Dissolved and Free Gases in the Amazon Deep-Sea Fan and Slope Sediments, Brazil

Luiz F. Rodrigues [1,*], João M. Ketzer [2], Rafael R. Oliveira [3], Victor H.J.M. dos Santos [1],
Adolpho H. Augustin [1], Jose A. Cupertino [1], Adriano R. Viana [4], Bruno Leonel [5]
and Wilhelm Dorle [5]

[1]   Institute of Petroleum and Natural Resources, PUCRS—Pontifícia Universidade Católica do Rio Grande do
       Sul, Porto Alegre 90619-900, Brazil; victor.santos@pucrs.br (V.H.J.M.d.S.);
       adolpho.augustin@pucrs.br (A.H.A.); jose.cupertino@pucrs.br (J.A.C.)
[2]   Department of Biology and Environmental Science, Linnaeus University, 391-82 Kalmar, Sweden;
       marcelo.ketzer@lnu.se
[3]   Neoprospecta, Florianópolis 88056-000, Brazil; rafael.oliveira@neoprospecta.com
[4]   PETROBRAS—E&P EXPLORATION, Rio de Janeiro 20031-170, Brazil; aviana@petrobras.com.br
[5]   Seaseep Dados de Petróleo, Rio de Janeiro 20031-144, Brazil; bleonel@seaseep.com.br (B.L.);
       wdorle@seaseep.com.br (W.D.)
*   Correspondence: frederico.rodrigues@pucrs.br; Tel.: +55-(51)-3320-3689

Received: 21 December 2018; Accepted: 22 January 2019; Published: 31 January 2019

check for
updates

Abstract: In this work, we investigated the molecular stable isotope compositions of hydrate-bound and dissolved gases in sediments of the Amazon deep-sea fan and adjacent continental slope, Foz do Amazonas Basin, Brazil. Some cores were obtained in places with active gas venting on the seafloor and, in one of the locations, the venting gas is probably associated with the dissociation of hydrates near the edge of their stability zone. Results of the methane stable isotopes (δ^{13}C and δD) of hydrate-bound and dissolved gases in sediments for the Amazon fan indicated the dominant microbial origin of methane via carbon dioxide reduction, in which ^{13}C and deuterium isotopes were highly depleted (δ^{13}C and δD of -102.2‰ to -74.2‰ V-PDB and -190 to -150‰ V-SMOW, respectively). The combination of C1/(C2+C3) versus δ^{13}C plot also suggested a biogenic origin for methane in all analysed samples (commonly >1000). However, a mixture of thermogenic and microbial gases was suggested for the hydrate-bound and dissolved gases in the continental slope adjacent to the Amazon fan, in which the combination of chemical and isotopic gas compositions in the C1/(C2+C3) versus δ^{13}C plot were <100 in one of the recovered cores. Moreover, the δ^{13}C-ethane of -30.0‰ indicates a thermogenic origin.

Keywords: Amazon fan; gas hydrate; thermogenic gas; biogenic gas; molecular composition; isotopic composition

---

1. Introduction

Energy production is one of the most important topics in our society nowadays owing to the increasing pressure on conventional sources based on petroleum. Global efforts encourage scientific programs to identify new alternative, cleaner, and unconventional energy sources [1]. One such alternative is gas hydrates, which are crystalline, ice-like structures that contain guest molecules, mainly methane [2]. Gas hydrates are widely distributed around the world and can be found in permafrost and submarine environments that present specific conditions of pressure and temperature for their formation [2–5]. Among the potential CH_4 sources, gas hydrates stand out as an important unconventional energy reserve, with quantities comparable to other known conventional sources [3–7]. The hydrocarbons trapped in hydrates can be derived from biological methanogenesis (biogenic origin) or formed by the thermal decomposition of organic matter (thermogenic origin) [4,8].

Besides being an important energy source, methane is an important player in the global biogeochemical carbon cycle and accounts for 20% of greenhouse effects [9–11]. In a global warming scenario, the large amount of methane may represent an environmental hazard, since destabilization of gas hydrates could result in a high volume of gas releasing into the atmosphere, potentially promoting catastrophic geological consequences [6,7,12]. Thus, the mapping of gas hydrate distribution, inventory assessment, and a comprehensive understanding of their role in energy and environmental issues should justify the establishment of a global agenda on this issue.

In Brazil, the Rio Grande Cone Gas Hydrate Province in the Pelotas Basin [13] is one of the main areas with gas hydrate deposits with associated chemosynthetic communities [14,15], and consists of the first discovery of gas hydrates in the western South Atlantic. The direct recovery of gas hydrate was performed through sediment sampling by piston coring and the results showed that the gas was mainly composed of methane of biogenic origin [13]. Recently, other works described the methane-sulphate interface dynamics [16] and the origin of organic matter in hydrate-bearing sediments [17] in the Pelotas Basin region.

The Ocean Drilling Program (ODP) 155 LEG, was performed in the Amazon fan area and resulted in a drilling campaign with 17 sites and covered an extensive area of the mid- and lower fan [18–20] The geochemical modelling conducted on the Amazon fan sediment concluded that the region is very susceptible to the production of biogenic methane. This CH_4 could be accumulated along the sediment column and, in regions with adequate pressure and temperature (waters deeper than 600 m), is likely to be trapped in gas hydrates [1]. The mass of organic carbon trapped in the Amazon deep-sea fan sediments is, for comparison, larger than the mass of carbon stored at the present-day Amazon forest [21]. In a previous work, from the seismic record of the Amazon fan, the presence of gas charged sediments was found [22].

From the ODP Leg 155 records and seismic investigations, a slope failure was detected that was associated with a massive sediment mass-transport deposits along the fan structure [18,23–25]. These catastrophic events were associated with glacial-interglacial cycles, which result in sea level (and pressure and bottom water temperature) changes and the destabilization of the gas hydrate layer. The subsequent gas release increased pore pressure and destabilization of the sediment package and resulted in the triggering of mass transport of deposits [23,24].

Another strong indicator of the existence of gas hydrates in the Amazon region is the presence of bottom simulating reflectors, observed within the upper fan [18,23–28]. These seismic records were described in water depths of about 2000 m, and are within the gas hydrate stability zone (GHSZ) [27]. The presence of gas hydrates together with gas venting sites at the seafloor in the Amazon deep-sea fan and Foz do Amazonas Basin was confirmed by the recovery of samples during a sea campaign in 2015. The gas compositions from hydrates recovered in vents indicate essentially biogenic sources with a possible thermogenic contribution, dominantly methane $\delta13C$ from −81.1% to −59.2% [21].

In the present work, additional chemistry tests were performed to improve the discussion of the origin of gas composition and isotopic analysis obtained through piston core sampling between hydrate-bound and dissolved gases in sediments collected in the Amazon fan and continental slope of the Foz do Amazonas Basin, Brazil. The gas composition results presented in this study indicate that the Amazon fan area represents a promising new energy frontier. The knowledge of the composition and origin of the hydrate-bound, dissolved and free gases in the Amazon deep-sea fan and slope sediments will help with assessing the potential of methane hydrates as an energy source and their relationship with the global carbon cycle and climate.

2. Study Area and Geological Setting

The Amazon fan is a deep-sea structure located in the Foz do Amazonas Basin, and extends seawards from the continental shelf, from the shelf break in front of the Amazon River to the Demerara Abyssal Plain (Figure 1). Sediments deposited in the fan were sourced mostly from the Andean belt [29] since the Late Miocene [18] and carried to the Atlantic Ocean and the Foz do Amazonas Basin by the Amazon River, which has one of the largest sediment [26] and organic matter [30] discharges in the world. In addition to the Amazon fan, we also included in this work data from the adjacent continental slope to the northwest. This area is formed by a series of submarine canyons, approximately 35 km wide and 130 km long, in which each canyon is formed by dozens of small submarine channels connected to a main canyon (Figure 1). These structures begin at the shelf break and develop down to the upper continental slope at water depths of more than 3000 m. Information about geological models and the hydrates' stability zone for the study area can be found in a previous manuscript [21] that discusses the new evidence on widespread gas venting from the gas hydrate stability zone on the upper Amazon deep-sea fan.

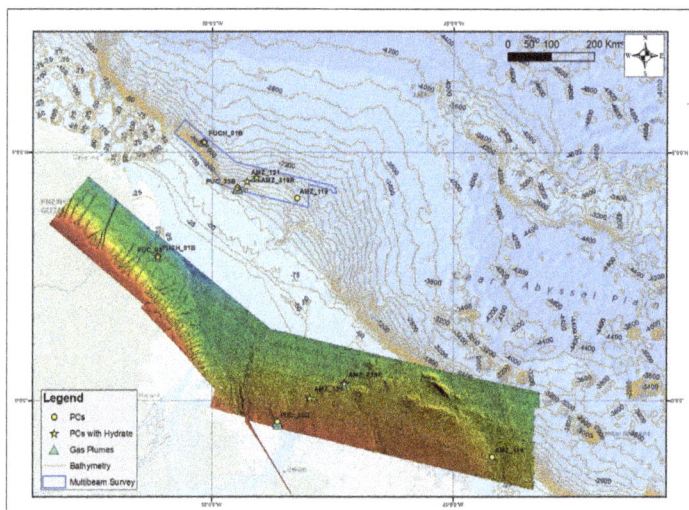

Figure 1. Location map showing the bathymetry (coloured) of the study area in the Amazon deep-sea fan, the adjacent continental slope area to the northwest with small canyons connected to a main canyon channel. The blue dots are the piston cores used in this study.

3. Samples and Methods

The samples and data used in this study were collected on board the research vessel Inspector II, equipped with a 6 m long piston corer unit. The precision of sea floor sampling was assured by an Ultra-short baseline (USBL) transponder/responder (Sonardyne, London, UK) fastened 5 m above the corer. A total of four oceanographic missions were carried out in July 2012, July 2015, August 2015 and September 2015. The first mission was dedicated to surveying the area with a multi-beam echo sounder (MBES) and covered an area of 18,490 km^2, generating high-resolution seafloor maps (15 × 15 m unit bathymetry cell and 10 × 10 m unit backscatter cell) and acoustic profiles of the water column. A high-resolution (CHIRP 3.5 kHz) seismic profile with average penetration of 10 m below the seafloor was acquired with a sub-bottom profiler, together with the MBES survey. The second, third and fourth missions were dedicated to core sediment sampling at specific targets (possible vent sites such as pockmarks and mounds with high backscatter signatures) on the seafloor, previously identified with the MBES data. Four cores were collected in the Amazon fan proper (cores PUC 35B,

AMZ 119, AMZ 121 and AMZ 319R), and two cores were collected in the adjacent continental slope area (cores PUC 03 and PUCH 01B). Cores PUC 35B, AMZ 319R, PUC 03 and PUCH 01B, were collected in active gas venting sites in the area [21]. The cores recovered were cut into sections of 1 m and the sediment samples were collected at the top and base of each core section.

Sediment samples for gas analysis (C1-C5 hydrocarbons) were collected from the top of each section (i.e., every 100 cm of the core). Samples were collected from the freshly exposed tops of the sections using a spatula and were immediately placed in gas-tight IsoJar containers from Isotech Laboratories Inc., Champaign (IL), EUA. The sediment sample filled one-third of gas-tight recipients (IsoJars), while another third was filled with distilled water, leaving the top third filled with air (headspace). The benzalkonium chloride bactericide were added into the IsoJars in order to eliminate microbial activity and the jars were kept refrigerated at 4 °C [31,32].

We also collected hydrate samples, in addition to sediment samples. The hydrate samples were collected using a spatula, placed in IsoJar containers and kept at room temperature. The headspace gases formed by the dissociation of hydrates inside the container were sampled and analysed. In this work, we present the carbon dioxide quantification and the respective values of carbon-stable isotope (δ^{13}C-CO$_2$), the deuterium isotope results of methane (δD-CH$_4$) and carbon-stable isotope results of ethane (δ^{13}C-C$_2$H$_6$) of three gas hydrate samples taken from three representative piston cores (PUCH 01B, AMZ 319R, and AMZ121). The results of δ^{13}C-CH$_4$ and gas composition (C1-C5 hydrocarbons) were previously reported [21].

Gas composition was determined using a gas chromatograph (Shimadzu Corporation, model GC-2014, from Kyoto, Japan) equipped with a VPPlot Alumina/KCl capillary column (Sigma-Aldrich Corporation., St. Louis, MO, USA), measuring 30 m × 0.53 mm with a 10.0 μm film. The analytical conditions used in this work are described in previous work [33]. Carbon-stable isotope analyses of the headspace gases were performed using a Thermo Fisher Scientific gas chromatograph (Waltham, MA, USA), coupled to a Thermo Scientific DELTAV Plus isotope ratio monitoring mass spectrometer via a Thermo GC IsoLink and Conflo IV interface (Thermo Fisher Scientific). The gas chromatograph contained a 30 m × 0.32 mm fused silica column, Carboxen Plot 1006 (Sigma-Aldrich Corporation., St. Louis, MO, EUA), and was operated at a heating ramp of 70–150 °C, over 30 min.

4. Results and Discussion

4.1. Amazon Fan

The gas composition of free gas from sediment samples collected in cores PUC 35B, AMZ 119, AMZ 121 and AMZ 319R were dominated by methane (81.05 to 99.97%), followed by CO$_2$ (0.76 to 12.62%) and ethane (0.01 to 0.3%). Traces of propane were found in AMZ 121(0.007 to 0.013%) and AMZ 319R (0.03 to 0.07%). The C1/(C2+C3) ratio varied from 330–31,049 whereas δ^{13}CH$_4$ and δD-CH$_4$ values ranged from −102.2% to −74.2% and −192% to −150%, respectively. It was not possible to analyse the carbon isotope compositions of ethane from hydrates, because the ethane concentrations were below the limit of detection of the technique (Table 1).

According to Bernard classification [34], the combination of gas, chemical and isotopic compositions of free gas from sediments in the C1/(C2+C3) versus δ^{13}C plot, suggested a biogenic origin for all analysed samples from the Amazon fan (Figure 2a). Microbial CO$_2$ reduction mainly generates methane, and there are traces of ethane and propane, so the relative C1/C2+ rations are very high (commonly >1000). Based on the classification scheme of Whiticar [35], δ^{13}CH$_4$ and δD-CH$_4$ are typically in the range for microbial methane production by CO$_2$ reduction (Figure 2b). This biogenic origin was also reported in other gas hydrate provinces around the world, like Rio Grande Cone [13], Blake Ridge [36], and the Black Sea [37,38].

The gas hydrate collected in PC AMZ 319R (Table 1) was essentially formed by methane (> 99%) with minor quantities of ethane, propane, propylene and low amounts of CO$_2$. The C1/(C2+C3) ratio for these samples ranges from 2406 to 2299 (Figure 2a), whereas δ^{13}C methane values range from

−79.5% to −77.3% [21], while for δD-CH$_4$, values range from −188% to −186% (Table 1). AMZ 121 was assayed by only one hydrate sample and the gas composition results showed a hydrate formed by methane (99.88%) and carbon dioxide (1.11%). The δ^{13}CH$_4$ /δD-CH$_4$ values were −81.1% [21] and −190%, respectively. According to the classification proposed by Bernard et al. [34], (Figure 2a), the samples obtained at the AMZ 319R and AMZ121 sites are consistent with the field of microbial gas and both samples collected would be of microbial origin produced by microbial CO$_2$ reduction (Figure 2b).

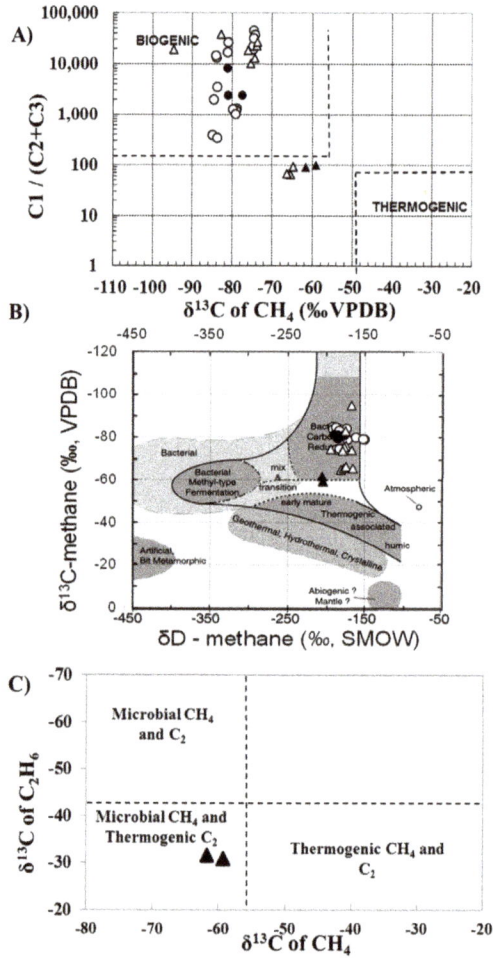

Figure 2. Diagram showing the relationship between molecular and isotopic compositions of hydrate-bound, free and dissolved gases in sediments of the Amazon fan and adjacent continental slope, based on the classifications of Whiticar [35], Milkov [39] and Taylor et al. [40]. (**A**) Chemical and isotopic compositions of free gas from sediments using C1/(C2+C3) versus ^{13}C plot; (**B**) Chemical and isotopic compositions of free gas from sediments using ^{13}C-CH$_4$ versus D-CH$_4$; (**C**) Chemical and isotopic compositions of free gas from sediments using ^{13}C$_2$H$_6$ versus ^{13}CH$_4$. Open and solid circles indicate hydrate-bound gases, and dissolved and free gases in pore water at the Amazon fan area, respectively, whereas open and solid triangles indicate hydrate-bound, free and dissolved gases in the adjacent continental slope study area northeast of the Amazon fan, respectively. The results of δ^{13}C-CH$_4$ and gas composition (C1–C5 hydrocarbons) for gas hydrate were previously reported [21].

4.2. Northwest Continental Slope Area

Average hydrocarbon compositions in gases from sediment samples of PUC 03 and PUCH 01B were, in general, strongly dominated by methane (91.06–98.88%) and CO_2 (0.20–8.94%). Only in PUCH 01B were ethane (1.12–1.47%), propane (0.005%), propylene (0.002–0.017%) and traces of butane and butene (1.05 and 1.33%, respectively) found.

The C1/(C2+C3) ratio for PUC 03 and PUCH 01B varied from 847–22,484 and 65.8–92.7, respectively, whereas $\delta^{13}CH_4$ values for PUC 03 and PUCH 01B ranged from −94.8‰ to −73.7‰ and −66.3‰ to −64.9‰, respectively. For δD-CH_4, the values range from −194‰ to −167‰ and −181‰ to −166‰ for PUC 03 and PUCH 01B, respectively. Regarding the isotope composition of carbon of ethane from hydrates, the values obtained were −30.8‰ and −31.5‰ (Table 1).

Figure 2a shows the relationship between the isotopic ratios and gas compositions of hydrate-bound gas, and free and dissolved gases in sediments. According to the classification proposed by Bernard et al. [34], the analysed methane gas obtained at the PUC 03 site plotted in the field of microbial gas, whereas some samples from the PUCH 01B site suggested a mix between microbial and thermogenic sources (Figure 2a). In the $\delta^{13}C$-δD diagram of methane shown in Figure 2b and based on the classification of Whiticar [35], C1 at all sites would be of microbial origin produced by microbial CO_2 reduction (Figures 1 and 2b).

The gas hydrate recovered at PUCH 01B revealed a mixture of biogenic and thermogenic gases, both in the chemical and isotopic compositions in the C1/(C2+C3) versus $\delta^{13}C$ plot (Figure 2a). Compared with the deep ocean sediment from the Gulf of Mexico, the gas hydrate recovered always contained >99% methane. The only exception was the gas hydrate obtained from the ODP Leg 96 with the low C1/(C2+C3) value of 160. Gas hydrate samples recovered from the Caspian Sea contained methane at 59.1–96.2% in the hydrocarbon gas mixtures, with C1/(C2+C3) values ranging from 1.7 to 45. This gas hydrate of the Caspian Sea was associated with fluid venting from the Buzdag and Elm mud volcanoes [41].

The gas hydrate recovered at PUCH 01B contained >98% methane, around 1% ethane, 0.02% propylene, and 0.07–0.5% of CO_2, and the C1/(C2+C3) value ranged between 89 and 98. Both samples collected would be of microbial origin, produced by microbial CO_2 reduction, as shown in Whiticar's diagram. However, Whiticar's diagrams [35] only correspond to results for the isotopic composition of methane. According to the isotopic analysis diagrams of ethane, the origin based on the classification of Milkov [39] is thermogenic. This suggests that a thermogenic gas is possibly present in the Northwest continental slope area, and constitutes a detectable gas constituent of the gas mix, together with methane resulting from microbial CO_2 reduction.

It is very important to explain that in some cores we found ethane without the presence of propane or butane (i.e., PUCH01B). Biodegradation of original gas composition can also produce "dry gas" due to the preferential removal of propane and butane [42]. Ethane is rarely or much less easily consumed, so the ethane/propane ratio increases with the biodegradation rate. Because propane amounts are much lower than those of ethane, its depletion does not significantly influence the Bernard ratio. In fact, there are examples of gas from biodegraded reservoirs with very low Bernard ratios, due to high ethane concentrations [42,43].

Light hydrocarbons and gases (C2–C5 hydrocarbons) can also be biodegraded; propane and n-butane are degraded the most rapidly. Biodegradation of C2–C5 hydrocarbons, increases the methane content. This may result in the production of a gas cap enriched in methane following the removal of wet gas components (C2–C5 alkanes), and probably also by direct production of methane during biodegradation [44–47]. However, even with the biodegradation of gas, it was possible to detect the presence of some amounts of other hydrocarbons distributed at different depths in the piston core. Although there were trace amounts of C3-C4, we suggest that the biodegradation process changed the original composition of gas. In addition, it was not possible to detect trace amounts of C3-C4 in some depths, because the concentrations were below the detection limit of the analytical technique.

Table 1. Molecular and stable isotope compositions of hydrate-bound gases (HG) and headspace gases at all sites. Methane (C1), ethane (C2), propane (C3), propene (=C3), butene (=C4), butane (C4), pentane (C5). The depth is in meters below sea floor (mbsf).

Piston Core	Depth (mbsf)	^{13}C-CH_4	D-CH_4	^{13}C-C_2H_6	^{13}C-CO_2	C1 (%)	C2 (%)	C3 (%)	=C3 (%)	=C4 (%)	C4 (%)	C5 (%)	CO_2 (%)	C1/C+
PUC 03	0.3	−76	−175	n.d.	−44.5	96.12	n.d.	n.d.	n.d.	n.d.	n.d.	n.d.	3.88	18,336
	1	−75.5	−177	n.d.	−41.1	96.52	n.d.	n.d.	n.d.	n.d.	n.d.	n.d.	3.48	9955
	2	−74.4	−195	n.d.	−45.2	96.91	n.d.	n.d.	n.d.	n.d.	n.d.	n.d.	3.09	13,030
	3	−73.9	−167	n.d.	−43.4	97.16	n.d.	n.d.	n.d.	n.d.	n.d.	n.d.	2.84	22,484
	4	−73.7	−179	n.d.	−48.6	92.19	n.d.	n.d.	n.d.	n.d.	n.d.	n.d.	7.81	1214
	5	−82.8	−176	n.d.	−46.2	94.22	n.d.	n.d.	n.d.	n.d.	n.d.	n.d.	5.78	847
	6	−94.8	−168	n.d.	−49.7	91.06	n.d.	n.d.	n.d.	n.d.	n.d.	n.d.	8.94	18,679
PUCH01B	0.3 (HG)	−59.2	−205	−30.8	−21.4	98.88	1.12	n.d.	0.002	n.d.	n.d.	n.d.	n.d.	98.1
	0.3 (HG)	−61.7	−206	−31.5	−21.8	98.75	1.23	n.d.	0.02	n.d.	n.d.	n.d.	n.d.	89.2
	0.7	−64.9	−182	n.d.	−41.2	98.54	1.42	n.d.	0.04	n.d.	n.d.	n.d.	n.d.	77.1
	1	−65.6	−178	n.d.	−41.5	98.49	1.29	0.005	0.017	n.d.	n.d.	n.d.	0.2	92.7
	1.4	−66.3	−175	n.d.	−38.7	96.77	1.47	n.d.	n.d.	n.d.	n.d.	n.d.	1.8	65.8
	1.9	−65.5	−166	n.d.	−45.3	94.49	1.43	n.d.	n.d.	1.33	1.05	n.d.	1.71	24.84
PUC35B	0.4	−81.1	−181	n.d.	−52.8	90.61	n.d.	n.d.	n.d.	n.d.	n.d.	n.d.	9.39	16,429
	1.4	−83.9	−183	n.d.	−49.7	91.88	n.d.	n.d.	n.d.	n.d.	n.d.	n.d.	8.12	13,010
	2.6	−84	−174	n.d.	−33.9	93.31	n.d.	n.d.	n.d.	n.d.	n.d.	n.d.	6.69	14,039
AMZ119	0.2	−74.6	−182	n.d.	−9.3	93	n.d.	n.d.	n.d.	n.d.	n.d.	n.d.	7	43,905
	1	−74.8	−182	n.d.	−7.7	81.05	n.d.	n.d.	n.d.	n.d.	n.d.	n.d.	18.5	19,017
	2.9	−74.2	−188	n.d.	−8.1	91.86	n.d.	n.d.	n.d.	n.d.	n.d.	n.d.	8.14	36,842
	4	−74.6	−179	n.d.	−10.4	87.38	n.d.	n.d.	n.d.	n.d.	n.d.	n.d.	12.62	22,283
	5.3	−74.6	−183	n.d.	−10.3	91.49	n.d.	n.d.	n.d.	n.d.	n.d.	n.d.	8.51	31,049
AMZ121	0.6	−102.2	−190	n.d.	−64.7	91.35	0.244	n.d.	n.d.	n.d.	n.d.	n.d.	8.41	375
	1.8	−84.9	−192	n.d.	−35.2	94.04	0.285	n.d.	n.d.	n.d.	n.d.	n.d.	5.68	330
	2.8	−84.6	−188	n.d.	−36.6	99.47	0.03	0.007	0.065	n.d.	n.d.	n.d.	0.49	3413
	3.0	−83.7	−192	n.d.	−33.1	97.17	n.d.	0.05	n.d.	n.d.	n.d.	n.d.	2.72	1974
	3.2	−83.7	−186	n.d.	−35.1	99.65	0.02	0.005	n.d.	n.d.	n.d.	n.d.	0.32	5536
	3.5 (HG)	−81.1	−190	n.d.	−35.0	98.88	0.01	n.d.	n.d.	n.d.	n.d.	n.d.	1.11	7936
AMZ319R	1.0	−78.8	−173	n.d.	−42.6	98.33	0.08	n.d.	n.d.	n.d.	n.d.	n.d.	1.59	1301
	1.5 (HG)	−80.9	−186	n.d.	−37.2	99.36	0.04	n.d.	n.d.	n.d.	n.d.	n.d.	0.60	2406
	2.4	−78.9	−150	n.d.	−42.0	98.02	0.08	0.03	n.d.	n.d.	n.d.	n.d.	1.82	1169
	3.5	−79.9	−162	n.d.	−39.4	97.87	0.08	n.d.	n.d.	n.d.	0.06	n.d.	1.99	1241
	3.7	−79.1	−152	n.d.	−43.0	97.14	0.10	0.05	n.d.	n.d.	n.d.	n.d.	2.64	1010
	3.4 (HG)	−77.3	−188	n.d.	−43.2	99.84	0.04	0.002	0.003	n.d.	n.d.	n.d.	0.11	2299

4.3. Alteration of Hydrocarbon Gas

Various marine sediments of εc values range from 49 to over 100, with values most commonly around 65–75 (Figure 3) [35,48,49]. Our data at AMZ 119 (Amazon fan proper) are consistent with methanogenesis by microbial CO_2 reduction in the literature, showing values ranging from 64 to 68. On the other hand, εc values of gases from the headspace from PUC 03, PUCH 01B, AMZ 121 and AMZ 319R range from 20 to 55, which is significantly smaller than that expected from CO_2 reduction. In the adjacent continental slope area, the εc values from PUC 03 range 25–35, while for PUCH 01B the range was 20–50. In the Amazon fan area, the AMZ 121, AMZ 319R and PUC 35B range was 37–50, 36–44 and 28–50, respectively.

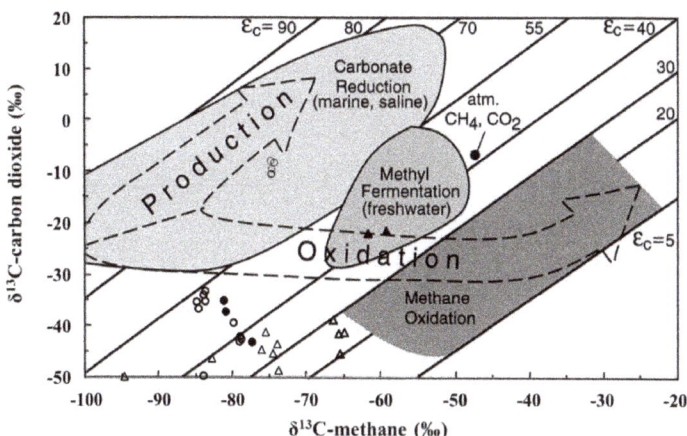

Figure 3. Combination plot of δ^{13}C-CO_2 and δ^{13}C-CH_4 with isotope fractionation lines (εc). Methanogenesis by carbonate reduction (saline, marine region) has a larger CO_2–CH_4 isotope separation than methane consumption. The figure also shows the observed CO_2–CH_4 carbon isotope partitioning trajectories resulting from both microbial methane formation and oxidation processes (adapted from Whiticar [35]).

At first glance, our εc results suggest that methane in gas hydrates in the PUC 01B sample originated from methyl fermentation in freshwater or the gas composition contained a fraction of ^{13}C enriched from a thermogenic source. However, this indication is not consistent with the molecular and isotopic data, which clearly document microbial methane generation via CO_2 reduction (Figure 2b).

There are some processes that may affect the molecular and isotopic composition of a gas after its formation/accumulation in a reservoir or during its migration to the surface: (i) mixing of several gas sources, and (ii) anaerobic oxidation of methane [35,50,51]. These processes are described below in the context of our samples.

4.3.1. Mixing of Several Gas Sources

Gases can be mixed in the reservoir or mixing can happen during migration through the sediments, where microbial gas pools may exist at shallower depths. A detectable thermogenic influence in the gas mix is possibly stronger, at the vicinity of gas conduits like faults, connecting deep gas sources to the surface [52]. The attribution of mixing is due to other secondary post-synthesis processes and the values of carbon isotopes of CH_4 alone can be misleading. The recognition of mixing should be based on δ^{13}C-CH_4 vs. C1/(C2+C3) plots, as can be seen in the continental slope area plots (Figure 2b).

If we look only at the δ^{13}C-CH_4 vs. C1/(C2+C3) plots, mixing of several sources can be partially excluded in the Amazon fan samples, according to the isotopic signatures of methane that show a clear microbial origin. However, in marine sediments, methanogenesis from microbial CO_2 reduction usually leads to an εc range of 49–100%. Furthermore, the carbon fractionation factor εc, which is the difference between δ^{13}C-CO$_2$ and δ^{13}C-CH$_4$, should remain constant upon depletion [35]. However, in this work, this factor varied widely between 28% and 67% in samples from the Amazon fan proper, and 20–50% in the adjacent continental slope area. The values obtained here fell below the normal range (49-100%), except for the gas sample PC AMZ 119. Low εc values may be partly explained by the fact that methane is transported by both diffusion as a dissolved phase and advection as gas bubbles, in which the latter is probably an important process owing to the presence of large quantities of gas venting on the water column [21]. The acoustic chimneys in the Amazon fan and adjacent slope area are usually associated with the presence of numerous fractures [53–55], which may act as potential fluid conduits for the upward migration of microbial gas or thermogenic gases [56]. Methane transport in the gas phase forms massive hydrates at the chimney sites in this basin [21].

The predominance of isotopically light methane in sampled sediment cores to depths of 380 mbsf, drilled in the Gulf of Mexico in 2005, and the minor amounts of clearly petroleum-related thermogenic gas, and the widespread occurrence of thermogenic gas in the northern Gulf Mexico was surprising. The exact origin of that deeper gas source was uncertain, but Lorenson et al. [57] based their results on the biodegradation of petroleum that originated in more deeply buried sediments. In the present work, we do not have the same interval depth as Lorenson's study; however, in a few meters we found: (1) trace amounts of heavy hydrocarbons in the gas hydrate composition; (2) enrichment of the heavier ^{13}C in methane of sediments with increasing depths; (3) enrichment of ^{13}C in methane comparing gas hydrate in different depths; and, (4) trace amounts of other low-molecular-weight hydrocarbons (propane, propylene and butane) in sediments.

Even with a biogenic origin related to microbial CO$_2$ reduction, it is important to mention that the gas composition of gas hydrate also showed trace amounts of propane and propylene in the deepest depth (3.4 mbsf) in AMZ 319R, which was typically indicative of a thermogenic contribution for the total gas composition. In Figure 4, even without apparent enrichment of gases dissolved in pore water, there was enrichment in ^{13}C in methane in the samples from gas hydrates, in relation to free gas and dissolved gas samples (i.e., δ^{13}CH$_4$ was −80.9% and −77.3% in 1.5 mbsf and 3.4 mbsf, respectively). Furthermore, the δ^{13}CH$_4$ values of free and dissolved gases showed enrichment of ^{13}C and there was a possible trend in which heavier carbon isotopes became more prominent with depth. These enrichments (both in gas hydrate as free and dissolved gasses) indicate that there was some possible thermogenic contribution to the mostly biogenic analysed gas mix. Finally, the low concentration of some hydrocarbons such as propane and butane, for example, is possibly due to high microbial degradation [44,47,58,59]. Therefore, the ratio C1/(C2+C3) could be modified to high values and could have masked the thermogenic origin. Figure 4 shows the relationship between the isotopic compositions and depth in sites AMZ 121 and AMZ 319R.

Amazon Fan

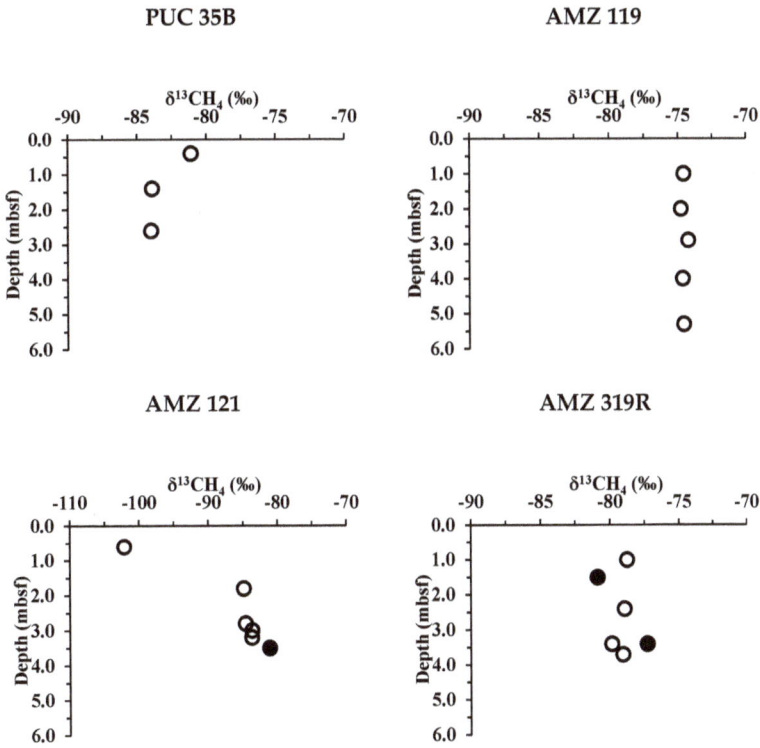

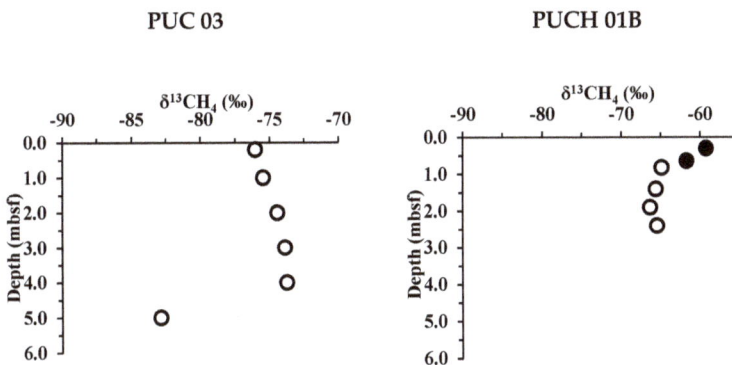

Figure 4. Diagrams showing the changes in the isotopic compositions with depth in sites of Amazon fan and northwest continental slope. Open and solid circles indicate dissolved gases in pore water and hydrate-bound gases, respectively.

Light isotopic values of CO_2, similar to the ones found in the Amazon deep-sea fan, were also described by Milkov [60]. Gases with CO_2 mostly enriched in ^{12}C have the most significant contribution of primary microbial and/or thermogenic methane and/or may represent the initial CO_2 from petroleum biodegradation, which had not yet been converted to methane [61] or was generated during biodegradation [62,63]. In gases with CO_2 mostly depleted in the lighter isotope ^{12}C, a significant portion of CO_2 was apparently converted into methane and most CO_2 was the residual gas from methanogenesis. Based on this discussion, the CO_2 produced during oil and gas biodegradation (without methanogenesis) should have d13C values ranging from −39% to −25% for oil [62–64] and <−40% for gas [65] which is generally more negative than the $\delta^{13}C$ of CO_2 liberated from dispersed organic matter during diagenesis.

Therefore, in this work, we suggest that the isotopic composition of the CO_2 is from gas oxidation (light hydrocarbons) and not from the methanogenic biodegradation of petroleum hydrocarbons through carbon dioxide reduction. Our $\delta^{13}C$ of CO_2 low values indicate, therefore, that CO_2 is not formed by methanogenic biodegradation of petroleum. The carbon isotope compositions of carbon dioxide reflect its genetic origins. However, as carbon dioxide has higher chemical reactivity and a wide range of potential genetic origins and sinks compared with hydrocarbons, it is usually difficult to determine its specific origin. Some main processes may affect molecular and isotopic composition of gas after its formation/accumulation in a reservoir or during its migration to the surface.

4.3.2. Anaerobic Oxidation of Methane

Another process that may affect the molecular and isotopic composition of a gas after its formation/accumulation in a reservoir or during its migration to the surface can be the anaerobic oxidation of methane. A possible explanation is the simplest occurrence of biogeochemical processes, specifically the anaerobic oxidation of methane coupled with carbonate precipitation within the sediment column. The latter process preferentially takes up methane with the lighter carbon isotope [66], therefore producing a ^{12}C-enriched CO_2 that leads to a decrease of εc, as shown in Figure 4 [35,67]. Methane is consumed by microbes at a greater rate than the heavier hydrocarbon gases. This reaction also results in a CO_2 pool that is enriched in ^{12}C [35], with reported εc values between 5 and 25 towards the latter stages of CH_4 consumption by AOM, which are due to the enrichment of ^{13}C in the residual methane, rather than to an increase in $^{12}CO_2$ from the oxidized methane. The larger pool of CO_2 relative to CH_4 obscures the oxidation trend of εc values in its initial stage [49]. Hence, AOM could be a potential factor contributing to the small εc observed in the studied samples.

Gases with CO_2 enriched in ^{12}C make the most significant contribution to primary microbial and/or thermogenic methane and/or may represent the initial CO_2 from petroleum biodegradation, which has not yet been converted to methane [61] or was generated during aerobic biodegradation [63]. In gases with CO_2 mostly depleted in ^{12}C, a significant portion of CO_2 was apparently converted into methane and most remaining CO_2 was the residual gas from methanogenesis [68]. Because ^{12}C–^{12}C bonds are easier to break than ^{12}C–^{13}C bonds, biodegraded residual gases become enriched in ^{13}C isotopes while produced CO_2 would be depleted in ^{13}C [44]. These gases are dissolved in biodegraded oils and contain high concentrations of ^{12}C-depleted CO_2, suggesting significant biodegradation of liquid hydrocarbons but perhaps limited biodegradation of gases [60].

In marine sediments, methanogenesis from microbial CO_2 reduction usually leads to an εc range of 49–100%. The values obtained for many samples in this work fell below this range. The most convincing explanation remains the occurrence of biogeochemical processes, more specifically the anaerobic oxidation of methane. This process preferentially takes up methane with the lighter carbon atom [66], therefore producing a ^{12}C-enriched CO_2, which leads to a decrease of εc, as shown in Figure 3. Ruffine et al. [56] determined the nature and origin of the gases bubbling at the Bay of Biscay. Their data were also plotted outside the appointed fields due to the very light carbon isotopic composition of CO_2. They attributed the very light carbon isotopic composition to anaerobic oxidation of methane, coupled with carbonate precipitation within the sedimentary column [56]. Recently, archaeal 16S rRNA gene sequences were recovered that were characteristic of anaerobic methane oxidizers from biodegraded oil reservoirs. This suggested that anaerobic methane oxidation occurred at some point during the subsurface degradation process [47]. As was mentioned, the gas biodegradation (without methanogenesis) should have d13C values <−40% for gas [65], which is generally more negative than the δ^{13}C of CO_2 liberated from dispersed organic matter during diagenesis.

5. Conclusions

Understanding the origin and composition of gases (methane, CO_2) in sediments and gases trapped in gas hydrates can help us to assess natural resources in sedimentary basins and also to better understand the global carbon cycle. Our findings suggest that, in addition to large quantities of methane formed by biogenic processes, the sediments and gas hydrates in the study area contain detectable quantities of gases of thermogenic origin. These gases are stored in the sediments and in gas hydrates, but also venting, and therefore, transferring mass (hydrocarbons of biogenic and thermogenic origins) to the ocean. The molecular and stable isotope compositions of hydrate-bound, free and dissolved gases in sediments from the Amazon fan and adjacent continental slope areas are different. For the Amazon fan there was a dominant microbial origin of methane via carbon dioxide reduction; however, a possible mixture of thermogenic and microbial gases resulted in relatively high methane δ^{13}C signatures in one of the sites in the adjacent continental slope. The values obtained for carbon fractionation factor (εc) between δ^{13}C-CO_2 and δ^{13}C-CH_4 were below the normal range expected for biogenic methane in marine settings. The low values of εc suggested that some processes may have affected the molecular and isotopic composition of gases after their formation/accumulation in a reservoir or during their migration due to: (i) the mixing of several gas sources or (ii) the anaerobic oxidation of methane.

Author Contributions: Conceptualization: L.F.R.; investigation: L.F.R., J.M.K., R.R.O., A.R.V., A.H.A. and J.A.C.; formal analysis: V.H.J.M.d.S., B.L., A.H.A. and L.F.R.; writing—original draft preparation: L.F.R.; writing—review and editing: L.F.R., J.M.K., J.A.C., A.R.V., B.L. and W.D.

Funding: This research received no external funding.

Acknowledgments: The authors thank SeaSeep Ltd. for collaboration in this research and for the permission to publish this paper, and the IHS Global Inc. Educational/Academic Grant Program for the use of the Kingdom Software. J.M.K. thanks the Brazilian National Research Council (CNPq —research grant No. 309915/2015-5), and the Brazilian Coordination for the Improvement of Higher Education Personnel (CAPES—research grant No. 0558/2015) for financial support.

Conflicts of Interest: The authors declare no conflict of interest.

References

1. Arning, E.T.; van Berk, W.; Vaz dos Santos Neto, E.; Naumann, R.; Schulz, H.M. The quantification of methane formation in Amazon Fan sediments (ODP Leg 155, Site 938) by hydrogeochemical modeling solid—Aqueous solution—Gas interactions. *J. S. Am. Earth Sci.* **2013**, *42*, 205–215. [CrossRef]
2. Sun, X.; Sun, C.; Xiang, J.; Jia, J.; Li, P.; Zhang, Z. Acidolysis hydrocarbon characteristics and significance of sediment samples from the ODP drilling legs of gas hydrate. *Geosci. Front.* **2012**, *3*, 515–521. [CrossRef]
3. Boswell, R.; Collett, T.S. Current perspectives on gas hydrate resources. *Energy Environ. Sci.* **2011**, *4*, 1206–1215. [CrossRef]
4. Chong, Z.R.; Hern, S.; Yang, B.; Babu, P.; Linga, P.; Li, X. Review of natural gas hydrates as an energy resource: Prospects and challenges. *Appl. Energy* **2016**, *162*, 1633–1652. [CrossRef]
5. Makogon, Y.F. Natural gas hydrates—A promising source of energy. *J. Nat. Gas Sci. Eng.* **2010**, *2*, 49–59. [CrossRef]
6. Archer, D. Methane hydrate stability and anthropogenic climate change. *Biogeosci. Discuss.* **2007**, *4*, 993–1057. [CrossRef]
7. Archer, D.; Buffett, B.; Brovkin, V. Ocean methane hydrates as a slow tipping point in the global carbon cycle. *Proc. Natl. Acad. Sci. USA* **2009**, *106*, 20596–20601. [CrossRef]
8. Sultan, N.; Cochonat, P.; Foucher, J.P.; Mienert, J. Effect of gas hydrates melting on seafloor slope instability. *Mar. Geol.* **2004**, *213*, 379–401. [CrossRef]
9. Reeburgh, W. Oceanic methane biogeochemistry. *Am. Chem. Soc.* **2007**, *107*, 486–513. [CrossRef]

10. Hunter, S.J.; Goldobin, D.S.; Haywood, A.M.; Ridgwell, A.; Rees, J.G. Sensitivity of the global submarine hydrate inventory to scenarios of future climate change. *Earth Planet. Sci. Lett.* **2013**, *367*, 105e115. [CrossRef]

11. Ribeiro, I.O.; de Souza, R.A.F.; Andreoli, R.V.; Kayano, M.T.; Costa, P. dos S. Spatiotemporal variability of methane over the Amazon from satellite observations. *Adv. Atmos. Sci.* **2016**, *33*, 852–864. [CrossRef]

12. Dickens, G.R.; Castillo, M.M.; Walker, J.C.G. A blast of gas in the latest Paleo-cene: Simulating first-order effects of massive dissociation of oceanic methane hydrate. *Geology* **1997**, *25*, 259e262. [CrossRef]

13. Miller, D.J.; Ketzer, J.M.; Viana, A.R.; Kowsmann, R.O.; Freire, A.F.M.; Oreiro, S.G.; Augustin, A.H.; Lourega, R.V.; Rodrigues, L.F.; Heemann, R.; et al. Natural gas hydrates in the Rio Grande Cone (Brazil): A new province in the western South Atlantic. *Mar. Pet. Geol.* **2015**, *67*, 187–196. [CrossRef]

14. Giongo, A.; Haag, T.; Simão, T.L.L.; Medina-Silva, R.; Utz, L.R.P.; Bogo, M.R.; Bonatto, S.L.; Zamberlan, P.M.; Augustin, A.H.; Lourega, R.V.; et al. Discovery of a chemosynthesis-based community in the western South Atlantic Ocean. *Deep Res. Part I Oceanogr. Res. Pap.* **2016**, *112*, 45–56. [CrossRef]

15. Medina-Silva, R.; Oliveira, R.R.; Trindade, F.J.; Borges, L.G.A.; Lopes, S.T.; Augustin, A.H.; Valdez, F.P.; Constant, M.J.; Simundi, C.L.; Eizirik, E.; et al. Microbiota associated with tubes of *Escarpia* sp. from cold seeps in the southwestern Atlantic Ocean constitutes a community distinct from that of surrounding marine sediment and water. *Antonie Van Leeuwenhoek* **2018**, *111*, 533–550. [CrossRef] [PubMed]

16. Rodrigues, L.F.; Ketzer, J.M.; Lourega, R.V.; Augustin, A.H.; Sbrissa, G.; Miller, D.; Heemann, R.; Viana, A.; Freire, A.F.M.; Morad, S. The influence of methane fluxes on the sulfate/methane interface in sediments from the Rio Grande Cone Gas Hydrate Province, Southern Brazil. *Braz. J. Geol.* **2017**, *47*, 369–381. [CrossRef]

17. Rodrigues, L.F.; Macario, K.D.; Anjos, R.M.; Ketzer, J.M.M.; Augustin, A.H.; Moreira, V.N.; dos Santos, V.H.J.M.; Muniz, M.C.; Cardoso, R.P.; Viana, A.R.; et al. Origin of organic matter in hydrate-bearing sediments of the Rio Grande Cone: Evidence from TOC, TN, δ13C and 14C isotopes. In Proceedings of the 9th International Conference on Gas Hydrates, Denver, CO, USA, 25–30 June 2017.

18. Flood, R.D.; Piper, D.J.W.; Klaus, A.; Burns, S.J.; Busch, W.H.; Cisowski, S.M.; Cramp, A.; Damuth, J.E.; Goñi, M.A.; Haberle, S.G.; et al. *Proceedings of the Ocean Drilling Program, Scientific Results*; Texas A & M University, Ocean Drilling Program: College Station, TX, USA, 1995; Volume 155, p. 695. ISSN 0884-5891.

19. Hinrichs, K.-U.; Rullkotter, J. Terrigenous and marine lipids in Amazon Fan sediments: Implications for sedimentological reconstructions. *Proc. Ocean Drill. Progr. Sci. Res.* **1997**, *155*, 539–553. [CrossRef]

20. Pirmez, C.; Flood, R.D.; Baptiste, J.; Yin, H.; Manley, P.L. Clay content, porosity and velocity of Amazon Fan sediments determined from ODP Leg 155 cores and wireline logs. *Geophys. Res. Lett.* **1997**, *24*, 317–320. [CrossRef]

21. Ketzer, J.M.; Augustin, A.; Rodrigues, L.F.; Oliveira, R.; Praeg, D.; Pivel, M.A.P.; Reis, A.T.; Silva, C.; Leonel, B. Gas hydrates and gas seepage in the Amazon deep-sea fan. *Geo-Mar. Lett.* **2018**, *38*, 1–10. [CrossRef]

22. Figueiredo, A.G.; Nittrouer, C.A.; de Alencar Costa, E. Gas-charged sediments in the Amazon submarine delta. *Geo-Mar. Lett.* **1996**, *16*, 31–35. [CrossRef]

23. Maslin, M.; Vilela, C.; Mikkelsen, N.; Grootes, P. Causes of catastrophic sediment failures of the Amazon Fan. *Quat. Sci. Rev.* **2005**, *24*, 2180–2193. [CrossRef]

24. Maslin, M.; Fleet, A.J. Equatorial western Atlantic Ocean circulation changes linked to the Heinrich events: deep-sea sediment evidence from the Amazon Fan. *Geol. Evol. Ocean Basins Res. Ocean Drill. Progr.* **1998**, *131*, 111–127. [CrossRef]

25. Reis, A.T.; Perovano, R.; Silva, C.G.; Vendeville, B.C.; Araujo, E.; Gorini, C.; Oliveira, V. Two-scale gravitational collapse in the Amazon Fan: A coupled system of gravity tectonics and mass-transport processes. *J. Geol. Soc. Lond.* **2010**, *167*, 593–604. [CrossRef]

26. Danmuth, J.E.; Kumar, N. Amazon Cone: Morphology, sedimentats, and growth pattern. *Geol. Soc. Am. Bull.* **1975**, *86*, 863–878. [CrossRef]

27. Manley, P.L.; Flood, R.D. Cyclic sediment deposition within Amazon deep-sea fan. *Am. Assoc. Pet. Geol. Bull.* **1988**, *72*, 912–925. [CrossRef]

28. Tanaka, M.D.; Silva, C.G.; Clennell, M.B. Gas Hydrates on the Amazon Submarine Fan, Foz do Amazonas Basin, Brazil. In Proceedings of the AAPG Annual Convention, Salt Lake, UT, USA, 11–14 May 2003; pp. 3–5.

29. Carvalho, G.C.R.d.; Gomes, C.J.S.; Martins Neto, M.A. O Cone do Amazonas, bacia da Foz do Amazonas: Uma nova discussão. *Rem Revista Escola de Minas* **2011**, *64*, 429–437. [CrossRef]

30. Richey, J.E.; Brock, J.T.; Naiman, R.J.; Wissmar, R.C.; Stallard, R.F. Organic Carbon: Oxidation and Transport in the Amazon River. *Science* **1980**, *207*, 1348–1351. [CrossRef]

31. IsoTech Labs Inc. Procedure for Taking Cuttings Samples in IsoJars®. Weatherford Laboratories. 2018. Available online: http://www.isotechlabs.com/customersupport/samplingprocedures/IsoJarSM.pdf (accessed on 14 June 2018).

32. Oung, J.N.; Lee, C.Y.; Lee, C.S.; Kuo, C.L. Geochemical Study on Hydrocarbon Gases in Seafloor Sediments, Southwestern Offshore Taiwan—Implications in the Potential Occurrence of Gas Hydrates. *Terr. Atmos. Ocean. Sci.* **2006**, *17*, 921–931. [CrossRef]

33. Rodrigues, L.F.; Goudinho, F.S.; Laroque, D.O.; Lourega, R.V.; Heemann, R.; Ketzer, J.M.M. An Alternative Gas Chromatography Setting for Geochemical Analysis. *J. Chem. Eng. Process Technol.* **2014**, *5*, 208. [CrossRef]

34. Bernard, B.B.; Brooks, J.M.; Sackett, W.M. Natural gas seepage in the Gulf of Mexico. *Earth Planet. Sci. Lett.* **1976**, *31*, 48–54. [CrossRef]

35. Whiticar, M.J. Carbon and hydrogen isotope systematics of bacterial formation and oxidation of methane. *Chem. Geol.* **1999**, *161*, 291–314. [CrossRef]

36. Paull, C.K.; Lorenson, T.D.; Borowski, W.S.; Ussler, W., III; Olsen, K.; Rodriguez, N.M. Isotopic Composition of CH4, CO2 Species, and Sedimentary Organic Matter Within Samples From the Blake Ridge: Gas Source Implications. Proc. Ocean Drill. Program. *Sci. Results* **2000**, *164*, 67–78. [CrossRef]

37. Pape, T.; Bahr, A.; Rethemeyer, J.; Kessler, J.D.; Sahling, H.; Hinrichs, K.U.; Klapp, S.A.; Reeburgh, W.S.; Bohrmann, G. Molecular and isotopic partitioning of low-molecular-weight hydrocarbons during migration and gas hydrate precipitation in deposits of a high-flux seepage site. *Chem. Geol.* **2010**, *269*, 350–363. [CrossRef]

38. Römer, M.; Sahling, H.; Pape, T.; Bahr, A.; Feseker, T.; Wintersteller, P.; Bohrmann, G. Geological control and magnitude of methane ebullition from a high-flux seep area in the Black Sea—The Kerch seep area. *Mar. Geol.* **2012**, *319*, 57–74. [CrossRef]

39. Milkov, A.V. Molecular and stable isotope compositions of natural gas hydrates: A revised global dataset and basic interpretations in the context of geological settings. *Org. Geochem.* **2005**, *36*, 681–702. [CrossRef]

40. Taylor, S.W.; Sherwood Lollar, B.; Wassenaar, I. Bacteriogenic Ethane in Near-Surface Aquifers: Implications for Leaking Hydrocarbon Well Bores. *Environ. Sci. Technol.* **2000**, *34*, 4727–4732. [CrossRef]

41. Kvenvolden, K.A. A review of the geochemistry of methane in gas hydrate. *Org. Geochem.* **1995**, *23*, 997–1008. [CrossRef]

42. Pallasser, R.J. Recognizing biodegradation in gas/oil accumulations through the d13C compositions of gas components. *Org. Geochem.* **2000**, *31*, 1363–1373. [CrossRef]

43. Waseda, A.; Iwano, H. Characterization of natural gases in Japan based on molecular and carbon isotope compositions. *Geofluids* **2008**, *8*, 286–292. [CrossRef]

44. James, A.T.; Burns, B.J. Microbial alteration of subsurface natural gas accumulations. *Am. Assoc. Pet. Geol. Bull.* **1984**, *68*, 957–960.

45. Horstad, I.; Larter, S.R. Petroleum migration, alteration, and remigration within Troll Field, Norwegian North Sea. *Bull. Am. Assoc. Petrol. Geol.* **1997**, *81*, 222–248. [CrossRef]

46. Boreham, C.J.; Hope, J.M.; Hartung-Kagi, B. Understanding source, distribution and preservation of Australian natural gas: A geochemical perspective. *Aust. Pet. Prod. Explor. Assoc. J.* **2001**, *41*, 523–547. [CrossRef]

47. Head, I.M.; Jones, D.M.; Larter, S.R. Biological activity in the deep subsurface and the origin of heavy oil. *Nature* **2003**, *426*, 344–352. [CrossRef] [PubMed]

48. Pohlman, J.W.; Bauer, J.E.; Canuel, E.A.; Grabowski, K.S.; Knies, D.L.; Mitchell, C.S.; Whiticar, M.J.; Coffin, R.B. Methane sources in gas hydrate-bearing cold seeps: Evidence from radiocarbon and stable isotopes. *Mar. Chem.* **2009**, *115*, 102–109. [CrossRef]

49. Whiticar, M.J.; Faber, E.; Schoell, M. Biogenic methane formation in marine and freshwater environments: CO2 reduction vs acetate fermentation—Isotope evidence. *Geochim. Cosmochim. Acta* **1986**, *50*, 693–709. [CrossRef]

50. Choi, J.; Kim, J.H.; Torres, M.E.; Hong, W.L.; Lee, J.W.; Yi, B.Y.; Bahk, J.J.; Lee, K.E. Gas origin and migration in the Ulleung Basin, East Sea: Results from the Second Ulleung Basin Gas Hydrate Drilling Expedition (UBGH2). *Mar. Pet. Geol.* **2013**, *47*, 113–124. [CrossRef]

51. Prinzhofer, A.; Pernaton, É. Isotopically light methane in natural gas: Bacterial imprint or diffusive fractionation? *Chem. Geol.* **1997**, *142*, 193–200. [CrossRef]

52. Wilson, R.M.; Macelloni, L.; Simonetti, A.; Lapham, L.; Lutken, C.; Sleeper, K.; D'Emidio, M.; Pizzi, M.; Knapp, J.; Chanton, J. Subsurface methane sources and migration pathways within a gas hydrate mound system, Gulf of Mexico. *Geochem. Geophys. Geosyst.* **2014**, *15*, 89–107. [CrossRef]
53. Bahk, J.J.; Kim, J.H.; Kong, G.S.; Park, Y.; Lee, H.; Park, Y.; Park, K.P. Occurrence of near-seafloor gas hydrates and associated cold vents in the Ulleung Basin, East Sea. *Geosci. J.* **2009**, *13*, 371–385. [CrossRef]
54. Haacke, R.R.; Hyndman, R.D.; Park, K.P.; Yoo, D.G.; Stoian, I.; Schmidt, U. Migration and venting of deep gases into the ocean through hydrate-choked chimneys offshore Korea. *Geology* **2009**, *37*, 531–534. [CrossRef]
55. Horozal, S.; Lee, G.H.; Yi, B.Y.; Yoo, D.G.; Park, K.P.; Lee, H.Y.; Kim, W.; Kim, H.J.; Lee, K. Seismic indicators of gas hydrate and associated gas in the Ulleung Basin, East Sea (Japan Sea) and implications of heat flows derived from depths of the bottom-simulating reflector. *Mar. Geol.* **2009**, *258*, 126–138. [CrossRef]
56. Ruffine, L.; Donval, J.P.; Croguennec, C.; Bignon, L.; Birot, D.; Battani, A.; Bayon, G.; Caprais, J.C.; Lantéri, N.; Levaché, D.; et al. Gas seepage along the edge of the Aquitaine shelf (France): Origin and local fluxes. *Geofluids* **2017**, *2017*. [CrossRef]
57. Lorenson, T.D.; Claypool, G.E.; Dougherty, J.A. Natural gas geochemistry of sediments drilled on the 2005 Gulf of Mexico JIP cruise. *Mar. Pet. Geol.* **2008**, *25*, 873–883. [CrossRef]
58. Horstad, I.; Larter, S.R.; Mills, N. A quantitative model of biological petroleum degradation within the Brent Group reservoir in the Gullfaks Field, Norwegian North Sea. *Org. Geochem.* **1992**, *19*, 107–117. [CrossRef]
59. Larter, S.; di Primio, R. Effects of biodegradation on oil and gas field PVT properties and the origin of oil rimmed gas accumulations. *Org. Geochem.* **2005**, *36*, 299–310. [CrossRef]
60. Milkov, A.V. Worldwide distribution and significance of secondary microbial methane formed during petroleum biodegradation in conventional reservoirs. *Org. Geochem.* **2011**, *42*, 184–207. [CrossRef]
61. Feisthauer, S.; Siegert, M.; Seidel, M.; Richnow, H.H.; Zengler, K.; Gründger, F.; Krüger, M. Isotopic fingerprinting of methane and CO2 formation from aliphatic and aromatic hydrocarbons. *Org. Geochem.* **2010**, *41*, 482–490. [CrossRef]
62. Aggarwal, P.K.; Fuller, M.E.; Gurgas, M.M.; Manning, J.F.; Dillon, M.A. Use of stable oxygen and carbon isotope analyses for monitoring the pathways and rates of intrinsic and enhanced in situ biodegradation. *Environ. Sci. Technol.* **1997**, *31*, 590–596. [CrossRef]
63. Zyakun, A.M.; Kosheleva, I.A.; Zakharchenko, V.N.; Kudryavtseva, A.I.; Peshenko, V.A.; Filonov, A.E.; Boronin, A.M. The use of the [13C]/[12C] ratio for the assay of the microbial oxidation of hydrocarbons. *Microbiology* **2003**, *72*, 592–596. [CrossRef]
64. Jones, D.M.; Head, I.M.; Gray, N.D.; Adams, J.J.; Rowan, A.K.; Aitken, C.M.; Bennett, B.; Huang, H.; Brown, A.; Bowler, B.F.J.; et al. Crude-oil biodegradation via methanogenesis in subsurface petroleum reservoirs. *Nature* **2008**, *451*, 176–180. [CrossRef] [PubMed]
65. Clayton, C.J.; Hay, S.J.; Baylis, S.A.; Dipper, B. Alteration of natural gas during leakage from a North Sea salt diapir field. *Mar. Geol.* **1997**, *137*, 69–80. [CrossRef]
66. Yoshinaga, M.Y.; Holler, T.; Goldhammer, T.; Wegener, G.; Pohlman, J.W.; Brunner, B.; Kuypers, M.M.M.; Hinrichs, K.U.; Elvert, M. Carbon isotope equilibration during sulphate-limited anaerobic oxidation of methane. *Nat. Geosci.* **2014**, *7*, 190–194. [CrossRef]
67. Boetius, A.; Ravenschlag, K.; Schubert, C.J.; Rickert, D.; Widdel, F.; Gieseke, A.; Amann, R.; Jørgensen, B.B.; Witte, U.; Pfannkuche, O. A marine microbial consortium apparently mediating anaerobic oxidation of methane. *Nature* **2000**, *407*, 623–626. [CrossRef] [PubMed]
68. Jeffrey, A.W.A.; Alimi, H.M.; Jenden, P.D. Geochemistry of Los Angeles Basin oil and gas systems. *Act. Margin Basins Aapg Mem.* **1991**, *52*, 197–219.

geosciences

MDPI

Article

Geotectonic Controls on CO_2 Formation and Distribution Processes in the Brazilian Pre-Salt Basins

Luiz Gamboa [1,*], André Ferraz [1], Rui Baptista [2] and Eugênio V. Santos Neto [3]

[1] Geology & Geophysical Department, Universidade Federal Fluminense, UFF, Niterói 2410-364, Brazil; andetienne@terra.com.br
[2] Geology Department, F. Ciências Universidade de Lisboa, 1749-016 Lisboa, Portugal; rjbaptista@fc.ul.pt
[3] Independent Consultant, Rio de Janeiro 22271-110, Brazil; santosnetoeugenio@gmail.com
* Correspondence: luizgamboa@gmail.com

Received: 7 March 2019; Accepted: 16 May 2019; Published: 5 June 2019

check for updates

Abstract: Exploratory work for hydrocarbons along the southeastern Brazilian Margin discovered high concentrations of CO_2 in several fields, setting scientific challenges to understand these accumulations. Despite significant progress in understanding the consequences of high CO_2 in these reservoirs, the role of several variables that may control such accumulations of CO_2 is still unclear. For example, significant differences in the percentages of CO_2 have been found in reservoirs of otherwise similar prospects lying close to each other. In this paper, we present a hypothesis on how the rifting geodynamics are related to these CO_2-rich accumulations. CO_2-rich mantle material may be intruded into the upper crustal levels through hyper-stretched continental crust during rifting. Gravimetric and magnetic potential methods were used to identify major intrusive bodies, crustal thinning and other geotectonic elements of the southeastern Brazilian Margin. Modeling based on magnetic, gravity, and seismic data suggests a major intrusive magmatic body just below the reservoir where a high CO_2 accumulation was found. Small faults connecting this magmatic body with the sedimentary section could be the fairway for the magmatic sourced gas rise to reservoirs. Mapping and understanding the crustal structure of sedimentary basins are shown to be important steps for "de-risking" the exploration process.

Keywords: carbon dioxide; Santos Basin; potential methods; gravimetric data; magnetic data; mantellic source; Bouguer anomaly; São Paulo Plateau

1. Introduction

Variations in the atmospheric composition and climatic changes throughout geologic time were crucial not only for the birth of life on Earth but also for acting as the trigger of several episodes of mass extinction and the development of new species after biological readjustments faced new environmental conditions. More recently over the historic human time-scale, the eruption of large carbon dioxide (CO_2) and dust emissions generated by recurrent volcanic eruptions affected the climate of the entire Earth.

At the present time, carbon dioxide is released not only from natural sources like biological and volcanic activities but also by a strong anthropogenic action including fossil fuel combustion for transport, industry, and energy generation.

Geology shows that the CO_2 natural emissions are commonly associated with Earth endogenic activity since the beginning of geological times. This endogenic sourced CO_2 could reach the Earth surface by natural sequestration in buried rock formations also occur in some specific regions depending on the geological framework. In some areas of the world, large accumulations of CO_2 have been found in geologic traps associated both with or without hydrocarbons.

In Brazil, significant CO_2 accumulations were not identified before the first decade of this century. Since 2006, large volumes of carbon dioxide and fluids with high CO_2 content have been discovered in some of the oil fields located in the pre-salt section of Santos Basin. In one of the drilled prospects, carbon dioxide was found to be up to 80% in the gas cap. This is an unusual fluid association for this petroleum province, as a CO_2-rich gas cap with condensate is on top of a heavy-oil (18° API) filled reservoir, whereas in nearby fields with apparently similar geological settings the CO_2 content is much lower, and oils are commonly lighter at 29° to 30° API [1].

High concentrations of CO_2 represent a major challenge to field developments not only because of technical difficulties related with the CO_2 separation from the hydrocarbons in case of large production rates but also because of the integrity risk of facilities as fast corrosion processes can reduce the lifetime or cause collapse of production pipes and other equipment [2]. Another important issue is related with the CO_2 disposal as international protocols towards the reduction of the emissions of gases linked to the greenhouse gas effect (GHG), especially after COP XXI, do not allow any CO_2 atmospheric release.

Disposal of CO_2 in the oil reservoirs during the production as an enhanced oil recovery (EOR) technique or after its depletion for CO_2 sequestration are solutions to minimize emissions. However, in both cases, dedicated handling facilities will be required, adding to the expense (capital and operating costs) of the project.

In addition to health, safety, and environment (HSE) implications, the CO_2 prediction in petroleum reservoirs is very important to enhance our knowledge of the petroleum system in exploration and de-risk new ventures. The interaction of CO_2 with other fluids and with host rocks can create important changes in the physical-chemical properties of the petroleum. Evaporative fractionation after a latter CO_2 arrival could explain the fluid stratification found in some petroleum reservoirs. Diagenetic processes due to this CO_2 arrival could also affect reservoir perm-porosity properties.

Although CO_2 occurrences in petroleum reservoirs are common worldwide, usually they are of low concentrations of less than 5%. Occurrences in the order of 20% correspond to less than 1% of all cases [3].

In the hydrocarbon prolific Santos Basin, the amount of CO_2 proved in several drilled prospects is very heterogeneous even in adjacent areas with very similar geologic contexts.

That is the case of the almost twin prospects of Tupi (Lula field) which has low CO_2 content and Jupiter where a concentration of 80% of CO_2 within the total gases is reported [1]. The origin of this gas from Earth's mantle has already been proven by the isotopic analyses of noble gases [4].

Several hypotheses have been proposed to explain this mantle sourced carbon dioxide in hydrocarbon reservoirs, namely regional crustal thinning, deep-seated faults, high fault density, igneous intrusions, among others [3–7].

The combined geological, geophysical setting, and gaseous geochemistry allow us to infer that the geodynamic evolution of São Paulo Plateau/Santos Basin area had several episodes of intense upwelling of magmatic material sourced from the mantle throughout late Cretaceous and Lower Eocene times as generally recognized in the regional geology.

In this paper, we present a hypothesis to explain the processes that control the geographic distribution of large volumes of CO_2 in the São Paulo Plateau region. This hypothesis can also provide a good explanation for the differences in CO_2 contents found in Jupiter structure and the neighbor discoveries.

2. Materials and Methods

As the area of this case study includes the Jupiter prospect, where high concentrations of carbon dioxide were reported [1], the reasons, source and carrier mechanisms for that abnormal CO_2 concentration are some of the project aims.

Seismic interpretation and geological analyses were undertaken but the most reliable hypothesis to explain the CO_2 concentrations in the reservoirs is from the analyses of the gravity and magnetic data obtained from global databases.

The public domain potential data set employed in this study - satellite data, ship track data, and others—were downloaded from websites: http://topex.ucsd.edu/WWW_html/mar_grav.html and http://www.geomag.us/models/wdmam.html. Our experience indicates that such a database is adequate to regional studies of continental margins and oceanic basins.

2.1. Gravimetric Data

As described by Sandwell and Smith [8], radar altimetry by satellite has enough accuracy to define the gravity field of oceanic basins due to the redundancy of data acquisition. Profiles collected over the years by satellites like ERS-1 and TOPEX-POSEIDON allowed the visualization of high-resolution gravity anomalies. The authors produced 4 mGal grids of concordance with data collected by ships. As the values of gravity over the oceans typically vary from 20 to 300 mGal, we conclude that the satellite gravity data are adequate for regional geophysical studies of the Earth's crust [9,10].

To visualize the gravity anomalous field in the region of the São Paulo Plateau the free air and Bouguer anomalies were initially determined. An infinite plate of 1750 Kg/m^3 was used to represent the density contrast between the water column and the marine substratum based on the work of Pawlosky [10]. This author obtained very satisfactory results in studies using gravity in the African continental margin of Namibia. Considering the similar nature of the conjugated margins, the same value of density in the south and southeast segments of the Brazilian continental margin was used with consistent results.

2.2. Magnetic Data

A global grid of the total magnetic field anomaly, named EMAG2 [11], was produced from satellite measurements, ship, and airborne surveys. This grid corresponds to a more accurate and up-to-date version of its predecessor WDMAM [12]. For their research, a compilation of airborne and ship surveys, available across the world, was meticulously performed by Maus and colleagues [11,12]. They minimized cross-errors found between the various surveys track lines. Finally, large wavelengths exceeding 330 km were subtracted from normalized data and replaced by the high accuracy magnetic measurements of the CHAMP satellite. This procedure allows a great improvement in the regional representation of the anomalous magnetic field because the long wavelengths are often corrupted in the marine surveys due to the lack of control of the temporal variations of the magnetic field.

3. Results

3.1. Regional Setting

The Santos Basin situated offshore southern Brazil between Campos and Pelotas Basins, offshore the states of Rio de Janeiro, São Paulo and Parana (Figure 1), is one of a series of basins located at the continental margin, the origin of which is connected to the early Cretaceous rifting of the South Atlantic.

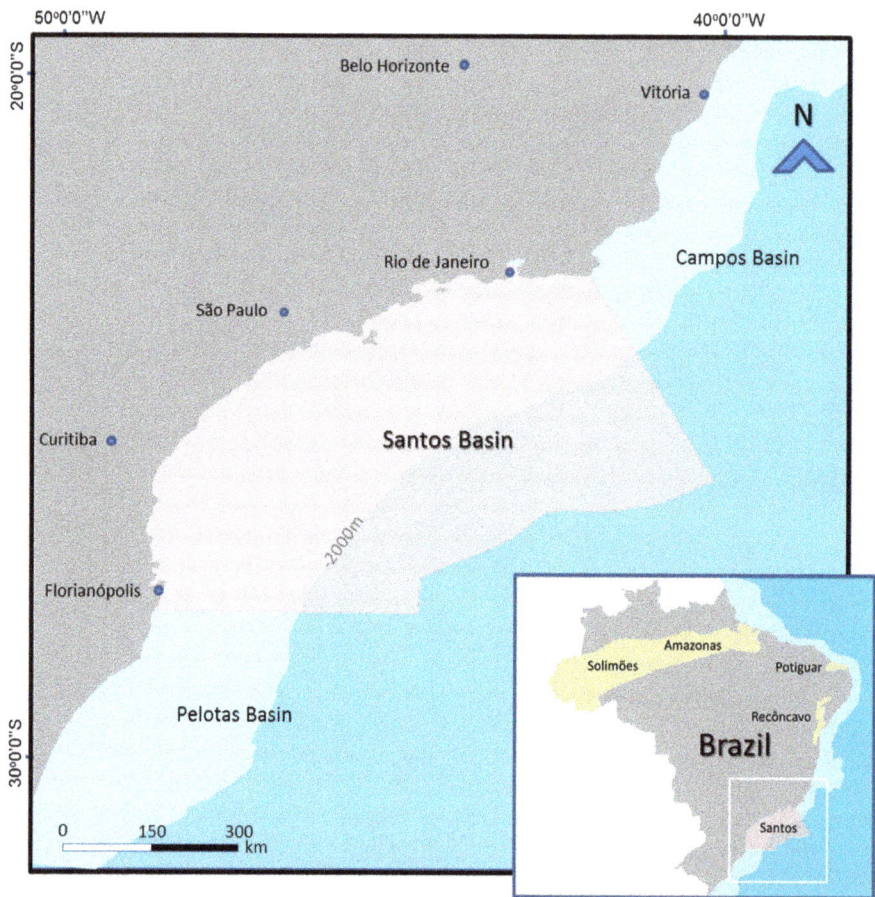

Figure 1. Santos Basin Location.

All these basins have identical sedimentary fills with similar depositional sequences triggered by the same main tectonic events. Moreira et al. [13] proposed the latest stratigraphic chart for Santos Basin (Figure 2).

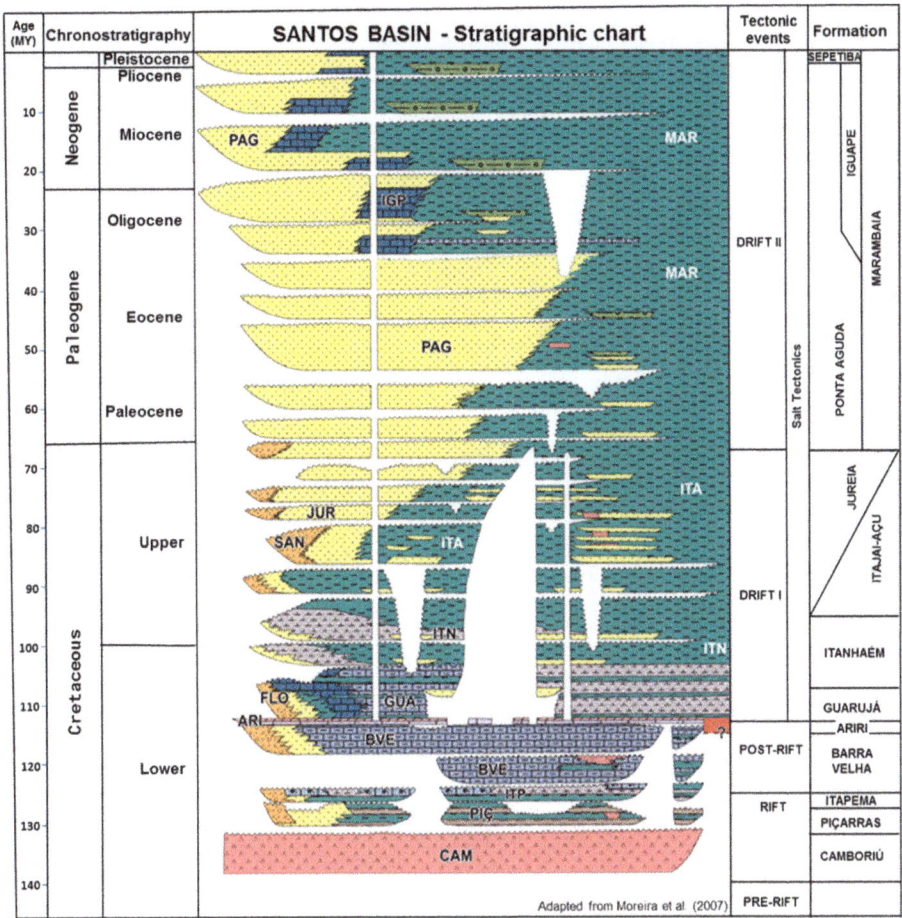

Figure 2. Simplified Santos Basin stratigraphic chart, modified from from [13].

During the process of separation of the African and South American plates, at the beginning of the rift phase, basaltic flows of the Camboriú Formation (Hauterivian) occurred, which is associated with the early stretching of the crust. Above the basalts, conglomerates and polimythic sandstones of the Piçarras Formation were deposited in proximal areas and organic-rich black shales in deeper parts of the basin, during a period of intense tectonic activity. Above the Piçarras Formation, still in the rift phase but in a less intense tectonic period, bounded by two prominent unconformities, the Itapema Formation was deposited with a great variety of limestones in proximal domains and layers of black shales in distal areas. In some areas, coquina reservoirs of great permo-porosity quality occur in this formation. Covering the Itapema Formation, the Barra Velha Formation occurs as a sequence of limestones, stromatolites, microbiolites, and shales deposited in a hypersaline shallow marine environment. Immediately above these reservoirs a 2000 m thick evaporitic sequence—the Ariri Formation—is a perfect seal. Above this evaporitic section several depositional sequences, mainly composed of open marine siliciclastics, mostly sandstones near shore and shales in greater depths, fill the basin since Albian to Quaternary.

The basin spans some 300 km along the continent and 700 km width in its dip direction. These dimensions indicate the unusual stretching suffered by the continental crust during the early stages of this basin formation [14]. The equivalent section at the conjugated margin, at the African side, is significantly narrower highlighting the asymmetry of the two continental passive margins.

This basin started to be explored for hydrocarbons in the early seventies and its first discovery, the Merluza field, was announced in 1979. More recently, after 2006, giant oil fields were discovered in its deep-water region, and Santos Basin became the most prolific basin of Brazil. However, in some of these significant hydrocarbon discoveries. abnormal amounts of CO_2 were identified. Due to the production problems created by these high CO_2 contents avoidance of such areas is desirable, and this can only be achieved by an understanding of the origin, migration paths, and trapping of this CO_2.

3.2. Structural Elements of Santos Basin-São Paulo Plateau

The main structural elements defined by the seismic data in this area can be highlighted and better defined using gravimetric and magnetic methods (Figures 3 and 4). A residual Bouguer anomaly map obtained by the removal of the regional trend, typically ascending towards the oceanic crust from its first vertical derivative or vertical gradient, reflects the basement structure of the basin and the intra crustal structures normally associated with mantle elevations (Figure 4). The western boundary of Santos Basin is defined by a hinge line highlighted by a prominent gravity positive anomaly.

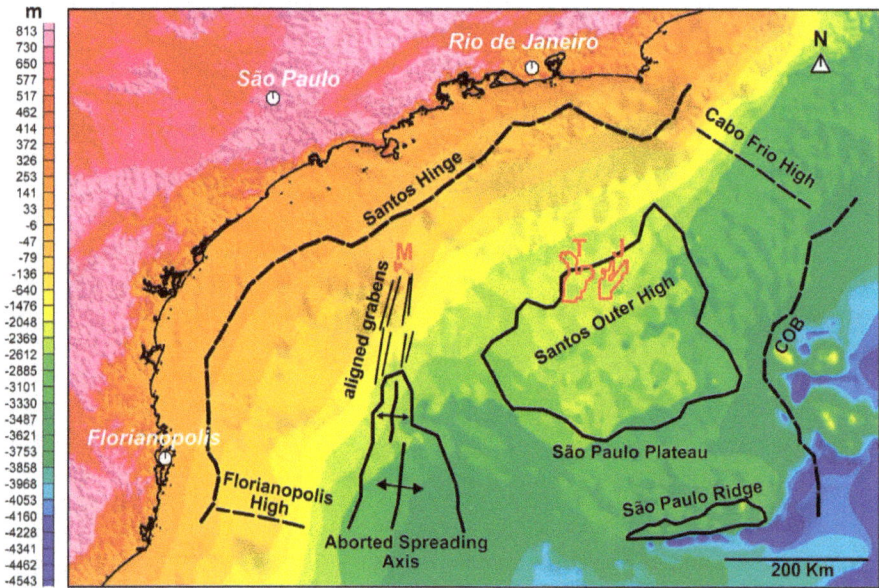

Figure 3. Santos Basin main structural features obtained from the potential methods interpretation, plotted on the bathymetric map. M and T are the locations of the Merluza and Tupi fields. J is the location of Jupiter prospect. COB corresponds to the continent–ocean boundary.

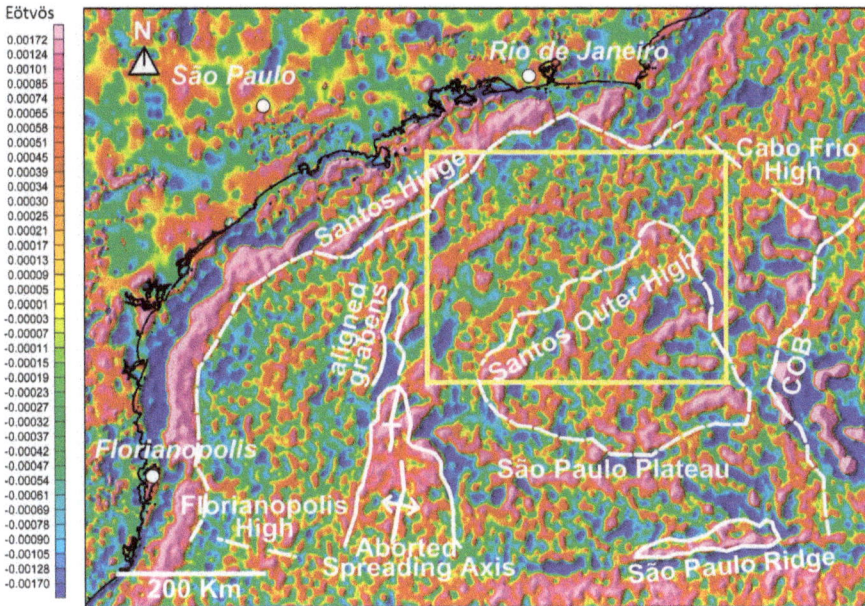

Figure 4. Residual Bouguer anomaly map. The blue areas correspond to the minimum gradients whereas the red and pink correspond to the maximum gradient areas of the residual anomalies. The yellow rectangle corresponds to the area studied in detail. COB corresponds to the continent–ocean boundary.

Another conspicuous positive anomaly of great magnitude is observed at the southwestern limit of the basin. This anomaly is associated with a mantle elevation and formation of oceanic crust in this region.

In map view, it has an arrow shape that narrows to the north (Figures 3 and 4). It extends northwards up to the region of the Merluza Field as a negative aligned anomaly. This arrow-shaped feature is interpreted as an early aborted spreading center related to an initial rifting process. A graben affecting the basement rocks is visible there on seismic data. No oceanic crust is interpreted within this northern graben. The oceanic crust was formed at its wider southern part whereas just a continental graben occurs further to the north (Figures 3 and 4).

A conspicuous large amplitude residual anomaly pattern delineates the Santos Outer High and further east the continent–ocean boundary (COB) is interpreted (Figure 4). The southern boundary of the São Paulo Plateau is clearly defined by a positive anomaly and by a prominent topographic feature named São Paulo Ridge, which is part of a major fracture zone—the Florianópolis Fracture Zone (Figure 4). The yellow rectangle in this figure defines the area where CO_2 anomalies were found in hydrocarbon reservoirs.

A good correlation is observed between the residual Bouguer map and the vertical gradient map of the total magnetic field (Figures 4 and 5). The hinge line of the basin is marked by a clear contrast between a high-frequency anomaly pattern, indicative of shallow basement and a smooth pattern of the deeper basement to the east.

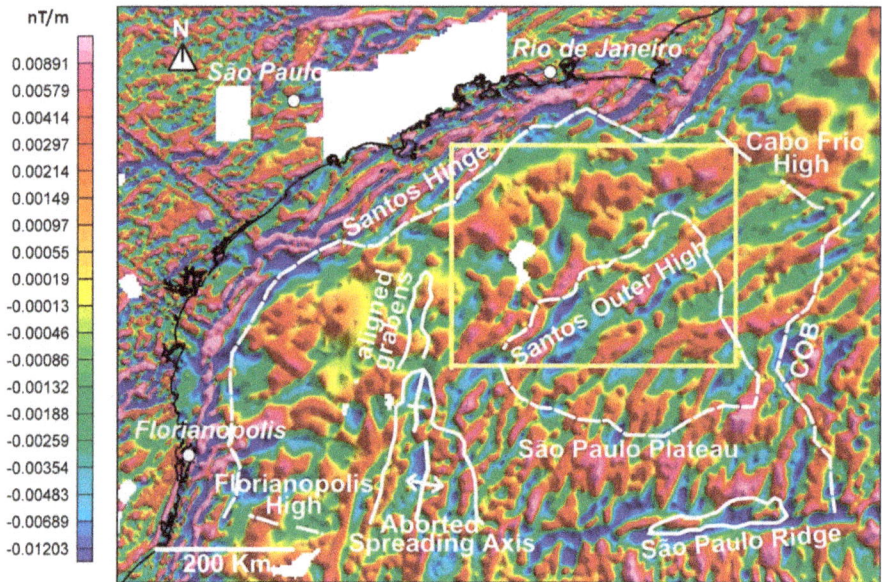

Figure 5. Vertical gradient of the total magnetic field map. The blue areas correspond to the minimum gradients, whereas the red and pink correspond to the maximum gradient areas. The yellow rectangle corresponds to the area studied in detail. COB corresponds to the continent–ocean boundary.

The Santos Basin Outer High identified on the gravimetric records (Figure 4) cannot be defined on the magnetic map (Figure 5). This is due to the predominance of intracrustal magnetic bodies as the source for such anomalies. Such source obliterates any contribution from the basement surface as a source of anomalies. Whereas in the gravimetric record is the opposite, the highs and lows related to the structure of the basement are responsible for the majority of the gravimetric contributions.

The abundant presence of such intracrustal sources generates a typical magnetic pattern that characterizes the crust under the São Paulo Plateau. The boundaries of this crustal domain are defined by a relative increase in the frequency content generated by linear anomalies controlled by the NE-SW structural trend of the basement rocks (Figure 5).

The distribution of these anomalies is similar to the one observed on a typical oceanic crust. The crust under the São Paulo Plateau can be interpreted as a highly stretched and magmatic injected continental crust or a transitional crust [15,16]. The western limit of this province is defined by the aborted spreading center axis, followed to the north by aligned grabens and by the NW boundary of Santos Outer High. On the other side, the eastern limit is the boundary between the continental and the oceanic crust (Figures 4 and 5).

Figure 6 shows the detailed studied area (yellow rectangle in Figures 4 and 5) where magmatic bodies injected into the crust were interpreted from the horizontal derivative in the direction of the magnetization induced by the current magnetic field. Additionally, a vertical derivative calculated from the horizontal one produced a view with more details and a better positioning of the interpreted magmatic bodies associated to the axes of maximum values of the derivative (Figure 6A).

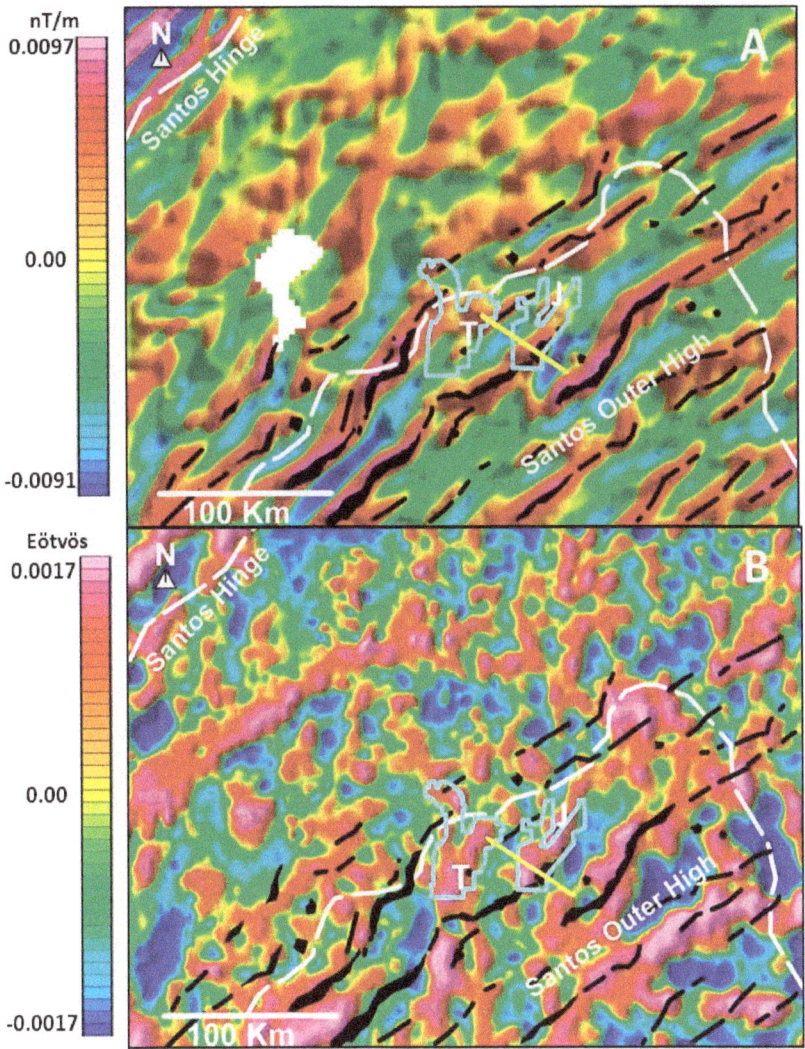

Figure 6. Interpretation map of the detail area highlighted in the Figures 4 and 5. (**A**) Magnetic map of derivative showing interpreted magmatic bodies injected in the crust, represented by black areas and black dashed lines, associated with axes of maximum values of the derivative (areas in warm tones). Observed declination and inclination of the total magnetic field (D = 23° W, I = −42°). (**B**) Residual Bouguer map showing the injected magmatic bodies underlying the basement structural highs (red and pink areas). The blue polygons indicate the areas of Tupi (T) and Jupiter (J) associated with their respective structural highs. The yellow straight line shows the location of the modeled section shown in Figure 7.

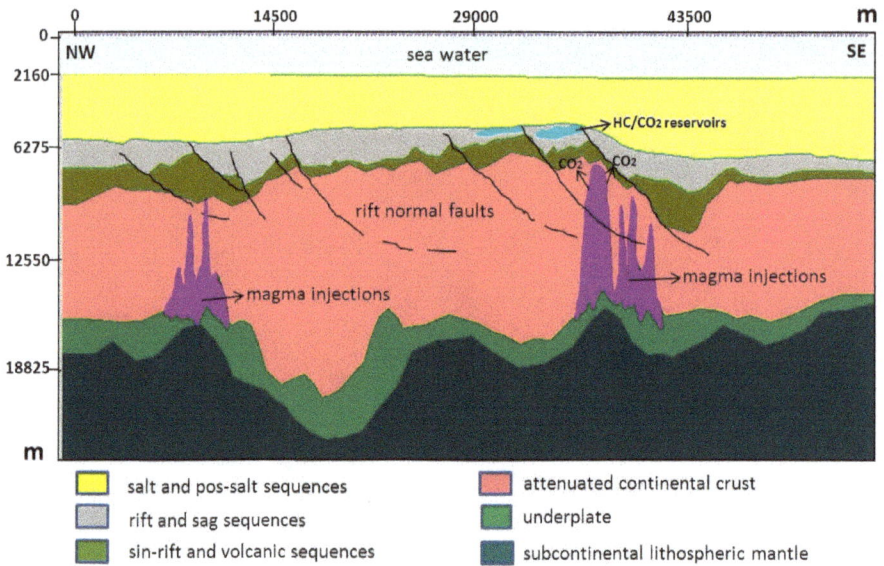

Figure 7. Modeled section based on magnetic, gravity and seismic information. The wells located on the region of the major intracrustal intrusion contain the highest amount of CO_2 found so far on the Brazilian Continental Margin. Section location in Figure 6.

The methodology for mapping the magmatic bodies was developed in this research based on the analysis of the derivative maps with the orientation of the modeled section of Figure 7. This kind of approach seems to be more effective than the most commonly used methodology in which the vertical derivative of the field reduced to the pole is determined. This is due to the characteristic positioning of the magmatic bodies, strongly aligned in the NE-SW direction of the rift structural trend (normal to the magnetic field) which greatly intensifies the induced magnetization component, thus favoring the horizontal derivative in the direction of the current magnetic field.

Figure 6B shows the residual Bouguer anomaly in relation to the underlying magmatic bodies. Positive residual anomalies (areas in red and pink) reflect the basement structural highs, bordered by normal faults that were active during the rifting stages of the basin.

The modeled section (Figure 7), perpendicular to the main trend of the faults in the area was obtained by the direct method using a 2.5 D model in which, to avoid ambiguities, restrictions were imposed using the 3D seismic data information, such as the base of the evaporitic section and the volcanic basement. The deeper levels of the crust were not defined by seismic data but from the gravity and magnetic data instead.

The attenuated continental crust under the volcanic sequence considered as the basement for this basin was also defined. The thickness of this crust is highly variable due to the great amounts of stretching experienced by this region.

These areas of crustal thinning are more susceptive to suffer magmatic injections with mantle-derived materials. Vertical magmatic intrusions could reach the higher crustal levels where they could be intercepted by rifting faults propagating at the higher levels of the basement or later on in younger formed grabens as well.

In the studied area, over the Jupiter structure, a strong total field magnetic anomaly is noticeable (Figure 8A). This anomaly is interpreted as produced by a robust intrusive body that reaches the upper crust close to the base of the sedimentary layers of the syn-rift section (Figure 7). This intrusive body is well evidenced by the transformation of the total magnetic field anomaly map by the methodology of the derivatives, developed in this research, described above (Figure 8B). The estimated position of the top of the intrusive body is, thus, mapped with improved accuracy.

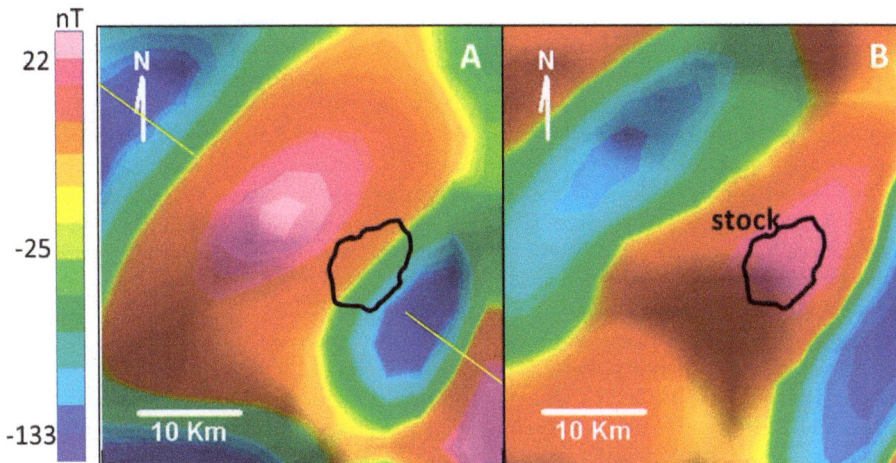

Figure 8. (**A**) Detailed map of the total field magnetic anomaly located at the SE of the Jupiter structure. The yellow straight line marks the map location of the modeled section (Figure 7). Note its high amplitude (greater than 150 nT) associated with a shallow localized source. (**B**) Derivative map of the total magnetic field anomaly showing the estimated location of the intrusive stock type body responsible for generating the total field magnetic anomaly. The top location of this body is defined by the black contour on the maximum values of the derivative. Note that the location of the intrusive body in the total field map is at the area between the maximum and minimum total field anomaly.

This is a typical anomaly of a localized body, associated to a probable large stock (Figure 8B). In the studied region, there are other similar anomalies, but these are much less prominent than the Jupiter anomaly that reaches more than 150 nT of amplitude (Figure 8A). The smaller anomalies verified in the area are probably associated with deeper or smaller intrusions and are therefore masked by regional total field anomalies and highlighted only by the gradients of the field (Figures 5 and 6A).

Numerous magmatic bodies intruded in the crust were defined extrapolating this model for the detailed anomaly in conjunction with the regional magnetic gradient map (Figure 6A). Plugs and stocks are shown on the map as black polygons and large dikes as black traces (Figure 6). These magmatic intrusions seem to be more abundant towards the distal parts of the basin in areas of the Outer High region controlled by the NE-SW structural trend of the rift.

4. Discussion

The large concentrations CO_2 in Santos Basin were unexpected during the early exploratory process of its deep-water areas. However, drilled pre-salt prospects have shown a wide range of CO_2 contents bringing together environmental and production complications. The mantellic origin of these CO_2 occurrences was established via noble gases isotopic analyses [4]. Nevertheless, processes and mechanisms responsible for the fate (introduction, migration, and preservation) of CO_2 in petroleum systems still remain unclear.

Many authors have suggested that the "CO_2-risk" in sedimentary basins could be related to the proximity of igneous intrusions and deep-seated faults, or to geothermal gradient higher than 30° C/km [3,5]. Other geoscientists suggested empirically that in some areas as, for example, in the Southeast Asia, CO_2 accumulations must be controlled by type and age of crustal basement, fault density, temperature, and pressure of reservoirs [6].

More recently, strong evidence has been gathered indicating that mantle helium and occurrences with higher percentages of CO_2 were related to areas of crustal thinning in depths of about 26–28 km, with thermal flux higher than 61 mW/m^2 [7].

Such conditions suggest rock melting due to the asthenosphere rising towards the crust. However, the occurrence of just one or more factors described above is not an unequivocal proxy for large accumulations of CO_2 in petroleum reservoirs. Such uncertainties are the main reason to investigate more thoroughly processes and mechanisms that generate, introduce, and accumulate CO_2 in petroleum systems. Independently of such myriad of details to be studied, a tool with the power to recognize deep related magmatic processes would be an excellent proxy to infer "CO_2-risk" in petroleum reservoirs.

In this way, our work focused on crustal studies is based mainly on potential method data associated with the geologic interpretation of the Jupiter Prospect data, where the highest CO_2 concentrations are reported up to date. The obtained results allow us to propose an association of the CO_2 and intracrustal intrusions of mantle-derived material.

Both gravity and magnetic data indicate the existence of a highly stretched continental crust under the São Paulo Plateau and a rather unique and conspicuous anomaly under the Jupiter Prospect. Modeling indicates that this anomaly corresponds to an intracrustal intrusion that reached almost to the top of the basement rocks in this area. We interpret this intrusion as the main responsible agent to transport CO_2 from the mantle into the reservoir levels in the pre-salt section of this area. Other occurrences of CO_2 in Santos Basin are all located in the stretched crust of the basin and are also interpreted as provenient from mantle material ascending along major fault segments.

The São Paulo Plateau, the distal portion of Santos Basin went to a complex rifting from its African counterparts and suffered extreme crustal stretching. That created a unique type of crust, which can almost be considered as a transition between continental to oceanic crust. Characteristics of both can be seen on the potential methods data. Thinning of the continental crust on the São Paulo Plateau facilitated mantle-derived material to rise to the upper crustal levels, bringing together CO_2 that eventually could migrate to upper levels and become trapped in reservoirs, with or without hydrocarbons or other gases like H_2S. The conspicuous magnetic anomaly found under the Jupiter Prospect indicates a shallow intrusion reaching almost the base of the sedimentary sections. The heavy oil found in the Prospect Jupiter can be explained by the selective extraction of light molecular weight of petroleum fraction due to the natural chromatographic effect produced by the CO_2 migration in a process similar to the one proposed by [17].

Very low CO_2 content is found in the petroleum fields located at the proximal areas of Santos Basin, neither on the fields of Recôncavo or Potiguar aborted rifts. In basins with expressive volcanism, like the Solimões Basin, where a thick continental crust exists, or even in Campos Basin, located to the north of Santos and with a much lesser amount of crustal stretching, almost no CO_2 is found. The lack of CO_2 in areas of lesser amounts of continental crustal stretching seems to be a rule.

Analyses of satellite potential data seem to work very well to characterize basins located on stretched continental where potentially mantle-derived material could rise to the upper crustal levels and bring large amounts of CO_2 with them. In this work, we have demonstrated the immense potential contained in these data. Potential data from regional satellite grid should thus be considered as an exploratory tool of great value to estimate "CO_2-risk".

5. Conclusions

Gravimetric and magnetic methods are reliable proxies for the identification of major intrusive bodies, crustal thinning, and other geotectonic elements that are related to the significant occurrences of CO_2 in the Southern offshore Brazilian basins, e.g., Santos and Campos.

Considering the unequivocal relationship between the abundance of CO_2 and the intensity of mantle signature of CO_2-associated helium [4], the approach used in this work can be an efficient tool to predict the "CO_2-risk" in those basins.

Despite the significant progress that has already been reached on the understanding of CO_2 fate, the role of other variables that could control the abundances of CO_2 in reservoirs is still unclear, sometimes twin prospects have significant differences in the percentages of such gas.

Author Contributions: Conceptualization, L.G. and R.B.; Formal analysis and supervision, L.G.; Grav-Mag Data curation and interpretation, A.F.; Visualization, L.G., A.F., R.B. and E.V.S.N.; Writing—original draft preparation, L.G., A.F. and R.B.; Writing—review and editing, L.G., A.F., R.B. and E.V.S.N.

Funding: The two first authors have initially developed this work as a small part of a broader spectrum R&D project sponsored by Petrogal Brasil, SA under ANP regulation. Complementary research and the elaboration of this paper received no external funding.

Acknowledgments: This work has been partially developed by two of the authors as a small part of a broader spectrum R&D project financed by Petrogal Brasil, SA under ANP regulation. These authors acknowledge Petrogal Brasil, SA, and ANP for that support. We thank Naresh Kumar and Peter Homewood for the revisions that much improved the paper. Authors extend their gratitude to the anonymous reviewers for their relevant and positive contributions.

Conflicts of Interest: The authors declare no conflict of interest. The funders had no role in the design of the study; in the collection, analyses, or interpretation of data; in the writing of the manuscript, or in the decision to publish the results.

References

1. Gaffney, C. *Exame e Avaliação de Dez Descobertas e Prospectos Selecionadas no Play do Pré-sal em Águas Profundas na Bacia de Santos, Brasil*; ANP-Report; ANP: Rio de Janeiro, Brasil, 2010.
2. Müller, N.; Elshahawi, H.; Dong, C.; Mullins, O.C.; Flannery, M.; Ardila, M.; Winheber, P.; McDade, E.C. Quantification of carbon dioxide using downhole Wireline formation tester measurements. In Proceedings of the SPE Annual Technical Conference and Exhibition, San Antonio, TX, USA, 24–27 September 2006.
3. Thrasher, J.; Fleet, A.J. Predicting the risk of carbon dioxide "pollution" in petroleum reservoirs. Organic Geochemistry: Developments and Applications to Energy, Climate, Environment and Human History. In Proceedings of the 17th International Meeting on Organic Geochemistry, Donostia-San Sebastian, Spain, 4–8 September 1995; pp. 1086–1088.
4. Santos, E.V.; Cerqueira, J.R.; Prinzhofer, A. Origin of CO_2 in Brazilian Basins. In Proceedings of the AAPG International Conference and Exhibition, Long Beach, CA, USA, 22–25 April 2012.
5. Clayton, C. Controls on the carbon isotope ratios of CO_2 in oil and gas fields. Organic Geochemistry: Developments and Applications to Energy, Climate, Environment and Human History. In Proceedings of the 17th International Meeting on Organic Geochemistry, Donostia-San Sebastian, Spain, 4–8 September 1995; pp. 1073–1074.
6. Imbus, S.W.; Katz, B.J.; Urwongse, T. Predicting CO_2 occurrence on a regional scale: Southeast Asia example. *Org. Geochem.* **1998**, *29*, 325–345. [CrossRef]
7. Li, M.; Wang, T.; Liu, J.; Lu, H.; Wu, W.; Gao, L. Occurrence and origin of carbon dioxide in the Fushan Depression, Beibuwan Basin, South China Sea. *Mar. Pet. Geol.* **2008**, *25*, 500–513. [CrossRef]
8. Sandwell, D.T.; Smith, W.H.F. Global marine gravity from retracked Geosat and ERS-1 altimetry: Ridge Segmentation versus spreading rate. *J. Geophys. Res.* **2009**, *114*, B01411. [CrossRef]
9. Maia, M. Comparing the use of ship and satellite data for geodynamic studies. In Proceedings of the ESA Symposium 15 years of Progress in Radar Altimetry, Venice, Italy, 12–18 March 2006.
10. Pawlowski, R. The use of gravity anomaly data for offshore continental margin demarcation. *Lead. Edge* **2008**, *27*, 722–727. [CrossRef]
11. Maus, S.; Barckhausen, U.; Berkenbosch, H.; Bournas, N.; Brozena, J.; Childers, V.; Dostaler, F.; Fairhead, J.D.; Finn, C.; Von Frese, R.R.B.; et al. EMAG2: A 2–arc min resolution Earth Magnetic Anomaly Grid compiled from satellite, airborne, and marine magnetic measurements. *Geochem. Geophys. Geosyst.* **2009**, *10*. [CrossRef]
12. Maus, S.; Sazonova, T.; Hemant, K.; Fairhead, J.D.; Dhananjay, R. World Digital Magnetic Anomaly Map (WDMAM) Candidate Version 1, National Geophysical Data Center candidate for the World Digital Magnetic Anomaly Map. *Geochem. Geophys. Geosyst.* **2007**, *6*. [CrossRef]
13. Moreira, J.; Madeira, C.; Gil, J.; Machado, M.A. Bacia de Santos. *Bol. Geoci. Petrobras* **2007**, *15*, 531–549.

14. Evain, M.; Afilhado, A.; Rigoti, C.; Loureiro, A.; Alves, D.; Klingelhoefer, F.; Schnurle, P.; Feld, A.; Fuck, R.; Soares, J.; et al. Deep structure of the Santos Basin-São Paulo Plateau System, SE Brazil. *J. Geophys. Res. Solid Earth* **2015**, *120*, 5401–5431. [CrossRef]
15. Kumar, N.; Gamboa, L.A.P. Evolution of the São Paulo Plateau (southeastern Brazilian margin) and implications for the early history of the South Atlantic. *Geol. Soc. Am. Bull.* **1979**, *90*, 281–293. [CrossRef]
16. Kumar, N.; Danfortth, A.; Nuttall, P.; Helwig, J.; Bird, D.E.; Venkatraman, S. From oceanic crust to exhumed mantle: A 40 year (1970–2010) perspective on the nature of crust under the Santos Basin, SE Brazil. *Geol. Soc. Lond. Spec. Publ.* **2012**, *369*, 147–165. [CrossRef]
17. Liu, Q.; Zhu, D.; Jin, Z.; Meng, Q.; Wu, X.; Yu, H. Effects of deep CO_2 on petroleum and thermal alteration: The case of the Huangqiao oil and gas field. *Chem. Geol.* **2017**, *469*, 214–229. [CrossRef]

geosciences

MDPI

Article

Gas Hydrate Estimate in an Area of Deformation and High Heat Flow at the Chile Triple Junction

Lucía Villar-Muñoz [1,*], Iván Vargas-Cordero [2], Joaquim P. Bento [3], Umberta Tinivella [4], Francisco Fernandoy [2,5], Michela Giustiniani [4], Jan H. Behrmann [1] and Sergio Calderón-Díaz [2]

[1] GEOMAR Helmholtz Centre for Ocean Research, Wischhofstr. 1-3, 24148 Kiel, Germany; jbehrmann@geomar.de

[2] Facultad de Ingeniería, Universidad Andrés Bello, Quillota 980, Viña del Mar 2531015, Chile; ivan.vargas@unab.cl (I.V.-C.); francisco.fernandoy@unab.cl (F.F.); sergio.calderon@unab.cl (S.C.-D.)

[3] Escuela de Ciencias del Mar, Pontificia Universidad Católica de Valparaíso, Av. Altamirano 1480, Valparaíso 2340000, Chile; jnettojunior@gmail.com

[4] Istituto Nazionale di Oceanografia e di Geofisica Sperimentale (OGS), Borgo grotta gigante 42/c, 34010 Sgonico, Italy; utinivella@inogs.it (U.T.); mgiustiniani@inogs.it (M.G.)

[5] Centro de Investigación Para la Sustentabilidad (CIS), Universidad Andrés Bello, República 252, Santiago 8370134, Chile

[*] Correspondence: lucia.villar@gmail.com; Tel.: +56-9-5226-4461

Received: 25 October 2018; Accepted: 31 December 2018; Published: 8 January 2019

check for
updates

Abstract: Large amounts of gas hydrate are present in marine sediments offshore Taitao Peninsula, near the Chile Triple Junction. Here, marine sediments on the forearc contain carbon that is converted to methane in a regime of very high heat flow and intense rock deformation above the downgoing oceanic spreading ridge separating the Nazca and Antarctic plates. This regime enables vigorous fluid migration. Here, we present an analysis of the spatial distribution, concentration, estimate of gas-phases (gas hydrate and free gas) and geothermal gradients in the accretionary prism, and forearc sediments offshore Taitao (45.5°–47° S). Velocity analysis of Seismic Profile RC2901-751 indicates gas hydrate concentration values <10% of the total rock volume and extremely high geothermal gradients (<190 °C·km^{-1}). Gas hydrates are located in shallow sediments (90–280 m below the seafloor). The large amount of hydrate and free gas estimated (7.21 × 10^{11} m^3 and 4.1 × 10^{10} m^3; respectively), the high seismicity, the mechanically unstable nature of the sediments, and the anomalous conditions of the geothermal gradient set the stage for potentially massive releases of methane to the ocean, mainly through hydrate dissociation and/or migration directly to the seabed through faults. We conclude that the Chile Triple Junction is an important methane seepage area and should be the focus of novel geological, oceanographic, and ecological research.

Keywords: BSR; gas hydrate; methane; seepage; active margin; Chile Triple Junction

1. Introduction

Gas hydrate is a crystalline ice-like solid formed by a mixture of water and gasses, mainly methane, giving place to a clathrate structure [1,2] that can be stored in the pore space of marine sediments under low temperature (<25 °C) and high pressure (>0.6 MPa) conditions. Methane gas may be produced biogenically at shallow depths or may migrate from a deeper source through advective transport along pathways such as fracture networks, faults, or shear zones (e.g., [3]). Since the gas hydrates are rich in methane, 1 m^3 of hydrate will yield 0.8 m^3 of water and 164 m^3 of methane at standard pressure and temperature (STP: 0 °C, 0.101325 Mpa) conditions [4], and a significant amount of hydrate represents unconventional and potential energy resources [5]. Moreover, gas hydrates play a part in global climate change, geo-hazards, and potential drilling hazards (e.g., [6–10]).

It is possible to identify gas hydrates in marine sediments using seismic profiles. The main indicator is the so-called Bottom Simulating Reflector (BSR), whose presence is related to the impedance contrast between high velocity gas hydrate-bearing and the underlying low velocity free gas layer [11–14]. Gas hydrate occurrences along the Chilean margin have been reported in many places by analysing the available seismic profiles (e.g., [11,14–27]), as well as more recently by direct identification of cold seeps emitting methane at the seafloor [28–34]. The first discovery of a seepage area was in 2004, offshore Concepción. Afterwards, other bathyal seep sites were identified, mainly by the presence of typical seep communities: (a) off the Limarí River at ~30° S (~1000 m water depth); (b) off El Quisco at 33° S (~340 m water depth); and c) off the Taitao Peninsula at ~46° S (~600 m water depth) [30–37].

Cold seeps sites are found in both active and passive margins and are related to the expulsion of methane-rich fluids. Chemosynthetic communities have been observed along active margins characterized by a well-developed accretionary prism, and along tectonically erosive margins [38]. The Chile Triple Junction (CTJ) area is a spectacular example of tectonic erosion (e.g., [39]). Even though many investigations are associated with seepage identification and gas expulsion quantification (gas bubbles) (e.g., [29,38,40–42]), there are few cases where the objective was to estimate the size of the gas source, as concentrations of gas hydrate and free gas [43].

Furthermore, the studies that report estimates of gas hydrates concentrations along the Chilean margin are scarce, even though, in the last decades, gas-phase concentrations have been estimated by fitting modelled velocity with theoretical velocity in the absence of gas [44]. These estimates reach an average of 15% and 1% of the total volume of gas hydrate and free gas concentrations, respectively [22,24,25,27]. A recent investigation of the southernmost Chilean continental margin showed that a regionally extensive methane hydrate reservoir, characterized by high gas hydrate and free gas concentrations, is present in the Patagonian marine sediments [27]. This could be an important natural resource for Chile, but because of the hydrate decomposition, this also potentially poses a great environmental threat.

On the other hand, the Chilean south-central margin is one of the tectonically most active regions on Earth, with very large and mega-scale earthquakes occurring every 130 and 300 years, respectively [45]. The margin segment close to the CTJ is characterized by high seismicity [46,47] that may trigger submarine sediments sliding and eventual gas hydrate dissociation. Some authors suggest that large subduction zone earthquakes have the potential to trigger hydrocarbon seepage to the ocean and possibly to the atmosphere (e.g., [29,48]). In this context, known gas hydrate quantities stored beneath marine sediments play an important role in the geohazard assessment. Besides, in subduction zones such as the Chilean margin, fluids play a key role in the nucleation and rupture propagation of earthquakes [49], and are a major agent of advective heat transfer from depth to the Earth's surface. For this reason, it is crucial to know the pathways where methane-rich fluids could migrate. The release of this methane stored in the forearc wedge could have consequences for the ocean and atmosphere systems, and the destabilized gas hydrate-bearing sediments are a formidable geohazard, in the form of submarine slumps, induced earthquakes, and tsunamis (e.g., [2,6,50–54]).

The particularity of the Chilean margin close to the CTJ, with anomalous heat flow and high seismicity, together with the presence of hydrothermal systems (e.g., [55]) and possible seafloor seeps, offers a unique scenario to study hydrate deposits. The aim of this study is to characterize and estimate the methane concentrations (hydrate and free gas phases) stored in the marine sediments in order to understand the potential amount of this gas that could be released through these natural pathways, likely affecting the geochemical properties of the seawater and, consequently, the marine ecology.

Geological Setting

The CTJ (Figure 1) is the site of the intersection of three tectonic plates: Nazca, Antarctic, and South America [39,56,57]. Here, the Chile Rise (CR), an active spreading centre, is being subducted beneath the South American continental margin. Ridge subduction began near Tierra del Fuego ~14 million years ago (Ma) and then migrated northwards to its current position north of the Taitao Peninsula (e.g., [15]). The Nazca plate subducts beneath South America in an ENE direction at a rate of about 70 km·Ma^{-1} north of the CTJ, and the Antarctic plate subducts in an ESE direction at about 20 km·Ma^{-1} south of the CTJ (e.g., [56]). The CR spreading rate has been estimated to have been about 70 km·Ma^{-1} over the past 5 Ma, but within the last 1 Ma, it has slowed down to about 60 km·Ma^{-1} (e.g., [58]).

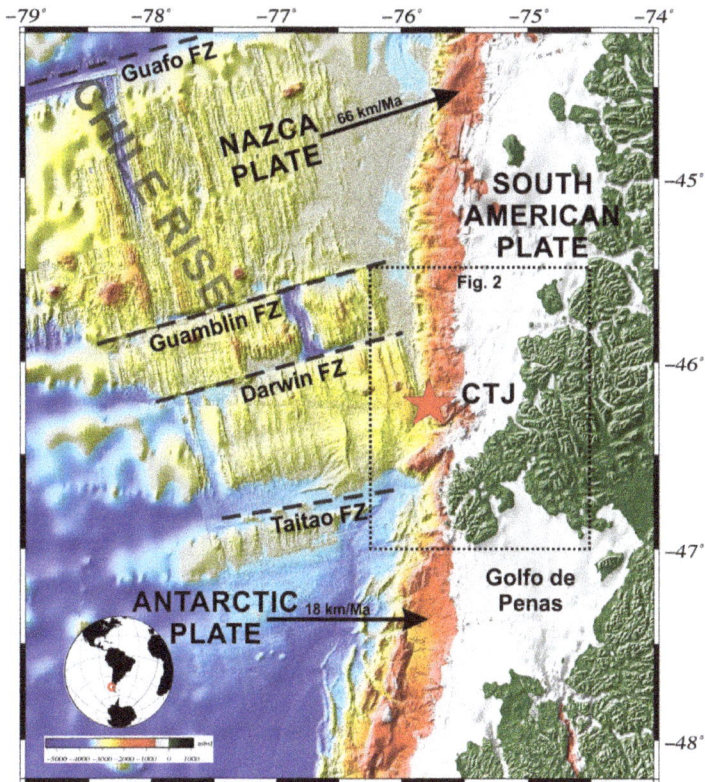

Figure 1. Location map of the study area offshore Taitao Peninsula. The bathymetry is based on GEBCO_08 Grid (version 20091120, http://www.gebco.net) and integrated with the IFREMER grid (cruise of the R/V L'Atalante, 1997). Tectonic setting of the Nazca, Antarctic, and South American plates: dashed black lines show the main Fracture Zones (FZ), red star marks a triple junction of the plates (CTJ), and dashed square corresponds to Figure 2.

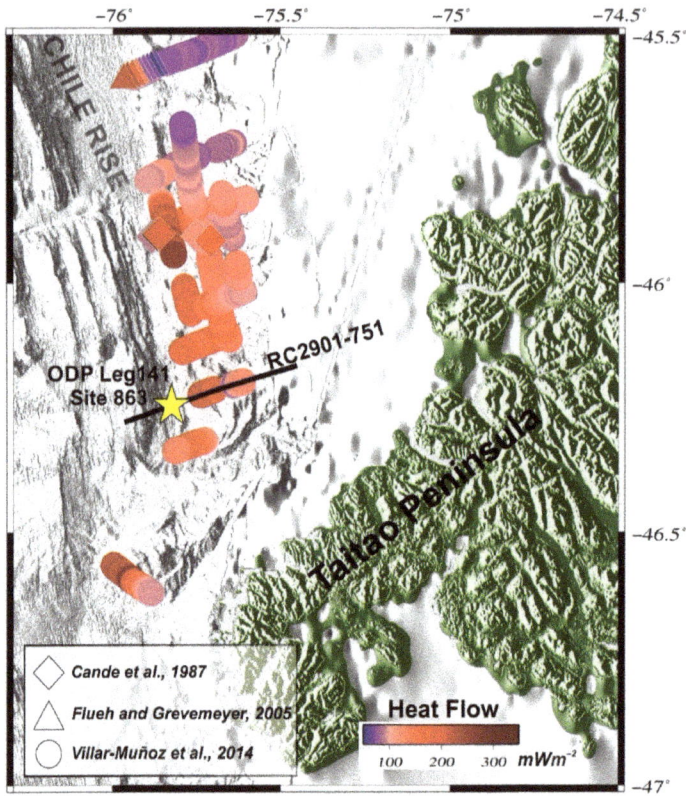

Figure 2. Heat flow (in m·Wm^{-2}) large-scale colour-coded based on BSR-derived heat flow and heat probes available for the area studied (after [26]). See text for description.

Close to the CTJ, the gas hydrate environment has peculiar characteristics relative to hydrate occurrences elsewhere. In fact, the ridge-trench collision perturbs pressure and temperature (PT) conditions within the sediment where hydrates have formed [11]. Excessively high heat flow, higher than 250 m·Wm^{-2}, was estimated above practically zero-age subducted crust (Figure 2). This is based on heat flow values derived from the depth of gas hydrate bottom-simulating reflectors [26,59] and direct measurements during the last decades [57,60].

The BSR-derived heat flow values are in general agreement with probe and borehole measurements [61]. Besides, high temperature gradients of 80–100 °C·km^{-1} were obtained at the toe of the continental wedge (e.g., Site 863 in Figure 2), just above the subducted zero-age crust [55]. The thermal anomaly in the region varies rapidly due to the presence of a strong convective circulation [62].

More recently, explorative work at the seafloor close to the CTJ has provided evidence for a sediment-hosted hydrothermal source near (~50 km) a methane-rich cold-seep area [63]. Advective methane transport operates within 5 km of the toe of the accretionary prism [59,64]. However, in the interior regions of the wedge, free gas migration and in situ gas production (within the hydrate stability region) build-up the hydrate [15], and BSR-depth towards the trench appears to rise in the sediments in proximity of the spreading ridge [15,26].

Moreover, gas at the base of the hydrate layer at the CTJ could also be produced from hydrate dissociation when changes in PT conditions shift the zone of hydrate stability upward, not only due to the accumulation of overburden, but also due to changes in PT conditions associated with active ridge subduction [11]. Increasing heat flow, associated with the approach of the CR, may have caused the base of the hydrate stability field to migrate ~300 m upwards in the sediments [15].

In this complex region, we find both active margin tectonic regimes: subduction erosion and subduction accretion occurring in close proximity (e.g., [65]). Bourgois et al [66] assumes that the tectonic evolution of the Chile margin in the area reflects the evolution of the tectonic regime at depth: subduction erosion from 5–5.3 to 1.5–1.6 Ma, followed by subduction accretion since 1.5–1.6 Ma. [67], indicates that subduction accretion occurring today along the pre-subduction segment is linked to a dramatic post-glacial increase in trench sediment supply. From evidence found by drilling at Ocean Drilling Program (ODP) Site 863 (Figure 2) at the CTJ proper, it was concluded that accretion ceased in late Pliocene, and presently, the small frontal accretionary prism is undergoing tectonic erosion [39,55].

2. Materials and Methods

2.1. Database

The analyzed seismic line was acquired in 1988 onboard the vessel R/V Robert Conrad within the framework of the project entitled "Paleogene geomagnetic polarity timescale" for Empresa Nacional del Petroleo (ENAP). The seismic profile was acquired using an air gun array with a size of 0.062 m^3. The shot spacing was approximately 50 m, and the streamer length was 3000 m and included 236 channels with an intertrace of 12.5 m. The seismic line RC2901-751 analyzed in this study was modelled to estimate gas hydrate and free gas concentrations.

During ODP Leg 141, the Site 863 located a few km south of the CTJ was drilled along the profile RC2901-751 in an area where the axis of the spreading ridge is subducting at 50 ka (Figure 2). Porosity and temperature data were obtained from this site.

2.2. Methods

The processing was performed using open source Seismic Unix software and codes ad-hoc [68] and includes a tested method reported in several studies [14,22,24,25,27,43,69]: (a) BSR identification, (b) seismic velocity modelling, (c) gas-phases estimates, and (d) geothermal gradient estimation.

(a) BSR identification: a stacking section was obtained by using standard processing (i.e. geometry arrangement, spherical divergence, velocity analysis, normal-moveout corrections, stacking, and filtering). The objective was to identify the BSR in a selected part of the stacking section. Once the BSR was recognized, the seismic velocity was modelled.

(b) Seismic velocity modelling: An in-depth velocity model was obtained using the Kirchhoff Pre-stack Depth migration (PreSDM) iteratively with a layer stripping approach (details in [70,71]). This approach uses the output of the PreSDM, the common image gathers (CIGs) [71]. In the seismic profile, three layers were modelled: the first between the seawater level and the seafloor reflector (SF layer); the second between the seafloor and the BSR (BSR layer); and the third between the BSR and the Base of Free Gas (BGR layer). It started with an initial constant velocity model equal to 1480 ms^{-1}. After four iterations, the SF reflector in the CIGs was flat, suggesting that the migration velocity was correct. The correct migration for BSR and BGR was reached after 25 and 15 iterations, respectively. Below the BGR, a velocity gradient was included and, to improve the migration result, the final velocity model was smoothed. Finally, band-pass filtering and mixing were applied to improve the final PreSDM image. The sensitivity was considered a depth error equal to 2.5% proposed by [22] after a sensitive test.

(c) Gas-phases estimates: Once the final velocity model had been built, it was converted into gas hydrate and free gas concentrations. At first, a qualitative estimate was performed, comparing the modelled velocity curves against theoretical curves in the absence of gas. Afterwards, positive anomalies were associated with gas hydrate presence, while negative anomalies were related to free gas presence. Modified Hamilton's curves [72] were adopted to estimate the theoretical velocity curves in the absence of hydrates and free gas or full water saturated sediments [73]. Gas hydrates and free gas concentrations were modified until the velocity model fitted the theoretical model, to obtain a quantitative estimate. The resultant is a concentration model in terms of total volume (for more details see [44]). Regarding the sensitivity, errors for gas hydrate and free gas estimates were assumed to be equal to 1.2% and 0.3% of volume, respectively. These errors were evaluated by [74], who performed a sensitive test to determine the influence of each parameter on the estimation of gas hydrate and free gas content. In fact, the main error was related to the assumptions of sediment properties.

(d) Geothermal gradient estimation: The geothermal gradient, indispensable to calculating the theoretical BSR-depth, was estimated using the following relation:

$$dT/dZ = (T_{BSR} - T_{SEA})/(Z_{BSR} - Z_{SEA}), \tag{1}$$

where BSR and seafloor depths (Z_{BSR}, Z_{SEA}) were extracted from the PreSDM section. Seafloor temperatures (T_{SEA}) were based on measurements from CTD data collected during ODP Leg 141 [75], while BSR temperatures (T_{BSR}) were based on the dissociation temperature-pressure function of gas hydrates [4]. Our estimation only considers methane because ethane concentration is negligible [22]. With regard to sensitivity, an error of depth equal to 2.5% was considered for seismic data [22].

3. Results

3.1. BSR Identification

The Kirchhoff PreSDM section (Figure 3) shows:

(a) A normal fault at a distance of 7 km representing the boundary between the lower and upper part of the continental rise and slope, respectively. Moreover, evidence of slip affecting the seafloor, as shallow faults and fractures, is registered from 8 to 15 km of distance;

(b) A strong and almost continuous BSR on the section that only gets weak or null where faults and fractures appear. Below the BSR, it is possible to recognize a weak but continuous reflector interpreted as BGR and, so, a free gas layer with a thickness of about 70 m;

(c) A variable depth of the BSR ranging between 80 and 150 m below seafloor (mbsf). The maximum depth of BSR was detected at about 2200 meters below sea level (mbsl) from 0 to 6 km, while the minimum depth (about 80 mbsf) was identified upwards (from 7 to 16 km). From 16 to 21 km of distance (in the "uplift part" of Figure 3), the BSR depth increases, reaching a depth of 150 mbsf.

3.2. Seismic Velocity Model

Above the BSR, a layer with a velocity ranging from 1650 to 1740 m/s was identified, while below the BSR, the velocity decreases from 1288 to 1550 m/s. Besides, below the BSR, the velocity decreases upwards (from 15 to 21 km of distance; see Figure 3 dark blue color), reaching its minimum value. An opposite velocity trend was observed above the BSR; in fact, when the velocity increases above the BSR (from 10 to 21 km of distance), the minimum velocity values are found below it. The BSR depth increases to the east, as shown by the velocity curves in Figure 3.

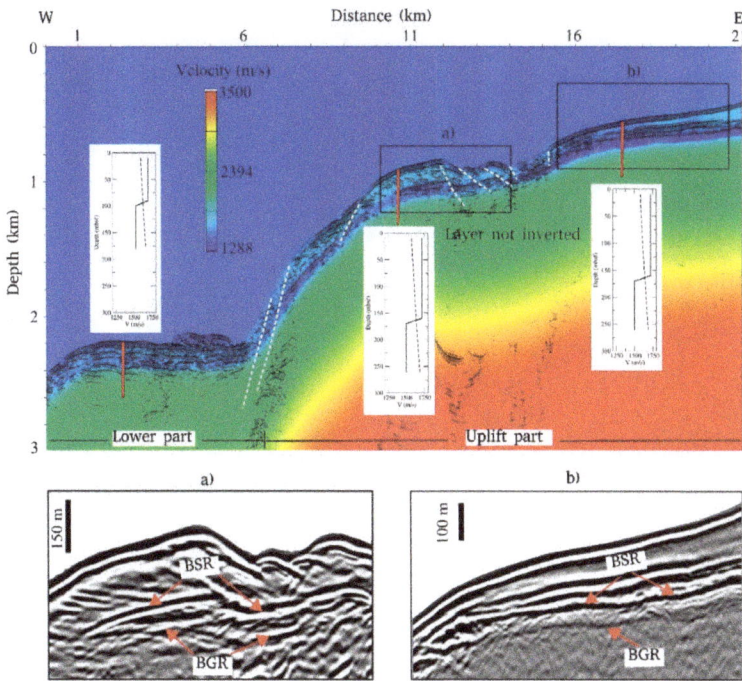

Figure 3. Velocity model superimposed in the Kirchhoff PreSDM section. The three inserts show the modelled velocity curves (solid black lines) and the theoretical curves in the absence of hydrates and free gas (dashed black lines) along the velocity model. Below, the rectangles indicate the position of the zooms in panel (a) and (b), in which red arrows indicate BSR and BGR (if present). The white dotted lines indicate faults and fractures.

3.3. Gas-Phases Estimates

High gas hydrates concentrations areas are located from 7 to 14 km of distance at approximately 1000 mbsl, reaching values ranging between 7 and 10% of total volume. Low gas hydrates concentrations regions (with values from 1 to 3% of total volume) are located from 1 to 6 km of distance at 2200 mbsl and from 15 to 20 km of distance at 600 mbsl (Figure 4). At shallow water depths, from 15 to 20 km of distance, high free gas concentrations were estimated, with values up to 0.8% of total volume. Note that hydrate and free gas concentrations show an opposite trend. In fact, from 7 to 14 km of distance, where gas hydrate concentrations increase (above the BSR), free gas concentrations decrease (see Top and Bottom panels in Figure 4). On the other hand, from 15 to 20 km of distance, where gas hydrate concentrations decrease, free gas concentrations increase.

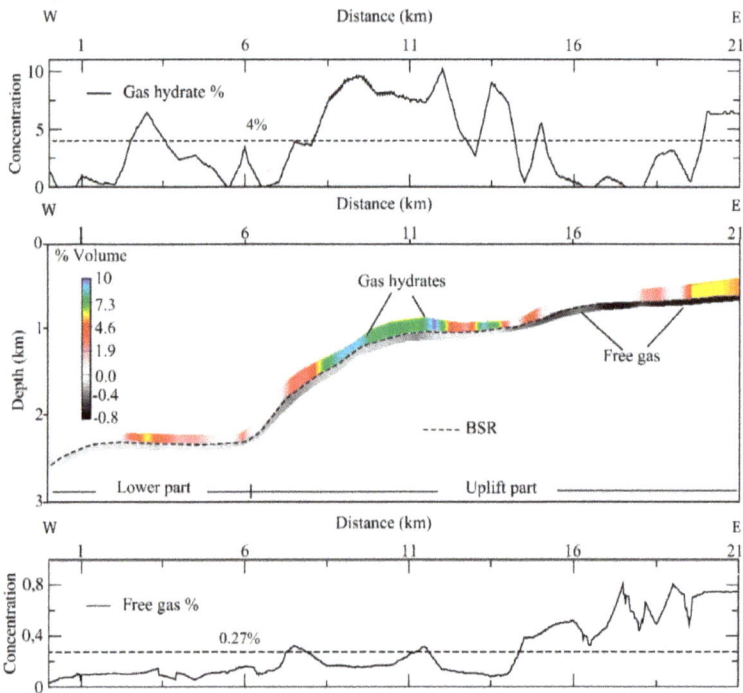

Figure 4. Gas hydrate and free gas concentration models and profiles relative to RC2901-751 seismic profile. Top panel: gas hydrate concentration values. Middle panel: gas-phase concentration model. Bottom panel: free gas concentration values. Dashed lines in the top and bottom panels correspond to the average gas hydrate and free gas concentrations, respectively.

3.4. Geothermal Gradient

The anomalous geothermal gradients calculated are variable in the seismic profile, ranging between 35 to 190 °C/km (Figure 5). The geothermal gradient increases towards the west (Figure 5), and the maximum values are at 2200 mbsl (see Figure 3). The minimum values were calculated on the east side of the profile (Figure 5) in correspondence of a water depth ranging from 600 to 1000 m. There are two isolated peaks (at ~9 and ~14 km of distance) of about 125 and 170 °C/km (Figure 5).

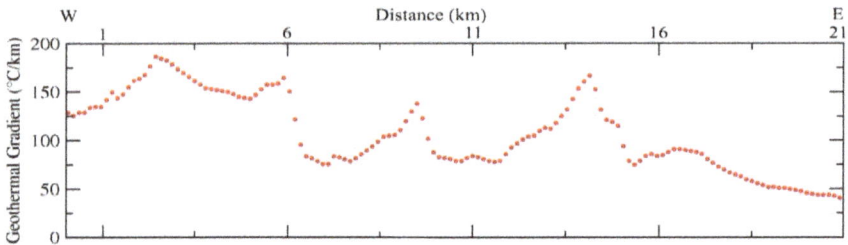

Figure 5. Geothermal gradient of the seismic profile RC2901-751. See text for details.

3.5. Gas Hydrate and Free Gas Volume at Standard Temperature and Pressure Conditions

In order to estimate the amount of methane stored in the marine sediments close to the CTJ region, bulk estimates of hydrate and free gas concentrations at standard temperature and pressure (STP) conditions were calculated using the following values:

- For gas hydrate: 4% of the total volume (dashed line in the upper panel of Figure 4), 50% porosity, thickness of the gas hydrate layer equal to 108 m, and a total projected area of about 2300 km². Considering these assumptions, the methane budget is 7.21×10^{11} m³ at STP conditions;
- For free gas: 0.27% of the total volume (dashed line in the lower section of Figure 4), 50% porosity, thickness of the free gas layer equal to 85 m, and a total projected area of about 2300 km². Considering these assumptions, the methane budget from gas hydrates is 4.1×10^{10} m³ at STP conditions.

The projected area was delimited based on the multi-resolution gridded Global Multi-Resolution Topography (GMRT) Synthesis [76] data and it comprises a part of the continental slope. The area was visually identified as the region that begins at the shelf break in the seaward edge of the shelf until it merges with the deep ocean floor at approximately 3000–3400 mbsl. All analyses were conducted with the open source Quantum Gis 3.4 (Qgis) and Generic Mapping Tools 5.4.4 (GMT) projects.

The free gas-volume expansion ratio was calculated using the Peng-Robinson equation of state [77], applying the methodology explained by [78]. Here, we assume that free gas is only composed of methane and it is located just below the gas hydrate stability zone. We divided the area containing free gas into five sub-areas to better assess the in-situ geothermal and pressure conditions and variations of the volume expansion ratios. Table 1 shows the free gas volume at in-situ and STP conditions. The rate of free gas volume expansion was calculated to estimate the volume of free gas content at STP conditions. The area was subdivided into five different regions, and at each one, pressure and temperature were calculated according to the corresponding geothermal gradient.

Table 1. Free gas volume at in-situ and STP conditions.

Interval (mbsl)	Area (m²)	Temperature (K)	Pressure (MPa)	Volume in-situ (m³)	Volume STP (m³)	Volume Expansion Ratio
500–1000	4.19×10^8	285.8	7.6	4.25×10^7	3.69×10^9	86.8
1000–1500	4.37×10^8	289.7	12.7	4.44×10^7	6.71×10^9	151.2
1500–2000	5.59×10^8	291.0	17.7	5.68×10^7	1.20×10^{10}	212.2
2000–2500	3.69×10^8	291.5	22.8	3.75×10^7	9.90×10^9	263.9
2500–3000	2.79×10^8	292.1	27.9	2.84×10^7	8.67×10^9	305.5
Total	2.06×10^9			2.10×10^8	4.10×10^{10}	

4. Discussion

The seismic section showed evidence of active tectonics; in fact, a large normal fault zone located at 7 km of distance represents the boundary between the western and eastern sectors (Lower and Uplift part in the Figure 3). Morphological features close to the normal fault can be associated with active tectonic extension and uplift processes above the subducting CR seafloor spreading centre [39]. Further upslope deformation is characterized by normal faults and fractures with small offsets affecting shallow sediments (Figure 3). The weak seismic character of BSR in the seaward (westward sector) is related to low free gas concentrations, while in the uplifted landward (eastward sector), a continuous and strong BSR can be related to high free gas concentrations up to 0.8% (Figure 4). These values are consistent with free gas concentrations reported by [11] along Seismic Line 745, located northward of this study area. A shallow BSR depth (average ~100 mbsf) can be explained by a high heat flow (average > 200 mW/m²) and geothermal gradient (average ~90 °C/km), as reported by [26] and in agreement with this study. In addition, vertical and lateral velocity variations above and below the BSR can be associated with gas hydrate and free gas presence and their concentration changes. Maximum velocity values above the BSR (up to 1740 m/s) can correspond to high gas hydrate concentrations, whereas low velocities below the BSR (around 1290 m/s) are related to high free gas concentrations (Figures 3 and 4). In fact, this low velocity can only be explained with free gas presence.

The gas-phase concentration distribution is in general agreement with heat flow reported by [26]. Moreover, low concentrations of gas hydrate and free gas coincide with high values of heat flow and geothermal gradients close to the Chile trench and the plate boundary (Figures 2 and 5), while high concentrations of gas hydrate and free gas are associated with a low heat flow and geothermal gradient further up the continental slope. A similar pattern was also recognized by [22] on the Chilean continental slope around 44° S.

The observation that both gas hydrate and free gas concentrations in the sediments have lower values close to the trench in the CTJ area could be explained as a result of gas hydrate dissociation and free gas migration in a regime of fluid advection under high heat flow conditions [59]. High heat flow is caused by the subduction of the Chile Rise [11,26,57,59,60], and geothermal fluids are supplied from deeper strata [67] that are undergoing deformation, anomalous compaction, and de-watering (e.g., [55,65]). The highest values of heat flow are located close to the heat source near the trench (Figures 2 and 5). We assume that in this area, the advective heat transfer in a regime of rising heat flow can change the pressure-temperature conditions, causing gas hydrate dissociation in the past and likely in the present. Low concentrations of free gas close to the trench (~0.1% of total volume), can be explained due to a variable production. Here, the dissociated hydrates is released as free gas and can migrate up into the hydrate stability zone, giving place to gas hydrate formation in higher areas (from 7 to 14 km of distance in Figure 4), increasing gas hydrate concentrations (~8% of total volume). However, active faults and fractures in the lower forearc can destroy stratigraphic seals and, consequently, impede free gas storage (e.g., [26,43]) above the subducting spreading ridge. This may explain the low concentrations of gas hydrate and free gas layers calculated close to the trench and high concentrations in shallow waters, where the lower values of heat flow were found, and deformation is less prevalent (Figures 3 and 4). Note, however, that low concentrations of gas hydrate and free gas were also found close to faults and fractures because of the enhancement of fluid-escape (Figures 3 and 4). Therefore, high heat flow due to spreading ridge subduction, tectonic faulting, and vigorous fluid advection at the leading edge of the overriding South American plate may indeed be a major factor for hydrate and gas reservoir distribution offshore Taitao Peninsula. Moreover, the highest value of free gas concentration, located in the shallower part of the accretionary wedge (~16 km of distance; Figure 4), can be explained by the upward migration of gas towards an impermeable hydrate layer, forming a structural trap [22]. Note also that this sector is characterized by the absence of faults that could act as pathways for upward fluid migration.

The anomalous heat flow close to the CTJ changes the stable PT conditions for the gas hydrate, promotes its dissociation and fluid escapes. The dissolved methane from gas hydrates could enter into the ocean through fluid ventings or as gas bubbles [79]. Some of the dissolved methane is diluted and oxidized as it rises through the ocean interior. However, an increase in gas methane entering the ocean above seawater saturation could lead to methane reaching the ocean surface mixed layer and being transported to the atmosphere via sea-air exchange [80].

A question worth discussing here is whether some of the methane in gas hydrates in the lower continental slope may in fact have been formed by abiotic processes (e.g., [81]) during the formation of serpentinite from ultramafic rocks. This can be valid for hydrates present in sediments just above the youngest crust of the CR subducted (near the trench), where active serpentinization and methane venting can initiate, develop, and survive, as was observed in similar regions (e.g., [82]). ODP Site 863 (see Figure 2 and [55]) is located on the seismic line presented in this study, right above the subducting oceanic spreading ridge. Pore waters squeezed from the drill cores recovered at ODP Site 863 show very high pH values up to 10.5, especially at drillhole depths greater than 600 meters below the sea floor. Along with the concentration profiles of F, B, Cl, and SO_4, this suggests that the pore fluids could be created from a sequence of reactions involving Mg-depleted fluids (see Figure 6 and description on p. 406 of [55]). This can be taken as an indication of metasomatic alteration in the serpentinized peridotite of the oceanic mantle (e.g., [83,84]) belonging to the downgoing plate at depth. Recently, Suess et al. [85] has shown that gas hydrates involving abiotically formed methane might be formed in sediment drifts overlying altered oceanic crust and mantle in slow-spreading environments. It is possible to envisage a similar scenario here, with the difference that the sediments of the lowermost continental slope are not directly sedimented above the spreading ridge, but are tectonically thrusted over the downgoing plate.

Finally, the estimated volume of gas hydrate calculated in the present study was lower than the values calculated in other regions along the Chilean margin (e.g., [27]). We hypothesize that this can be explained by the following reasons: (a) limited sediment accumulation due to the shortening of the wedge close to the CTJ, which causes unfavourable conditions for the formation of gas hydrates [11,39]; (b) the presence of faults and fractures that can locally promote fluid escape and prevent gas hydrate formation (e.g., [43,85]); (c) faults identified in the seismic profile (Figure 3) cross the transition layer of the gas hydrate phase and serve as pipes that drain water and methane to the seafloor (e.g., [85]); (d) the CTJ is characterised by an anomalous thermal state (e.g., [26]) that inhibits the formation of gas hydrates, by changing the gas hydrate stability zone.

5. Conclusions

The results of this research for the gas hydrate in the margin close to the Chile Triple Junction lead us to conclude that:

- The values for gas hydrate concentration are lower than 10% of the total rock volume. The highest concentrations are calculated in shallower waters, where the geothermal gradient is low and deformation is less prevalent;
- The amount of hydrate and free gas estimated over the studied area were 7.21×10^{11} m^3 and 4.1×10^{10} m^3, respectively;
- An inverse correlation between gas-phase concentrations and geothermal gradient is recognized. Low gas hydrate and free gas concentrations coincide with high values of geothermal gradients over the studied area;
- An extremely high geothermal gradient close to the trench was calculated, reaching values up to 190 °C·km^{-1}, caused by the subduction of the CR at the CTJ, altering the stable PT conditions for the gas hydrate, which promotes its dissociation and upward migration, and fluid escapes;
- High heat flow, tectonic faulting, and vigorous fluid advection may be important factors for hydrate and gas reservoir distribution offshore Taitao Peninsula;
- The CTJ is an important methane seepage area and should be the focus of novel geological, oceanographic, and ecological research.

Author Contributions: Conceptualization, L.V.-M. and I.V.-C.; formal analysis, L.V.-M. and I.V.-C.; funding acquisition, I.V.-C.; investigation, L.V.-M. and I.V.-C.; methodology, L.V.-M., I.V.-C., J.P.B., U.T., F.F., M.G., and S.C.; software, L.V.-M., I.V.-C., J.P.B., and U.T.; supervision, J.H.B.; visualization, L.V.-M. and J.P.B.; writing—original draft, L.V.-M.; writing—review & editing, I.V.-C., J.P.B., U.T., F.F., M.G., J.H.B., and S.C.

Funding: This research was funded by CONICYT- Fondecyt de Iniciación, 11140216.

Acknowledgments: Special thanks are due to Steven Cande and Stephen Lewis, who acquired the openly available data (http://www.ig.utexas.edu/) of R/V Robert Conrad Cruise RC2901. Lucía Villar-Muñoz acknowledges tenure of a DAAD scholarship for her postgraduate research and is grateful to the founders of GMT (Wessel and Smith). We are very grateful to Daniela Lazo and Rafael Santana, who contributed to the writing process.

Conflicts of Interest: The authors declare no conflict of interest. The funders had no role in the design of the study; in the collection, analyses, or interpretation of data; in the writing of the manuscript, or in the decision to publish the results.

References

1. Sloan, E.D. Fundamental principles and applications of natural gas hydrates. *Nature* **2003**, *426*, 353–363. [CrossRef] [PubMed]
2. Sloan, E.D.; Koh, C. *Clathrate Hydrates of Natural Gases*, 3rd ed.; CRC Press: Boca Raton, FL, USA, 2007; 752p.
3. Schmidt, M.; Hensen, C.; Morz, T.; Müller, C.; Grevemeyer, I.; Wallmann, K.; Mau, S.; Kaul, N. Methane hydrate accumulation in Mound 11 mud volcano, Costa Rica forearc. *Mar. Geol.* **2005**, *216*, 77–94. [CrossRef]
4. Sloan, E.D. *Clathrate Hydrates of Natural Gases*, 2nd ed.; CRC Press: Boca Raton, FL, USA, 1998; 705p.

5.   Milkov, A.V. Global estimates of hydrate-bound gas in marine sediments: How much is really out there? *Earth-Sci. Rev.* **2004**, *66*, 183–197. [CrossRef]

6.   Crutchley, G.J.; Mountjoy, J.J.; Pecher, I.A.; Gorman, A.R.; Henrys, S.A. Submarine Slope Instabilities Coincident with Shallow Gas Hydrate Systems: Insights from New Zealand Examples. In *Submarine Mass Movements and their Consequences*; Advances in Natural and Technological Hazards Research; Lamarche, G., Mountjoy, J., Bull, S., Hubble, T., Krastel, S., Lane, E., Micallef, A., Moscardelli, L., Mueller, C., Pecher, I., Eds.; Springer: Cham, Switzerland, 2016; 41p.

7.   Hovland, M.; Orange, D.; Bjorkum, P.A.; Gudmestad, O.T. Gas hydrate and seeps-effects on slope stability: The "hydraulic model". In Proceedings of the Eleventh International Offshore and Polar Engineering Conference, Stavanger, Norway, 17–22 June 2001; Volume 1, pp. 471–476.

8.   Kretschmer, K.; Biastoch, A.; Rüpke, L.; Burwicz, E. Modeling the fate of methane hydrates under global warming. *Glob. Biogeochem. Cycles* **2015**, *29*, 610–625. [CrossRef]

9.   Mountjoy, J.J.; Pecher, I.; Henrys, S.; Crutchley, G.; Barnes, P.M.; Plaza-Faverola, A. Shallow methane hydrate system controls ongoing, downslope sediment transport in a low-velocity active submarine landslide complex, Hikurangi Margin, New Zealand. *Geochem. Geophys. Geosyst.* **2014**, *15*, 4137–4156. [CrossRef]

10. Ruppel, C.D.; Kessler, J.D. The interaction of climate change and methane hydrates. *Rev. Geophys.* **2017**, *55*, 126–168. [CrossRef]

11. Bangs, N.L.; Sawyer, D.S.; Golovchenko, X. Free gas at the base of the gas hydrate zone in the vicinity of the Chile triple Junction. *Geology* **1993**, *21*, 905–908. [CrossRef]

12. Hyndman, R.D.; Spence, G.D. A seismic study of methane hydrate marine bottom-simulating-reflectors. *J. Geophys. Res.* **1992**, *97*, 6683–6698. [CrossRef]

13. Kvenvolden, K.A. Comparison of marine gas hydrates in sediments of an active and passive continental margin. *Mar. Pet. Geol.* **1985**, *2*, 65–70. [CrossRef]

14. Vargas-Cordero, I.; Tinivella, U.; Accaino, F.; Loreto, M.F.; Fanucci, F.; Reichert, C. Analyses of bottom simulating reflections offshore Arauco and Coyhaique (Chile). *Geo-Mar. Lett.* **2010**, *30*, 271–281. [CrossRef]

15. Brown, K.M.; Bangs, N.L.; Froelich, P.N.; Kvenvolden, K.A. The nature, distribution, and origin of gas hydrate in the Chile Triple Junction region. *Earth Planet. Sci. Lett.* **1996**, *139*, 471–483. [CrossRef]

16. Grevemeyer, I.; Kaul, N.; Díaz-Naveas, J.L. Geothermal evidence for fluid flow through the gas hydrate stability field off Central Chile-transient flow related to large subduction zone earthquakes? *Geophys. J. Int.* **2006**, *166*, 461–468. [CrossRef]

17. Loreto, M.F.; Tinivella, U.; Ranero, C. Evidence for fluid circulation, overpressure and tectonic style along the Southern Chilean margin. *Tectonophysics* **2007**, *429*, 183–200. [CrossRef]

18. Polonia, A.; Brancolini, G.; Torelli, L.; Vera, E. Structural variability at the active continental margin off southernmost Chile. *J. Geodyn.* **1999**, *27*, 289–307. [CrossRef]

19. Polonia, A.; Brancolini, G.; Torelli, L. The accretionary complex of southernmost Chile from the strait of Magellan to the Drake passage. *Terra Antarct.* **2001**, *8*, 87–98.

20. Polonia, A.; Torelli, L. Antarctic/Scotia plate convergence off southernmost Chile. *Geol. Acta* **2007**, *5*, 295–306.

21. Polonia, A.; Torelli, L.; Brancolini, G.; Loreto, M.F. Tectonic accretion versus erosion along the southern Chile trench: Oblique subduction and margin segmentation. *Tectonics* **2007**, *26*, TC3005. [CrossRef]

22. Vargas-Cordero, I.; Tinivella, U.; Accaino, F.; Loreto, M.F.; Fanucci, F. Thermal state and concentration of gas hydrate and free gas of Coyhaique, Chilean Margin (44° 30' S). *Mar. Pet. Geol.* **2010**, *27*, 1148–1156. [CrossRef]

23. Vargas-Cordero, I.; Tinivella, U.; Accaino, F.; Fanucci, F.; Loreto, M.F.; Lascano, M.E.; Reichert, C. Basal and Frontal Accretion Processes versus BSR Characteristics along the Chilean Margin. *J. Geol. Res.* **2011**, *2011*, 1–10. [CrossRef]

24. Vargas-Cordero, I.; Tinivella, U.; Villar-Muñoz, L.; Giustiniani, M. Gas hydrate and free gas estimation from seismic analysis offshore Chiloé island (Chile). *Andean Geol.* **2016**, *43*, 263–274. [CrossRef]

25. Vargas-Cordero, I.; Tinivella, U.; Villar-Muñoz, L. Gas Hydrate and Free Gas Concentrations in Two Sites inside the Chilean Margin (Itata and Valdivia Offshores). *Energies* **2017**, *10*, 2154. [CrossRef]

26. Villar-Muñoz, L.; Behrmann, J.H.; Diaz-Naveas, J.; Klaeschen, D.; Karstens, J. Heat flow in the southern Chile forearc controlled by large-scale tectonic processes. *Geo-Mar. Lett.* **2014**, *34*, 185–198. [CrossRef]

27. Villar-Muñoz, L.; Bento, J.P.; Klaeschen, D.; Tinivella, U.; Vargas-Cordero, I.; Behrmann, J.H. A first estimation of gas hydrates offshore Patagonia (Chile). *Mar. Pet. Geol.* **2018**, *96*, 232–239. [CrossRef]

28. Coffin, R.; Pohlman, J.; Gardner, J.; Downer, R.; Wood, W.; Hamdan, L.; Walker, S.; Plummerg, R.; Gettrus, J.; Diaz, J. Methane hydrate exploration on the mid Chilean coast: A geochemical and geophysical survey. *J. Pet. Sci. Eng.* **2007**, *56*, 32–41. [CrossRef]

29. Geersen, J.; Scholz, F.; Linke, P.; Schmidt, M.; Lange, D.; Behrmann, J.H.; Volker, D.; Hensen, C. Fault zone controlled seafloor methane seepage in the rupture area of the 2010 Maule earthquake, Central Chile. *Geochem. Geophys. Geosyst.* **2016**, *17*, 4802–4813. [CrossRef]

30. Jessen, G.L.; Pantoja, S.; Gutierrez, M.A.; Quinones, R.A.; Gonzalez, R.R.; Sellanes, J.; Kellermann, M.Y.; Hinrichs, K.U. Methane in shallow cold seeps at Mocha Island off central Chile. *Cont. Shelf Res.* **2011**, *31*, 574–581. [CrossRef]

31. Sellanes, J.; Quiroga, E.; Gallardo, V. First direct evidence of methane seepage and associated chemosynthetic communities in the bathyal zone off Chile. *J. Mar. Biol. Assoc. UK* **2004**, *84*, 1065–1066. [CrossRef]

32. Sellanes, J.; Krylova, E. A new species of Calyptogena (Bivalvia, Vesicomyidae) from a recently discovered methane seepage area off Concepción Bay, Chile (36S). *J. Mar. Biol. Assoc. UK* **2005**, *85*, 969–976. [CrossRef]

33. Sellanes, J.; Quiroga, E.; Neira, C. Megafaunal community structure and trophic relationships of the recently discovered Concepción Methane Seep Area (Chile, 36S). *ICES J. Mar. Sci.* **2008**, *65*, 1102–1111. [CrossRef]

34. Scholz, F.; Hensen, C.; Schmidt, M.; Geersen, J. Submarine weathering of silicate minerals and the extent of pore water freshening at active continental margins. *Geochim. Cosmochim. Acta* **2013**, *100*, 200–216. [CrossRef]

35. German, C.R.; Shank, T.M.; Lilley, M.D.; Lupton, J.E.; Blackman, D.K.; Brown, K.M.; Baumberger, T.; FrühGreen, G.; Greene, R.; Saito, M.A.; et al. Hydrothermal Exploration at the Chile Triple Junction—ABE's Last Adventure. In *AGU Fall Meeting Abstracts*; American Geophysical Union: Washington, DC, USA, 2010.

36. Oliver, P.G.; Sellanes, J. New species of Thyasiridae from a methane seepage area off Concepción, Chile. *Zootaxa* **2005**, *1092*, 1–20. [CrossRef]

37. Völker, D.; Geersen, J.; Contreras-Reyes, E.; Sellanes, J.; Pantoja, S.; Rabbel, W.; Thorwart, M.; Reichert, C.; Block, M.; Weinrebe, W.R. Morphology and geology of the continental shelf and upper slope of southern Central Chile (33S–43S). *Int. J. Earth Sci. (Geol. Rundsch.)* **2014**, *103*, 1765. [CrossRef]

38. Olu, K.; Duperret, A.; Sibuet, M.; Foucher, J.P.; Fiala-Medioni, A. Structure and distribution of cold seep communities along the Peruvian active margin: Relationship to geological and fluid patterns. *Mar. Ecol. Prog. Ser.* **1996**, *132*, 109–125. [CrossRef]

39. Behrmann, J.H.; Lewis, S.D.; Cande, S.C. Tectonics and geology of spreading ridge subduction at the Chile Triple Junction: A synthesis of results from Leg 141 of the Ocean Drilling Program. *Geol. Rundsch.* **1994**, *83*, 832–852. [CrossRef]

40. Hillman, J.I.T.; Klaucke, I.; Bialas, J.; Feldman, H.; Drexler, T.; Awwiller, D.; Atgin, O.; Çifçi, G. Gas migration pathways and slope failures in the Danube Fan, Black Sea. *Mar. Pet. Geol.* **2018**, *92*, 1069–1084. [CrossRef]

41. Hovland, M.; Svensen, H.; Forsberg, C.F.; Johansen, H.; Fichler, C.; Fosså, J.H.; Jonsson, R.; Rueslåtten, H. Complex pockmarks with carbonate-ridges off mid-Norway: Products of sediment degassing. *Mar. Geol.* **2005**, *218*, 191–206. [CrossRef]

42. Römer, M.; Sahling, H.; Pape, T.; Bohrmann, G.; Spieß, V. Quantification of gas bubble emissions from submarine hydrocarbon seeps at the Makran continental margin (offshore Pakistan). *J. Geophys. Res.* **2012**, *117*, C10015. [CrossRef]

43. Vargas-Cordero, I.; Tinivella, U.; Villar-Muñoz, L.; Bento, J.P. High Gas Hydrate and Free Gas Concentrations: An Explanation for Seeps Offshore South Mocha Island. *Energies* **2018**, *11*, 3062. [CrossRef]

44. Tinivella, U.; Carcione, J.M. Estimation of gas-hydrate concentration and free-gas saturation from log and seismic data. *Lead Edge* **2001**, *20*, 200–203. [CrossRef]

45. Cisternas, M.; Atwater, B.F.; Torrejón, F.; Sawai, Y.; Machuca, G.; Lagos, M.; Eipert, A.; Youlton, C.; Salgado, I.; Kamataki, T.; et al. Predecessors of the giant 1960 Chile earthquake. *Nature* **2005**, *437*, 404–407. [CrossRef]

46. Agurto-Detzel, H.; Rietbrock, A.; Bataille, K.; Miller, M.; Iwamori, H.; Priestley, K. Seismicity distribution in the vicinity of the Chile Triple Junction, Aysén Region, southern Chile. *J. S. Am. Earth Sci.* **2014**, *51*, 1–11. [CrossRef]

47. Murdie, R.E.; Prior, D.J.; Styles, P.; Flint, S.S.; Pearce, R.G.; Agar, S.M. Seismic responses to ridge-transform subduction: Chile triple junction. *Geology* **1993**, *21*, 1095–1098. [CrossRef]

48. Fischer, D.; Mogollón, J.M.; Strasser, M.; Pape, T.; Bohrmann, G.; Fekete, N.; Spiess, V.; Kasten, S. Subduction zone earthquake as potential trigger of submarine hydrocarbon seepage. *Nat. Geosci.* **2013**, *6*, 647–651. [CrossRef]

49. Sibson, R.H. Interactions between temperature and pore fluid pressure during earthquake faulting—A mechanism for partial or total stress relief. *Nat. Phys. Sci.* **1973**, *243*, 66–68. [CrossRef]

50. Boobalan, A.J.; Ramanujam, N. Triggering mechanism of gas hydrate dissociation and subsequent submarine landslide and ocean wide Tsunami after Great Sumatra-Andaman 2004 earthquake. *Arch. Appl. Sci. Res.* **2013**, *5*, 105–110.

51. Elger, J.; Berndt, C.; Rüpke, L.H.; Krastel, S.; Gross, F.; Geissler, W.H. Submarine slope failures due to pipe structure formation. *Nat. Commun.* **2018**, *9*, 715. [CrossRef] [PubMed]

52. Kvenvolden, K.A. Gas hydrates-geological perspective and global change. *Rev. Geophys.* **1993**, *31*, 173–187. [CrossRef]

53. Waite, W.F.; Santamarina, J.C.; Cortes, D.D.; Dugan, B.; Espinoza, D.N.; Germaine, J.; Jang, J.; Jung, J.W.; Kneafsey, T.J.; Shin, H.; et al. Physical properties of hydrate-bearing sediments. *Rev. Geophys.* **2009**, *47*, RG4003. [CrossRef]

54. Xu, W.; Germanovich, L.N. Excess pore pressure resulting from methane hydrate dissociation in marine sediments: A theoretical approach. *J. Geophys. Res.* **2006**, *111*. [CrossRef]

55. Behrmann, J.H.; Lewis, S.D.; Musgrave, R.; Bangs, N.; Bodén, P.; Brown, K.; Collombat, H.; Didenko, A.N.; Didyk, B.M.; Froelich, P.N.; et al. Chile Triple Junction. In Proc. ODP. *Init. Repts. (Pt. A)* **1992**, *141*, 1–708.

56. Cande, S.C.; Leslie, R.B. Late Cenozoic tectonics of the southern Chile trench. *J. Geophys. Res.* **1986**, *91*, 471–496. [CrossRef]

57. Cande, S.C.; Leslie, R.B.; Parra, J.C.; Hodbart, M. Interaction between the Chile ridge and Chile trench: Geophysical and geothermal evidences. *J. Geophys. Res.* **1987**, *92*, 495–520. [CrossRef]

58. Herron, E.M.; Cande, S.C.; Hall, B.R. An active spreading center collides with a subduction zone: A geophysical survey of the Chile margin triple Junction. *Mem. Geol. Soc. Am.* **1981**, *154*, 683–701.

59. Brown, K.M.; Bangs, N.L.; Marsaglia, K.; Froleich, P.N.; Zheng, Y.; Didyk, B.M.; Prior, D.; Richford, E.L.; Torres, M.; Kumsov, V.B.; et al. A summary of ODP 141 hydrogeologic, geochemical, and thermal results. *Proc. ODP Sci. Results* **1995**, *141*, 363–373.

60. Flueh, E.; Grevemeyer, I. *FS SONNE Cruise Report SO 181 TIPTEQ-from the Incoming Plate to Megathrust Earthquakes. Rep. 06.12.2004–26.02.2005*; Leibniz-Institut für Meereswissenschaften an der University Kiel: Kiel, Germany, 2005; 533p.

61. Völker, D.; Grevemeyer, I.; Stipp, M.; Wang, K.; He, J. Thermal control of the seismogenic zone of southern central Chile. *J. Geophys. Res.* **2011**, *116*, B10305. [CrossRef]

62. Lagabrielle, Y.; Guivel, C.; Maury, R.; Bourgois, J.; Fourcade, S.; Martin, H. Magmatic-tectonic effects of high thermal regime at the site of active ridge subduction: The Chile Triple Junction model. *Tectonophysics* **2000**, *326*, 255–268. [CrossRef]

63. German, C.R.; Ramirez-Llodra, E.; Baker, M.C.; Tyler, P.A. ChEss Scientific Steering Committee. Deep-Water Chemosynthetic Ecosystem Research during the Census of Marine Life Decade and Beyond: A Proposed Deep-Ocean Road Map. *PLoS ONE* **2011**, *6*, e23259. [CrossRef]

64. Bangs, N.L.; Brown, K.M. Regional heat flow in the vicinity of the Chile Triple Junction constrained by the depth of the bottom simulating reflection. *Proc. ODP Sci. Results* **1995**, *141*, 253–259.

65. Behrmann, J.H.; Kopf, A. Balance of tectonically accreted and subducted sediment at the Chile Triple Junction. *Int. J. Earth Sci. (Geol. Rundsch.)* **2001**, *90*, 753–768. [CrossRef]

66. Bourgois, J.; Martin, H.; Lagabrielle, Y.; Le Moigne, J.; Frutos Jara, J. (Chile margin triple junction area) Subduction erosion related to spreading-ridge subduction: Taitao peninsula. *Geology* **1996**, *24*, 723–726. [CrossRef]

67. Bangs, N.L.; Cande, S.C. Episodic development of a convergent margin inferred from structures and processes along the southern Chile margin. *Tectonics* **1997**, *16*, 489–503. [CrossRef]

68. Cohen, J.K.; Stockwell, J.W. *CWP/SU: Seismic Unix Release 4.0: A Free Package for Seismic Research and Processing*; Center for Wave Phenomena, Colorado School of Mines: Golden, CO, USA, 2008; pp. 1–153.

69. Loreto, M.F.; Tinivella, U.; Accaino, F.; Giustiniani, M. Offshore Antarctic Peninsula gas hydrate reservoir characterization by geophysical data analysis. *Energies* **2011**, *4*, 39–56. [CrossRef]

70. Yilmaz, O. *Seismic Data Analysis: Processing, Inversion and Interpretation of Seismic Data*, 2nd ed.; Society of Exploration Geophysicists: Oklahoma, OK, USA, 2001; 2027p.

71. Liu, Z.; Bleistein, N. Migration velocity analysis: Theory and an iterative algorithm. *Geophysics* **1995**, *60*, 142–153. [CrossRef]

72. Hamilton, E.L. Sound velocity gradients in marine sediments. *J. Acoust. Soc. Am.* **1979**, *65*, 909–922. [CrossRef]
73. Tinivella, U. A method for estimating gas hydrate and free gas concentrations in marine sediments. *Bollettino di Geofisica Teorica ed Applicata* **1999**, *40*, 19–30.
74. Tinivella, U. The seismic response to overpressure versus gas 638 hydrate and free gas concentration. *J. Seism. Explor.* **2002**, *11*, 283–305.
75. Grevemeyer, I.; Villinger, H. Gas hydrate stability and the assessment of heat flow through continental margins. *Geophys. J. Int.* **2001**, *145*, 647–660. [CrossRef]
76. Ryan, W.B.F.; Carbotte, S.M.; Coplan, J.O.; O'Hara, S.; Melkonian, A.; Arko, R.; Weissel, R.A.; Ferrini, V.; Goodwillie, A.; Nitsche, F.; et al. Global Multi-Resolution Topography synthesis. *Geochem. Geophys. Geosyst.* **2009**, *10*, Q03014. [CrossRef]
77. Peng, D.; Robinson, D.B. A new two-constant equation of state. *Ind. Eng. Chem. Fundam.* **1976**, *15*, 59–64. [CrossRef]
78. Barth, G. *Methane Gas Volume Expansion Ratios and Ideal Gas Deviation Factors for the Deep Water Bering Sea Basins*; USGS Open-File Report; USGS: Reston, VA, USA, 2005; 1451p.
79. Reeburgh, W.S. Oceanic Methane Biogeochemistry. *Chem. Rev.* **2007**, *107*, 486–513. [CrossRef]
80. Cynar, F.J.; Yayanos, A.A. *Biogeochemistry of Global Change: Radiatively Active Trace Gases*; Oremland, R.S., Ed.; Chapman-Hall: Atlanta, GA, USA, 1993; pp. 551–573.
81. McCollom, T.M. Abiotic methane formation during experimental serpentinization of olivine. *PNAS* **2016**, *113*, 13965–13970. [CrossRef] [PubMed]
82. Johnson, J.E.; Mienert, J.; Plaza-Faverola, A.; Vadakkepuliyambatta, S.; Knies, J.; Bünz, S.; Andreassen, K.; Ferré, B. Abiotic methane from ultraslow-spreading ridges can charge Arctic gas hydrates. *Geology* **2015**, *43*, 371–374. [CrossRef]
83. Bach, W.; Paulick, H.; Garrido, C.J.; Ildefonse, B.; Meurer, W.P.; Humphris, S.E. Unraveling the sequence of serpentinization reactions: Petrography, mineral chemistry, and petrophysics of serpentinites from MAR 15°N (ODP Leg 209, Site 1274). *Geophys. Res. Lett.* **2006**, *33*, 4–7. [CrossRef]
84. Kelley, D.S.; Karson, J.A.; Früh-Green, G.L.; Dana, R.; Yoerger, D.R.; Shank, T.M.; Butterfield, D.A.; Hayes, J.M.; Schrenk, M.O.; Olson, E.J.; et al. A serpentinite-hosted ecosystem: The Lost City hydrothermal vent field. *Science* **2005**, *307*, 1428–1434. [CrossRef] [PubMed]
85. Suess, E.; Torres, M.; Bohrmann, G.; Collier, R.W.; Greinert, J.; Linke, P.; Rehder, G.; Tréhu, A.; Wallmann, K.; Winckler, G.; et al. Gas hydrate destabilization: Enhanced dewatering, benthic material turnover and large methane plumes at the Cascadia convergent margin. *Earth Planet. Sci. Lett.* **1999**, *170*, 1–15. [CrossRef]

geosciences

MDPI

Article

Potential Instability of Gas Hydrates along the Chilean Margin Due to Ocean Warming

Giulia Alessandrini [1,*] , Umberta Tinivella [2] , Michela Giustiniani [2] ,
Iván de la Cruz Vargas-Cordero [3] and Silvia Castellaro [1]

[1]   Dipartimento di Fisica e Astronomia, Sezione Geofisica, Università di Bologna, 40127 Bologna, Italy;
     silvia.castellaro@unibo.it
[2]   Istituto Nazionale di Oceanografia e di Geofisica Sperimentale (OGS), Borgo grotta gigante 42/c,
     34010 Sgonico, Italy; utinivella@inogs.it (U.T.); mgiustiniani@inogs.it (M.G.)
[3]   Facultad de Ingeniería, Universidad Andres Bello, Quillota 980, Viña del Mar 2531015, Chile;
     ivan.vargas@unab.cl
*   Correspondence: giulia.alessandrin11@unibo.it; Tel.: +39-085-2095169

Received: 22 March 2019; Accepted: 16 May 2019; Published: 21 May 2019

check for updates

Abstract: In the last few years, interest in the offshore Chilean margin has increased rapidly due to the presence of gas hydrates. We have modelled the gas hydrate stability zone off Chilean shores (from 33° S to 46° S) using a steady state approach to evaluate the effects of climate change on gas hydrate stability. Present day conditions were modelled using published literature and compared with available measurements. Then, we simulated the effects of climate change on gas hydrate stability in 50 and 100 years on the basis of Intergovernmental Panel on Climate Change and National Aeronautics and Space Administration forecasts. An increase in temperature might cause the dissociation of gas hydrate that could strongly affect gas hydrate stability. Moreover, we found that the high seismicity of this area could have a strong effect on gas hydrate stability. Clearly, the Chilean margin should be considered as a natural laboratory for understanding the relationship between gas hydrate systems and complex natural phenomena, such as climate change, slope stability and earthquakes.

Keywords: gas hydrate; modelling; climate change; Chilean margin; slope stability; earthquake

1. Introduction

Many scientists worldwide have been working to better understand the onshore and offshore distribution of gas hydrate and its stability conditions. Natural gas hydrate is studied for a number of reasons, e.g., hydrate accumulations can store large amounts of natural gas, which could represent a potential energy resource (i.e., [1,2]).

Gas hydrates play an important role in the Chilean margin, mainly on account of critical issues concerning their potential dissociation. In fact, any variation in pressure and/or temperature conditions can lead to gas hydrate dissociation [3]. This may occur in the near future, as modelled by several previous studies (e.g., [4]), since the most recent assessment made by International Panel on Climate Change [5] confirmed that climate change may result in rising ocean temperatures and sea level. The release of huge quantities of natural gas in the water column could affect the marine ecosystem resulting in significant impact to benthic organisms [6]. Furthermore, methane is an important greenhouse gas, so after its release into the ocean, it could reach the atmosphere, resulting in positive feedback for global warming, as underlined by previous studies [7–9], although this is still the subject of debate among the scientific community (i.e., [10–13]). In fact, many factors prevent the methane from gas hydrate from reaching the atmosphere, such as methane release velocities and rates from the subsurface, and methane oxidation to carbon dioxide by microbial and chemical processes [14].

In recent years, the relationship between gas hydrate and submarine slides has been widely studied (e.g., [15]). Excess pore pressure has been identified as a key parameter in assessing slope instability [16]. Shear strain resistance significantly increases in hydrate-bearing sediments compared to hydrate-free sediments (e.g., [17]); during gas hydrate dissociation, released gas could increase the local pore fluid pressure in the sediment. This leads to a decrease of normal stress and, as a consequence, the formation of weak layers, in which less shear stress is needed to trigger failure (e.g., [18]). For this reason, the change in mechanical characteristics of marine sediments due to gas hydrate dissociation could lead to slope instability (e.g., [15]). Submarine slides would affect (i) gas hydrate stability itself, (ii) marine ecosystems, (iii) seafloor infrastructure and (iv) coastal areas, due to tsunamis that could be triggered [19].

The presence of gas hydrate in the Chilean margin has been confirmed by several geophysical cruises, in particular along the accretionary prism [20–22]. Gas hydrate has been detected in water depths up to 4 km with a depth range of 100–600 m below the seafloor (m b.s.f.) [13,23–28]. During ODP Leg 141, drill holes near the Chile Triple Junction sampled gas hydrates [29]. Mean volume concentrations of 18% and 1% have been observed for gas hydrate and free gas, respectively [20]. Recently, seismic data analysis confirmed the presence of gas hydrate and free gas, estimating concentration values in agreement with direct measurements [13,27,28,30,31]. In addition, studies off Valdivia estimated about 3.5% gas hydrate saturation in the pore space of marine sediments [21,30].

The study area is located in the southern segment of the central Peru-Chile margin (Figure 1) from 33° S to 46° S, covering about 1500 km from North to South. It includes the offshore regions from Valparaíso to Península de Taitao. In this area, the oceanic Nazca Plate subducts eastward, under the continental South American Plate, with an average convergence rate of about 8 cm/y [32–38].

The Juan Fernández and Chile ridges are the main bathymetric anomalies and represent the northern and southern boundaries for the study area, respectively [39]. The study area is characterized by a high sedimentation rate due to rapid erosion of the Andes and efficient fluvial transport, resulting in turbidite trench infill greater than 2 km [40].

Methane hydrate is stable in seafloor sediments at depths greater than 500 m below sea level (m b.s.l.), considering the equations of Sloan [3], as modelled by Tinivella et al. [18]. For the above considerations, this work aims to make a preliminary evaluation of the stability of gas hydrate and the possible effect of climate change on its stability by using a steady state approach along the Chilean margin.

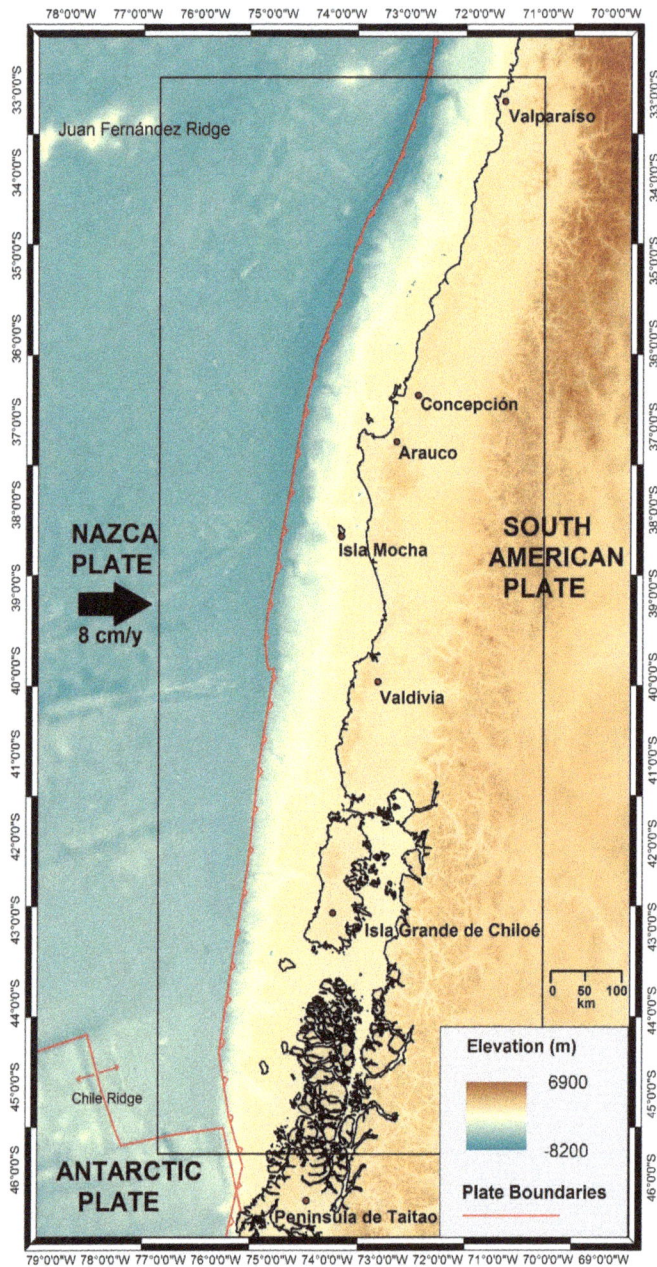

Figure 1. Topography and major tectonic elements of the study area. The black rectangle shows a sub-set of the study area selected for this work. The modelled area is the shelf-slope system, in which seafloor depths reach more than 500 m b.s.l.

2. Materials and Methods

2.1. Data Collection and Analysis

Gas hydrate stability on continental margins is a function of hydrostatic pressure, seafloor temperature, salinity, geothermal gradient and gas composition [3]. To model the Gas Hydrate Stability Zone (GHSZ) on the Chilean margin, the above-mentioned data have been gathered through bibliographic sources and available databases. Data analysis was performed using Geographic Information System (GIS) methods.

Bathymetry. In order to consider the hydrostatic pressure in our modelling, a bathymetric model (Figure 1) was downloaded from the GMRT website [41]. For this work, a WGS84 projection was chosen. The selected area from 33° S to 46° S and from 77° W to 71.5° W was imported and managed in GIS.

Seafloor temperature. Seafloor temperature data are available on the National Oceanographic Data Center website [42]. The selected data for the modelling area were interpolated in order to create a 500 × 500 m cell seafloor temperature grid (Figure 2A). A comparison between the bathymetric model and the seafloor temperature distribution suggests that seafloor topography strongly affects water mass temperatures as expected. In fact, colder waters (1 °C) fill deeper basins, while eastward, close to the Chilean shoreline, water temperatures are higher (13 °C).

Water column salinity. Water column salinity data were downloaded from the National Oceanographic Data Center website [42]. The selected data for the modelling area were imported in GIS and interpolated in a 500 × 500 m cell salinity grid. Figure 2B shows a water column salinity that varies between 33 and 34.

Geothermal gradient. Geothermal gradient data have been derived from the heat flow/thermal conductivity ratio. First, the heat flow and the thermal conductivity grids were built. To do this, heat flow data have been gathered, merged together and interpolated in a 500 × 500 m cell grid. They come from direct and indirect measurements. Heat flow data were collected during ODP Legs 141 and 202 [29,34,43]. The indirect data have been obtained from seismic data acquired along the Chilean margin, using the heat flow calculation method of Cande et al. [23] and reported in Villar-Muñoz et al. [44]. They show a progressive regional increase of the heat flow values from North to South: from about 24 mW/m^2, off Valparaíso, to 250 mW/m^2, close to the Chile Triple Junction.

Conductivity data come from a regional estimate, based on data collected during ODP Legs 141 and 202 [29,34,43,44]. Based on the average measured values, in the northern region the thermal conductivity was assumed equal to 0.85 W mK^{-1} [34,44], whereas in the southern area it decreases to 1.25 W mK^{-1} [29,43]. These data were interpolated through a linear regression algorithm in order to build a thermal conductivity grid. Combining the heat flow and the thermal conductivity grids, it was possible to calculate the geothermal gradient grid. Nevertheless, it was not possible to display the geothermal gradient in a map, because the number of data available for the interpolation was too low if compared to the previous two grids. The average geothermal gradient is around 49°C/km, which is consistent with previous observations [43].

Gas composition. Based on log and downhole temperature measurements carried out during ODP Legs 141 and 202 [43,45], the composition of gas hydrate in these sites is mainly methane. Unfortunately, no other direct measurements are available regarding the gas composition along the entire margin. So, we considered a pure methane hydrate that is more sensitive to climate change, as demonstrated by several authors (e.g., [46]).

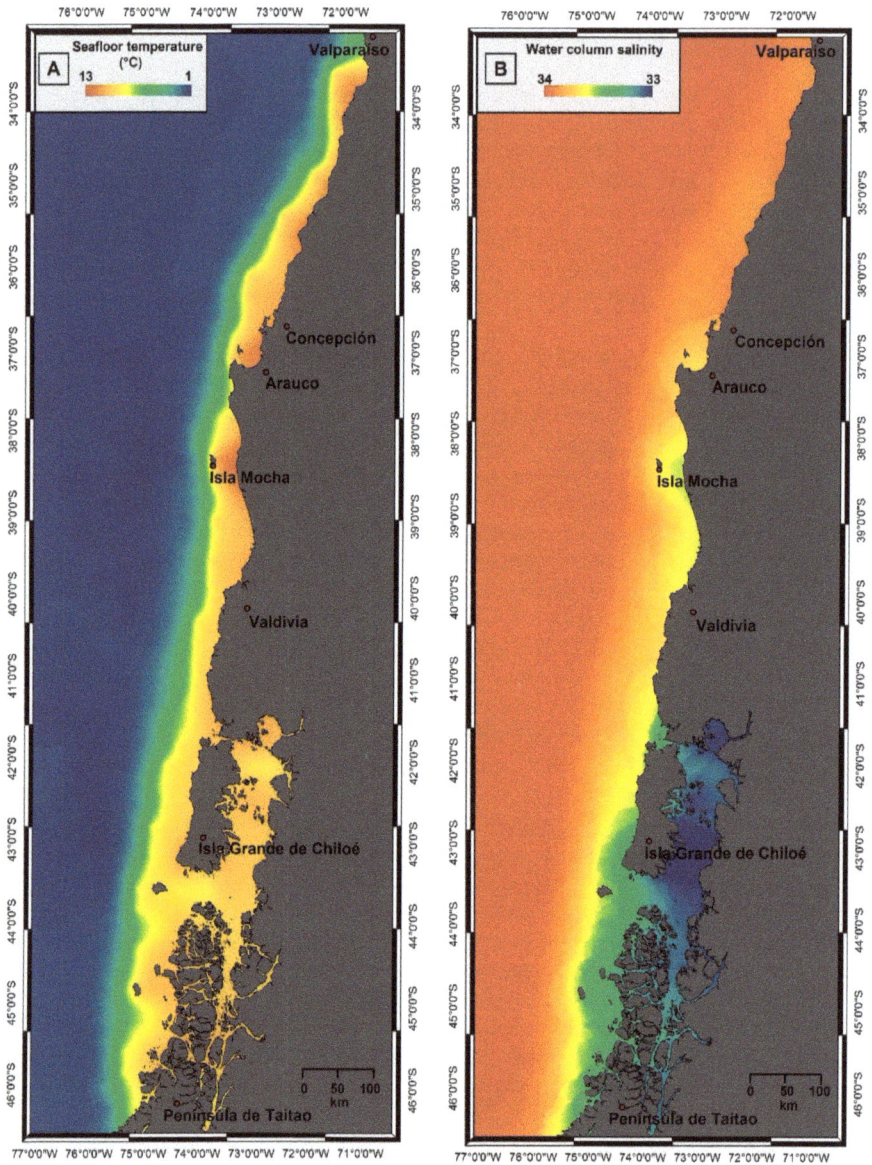

Figure 2. (A) Seafloor temperature grid. **(B)** Water column salinity grid.

2.2. Modelling

We adopted a steady state approach to model the base of the GHSZ assuming that: (a) a seabed temperature perturbation drives heat and has sufficient time to diffuse through the entire GHSZ; (b) during that time gas hydrate does not form or dissociate within the GHSZ; and (c) there is no latent heat. First, we simulated the present-day conditions in order to evaluate the zones where gas hydrate stability is verified (hereafter called "Scenario *S0*") using the above described data. The base of the GHSZ was defined as the intersection between the gas hydrate stability curve and the geothermal curve (i.e., [18,47]). The modelling considers Sloan equations [3], concerning gas hydrates equilibrium phases, and Dickens & Quinby-Hunt formula [48], which considers different salinity values. Applying this approach, the depth of the base of GHSZ (z_{GHSZ}) was calculated by solving the following equation:

$$\{7.054 \times 10^{-3} - 2.83 \times 10^{-4} \times [\log \cdot \varrho_w + \log (z_w + z_{GHSZ})]\} \times (T_0 + 273.15 + 1 \times 10^{-3} \, GG \, z_{GHSZ}) = 1,$$

where z_w is the water depth (m), T_0 is the seafloor temperature (°C), GG is the geothermal gradient (°C/km) and ϱ_w is the water density (i.e., [46]). The input data for this calculation were the above described manipulated dataset, described in Section 2.1.

Considering the Intergovernmental Panel on Climate Change (IPCC) [5] and National Aeronautics and Space Administration (NASA) [49] forecasts for future global warming, we simulated the effects of climate change on the GHSZ. More precisely, we considered different seafloor temperature increases (ΔT) and different sea level rises ($\Delta s.l.$), for 50- and 100-year-long terms. Based on the above cited forecasts, the following scenarios were modelled:

Scenarios in 50 years:
- *S1*: $\Delta T = 2 \, °C$,
- *S2*: $\Delta s.l. = 1.6 \, m$,
- *S3*: $\Delta T = 2 \, °C$ and $\Delta s.l. = 1.6 \, m$.

Scenarios in 100 years:
- *S4*: $\Delta T = 4 \, °C$,
- *S5*: $\Delta s.l. = 3.2 \, m$,
- *S6*: $\Delta T = 4 \, °C$ and $\Delta s.l. = 3.2 \, m$.

3. Results

The depth of the base of GHSZ for the Scenario *S0* is mapped in Figure 3. With sufficient methane, gas hydrate could form from the seafloor down to 580 m b.s.f. along the lower slope, in which bathymetric depths are as great as 6 km. The results of our modelling are in agreement with geophysical data, as reported in Coffin et al. [50]. Travelling to the east, along the upper slope, the base of the GHSZ becomes shallower until it intersects the seafloor at about 500 m b.s.l. (Figure 3). The area of this intersection is the most sensitive to seafloor temperature variations due to its smaller thickness, as observed by Marín-Moreno et al. [51]. To assess the error in modelling, we consider the error estimated by seismic velocity model perturbation to be 5% because the geothermal gradients were obtained from seismic data [22], and the error in bathymetric data as 1.5% in agreement with Tinivella et al. [18].

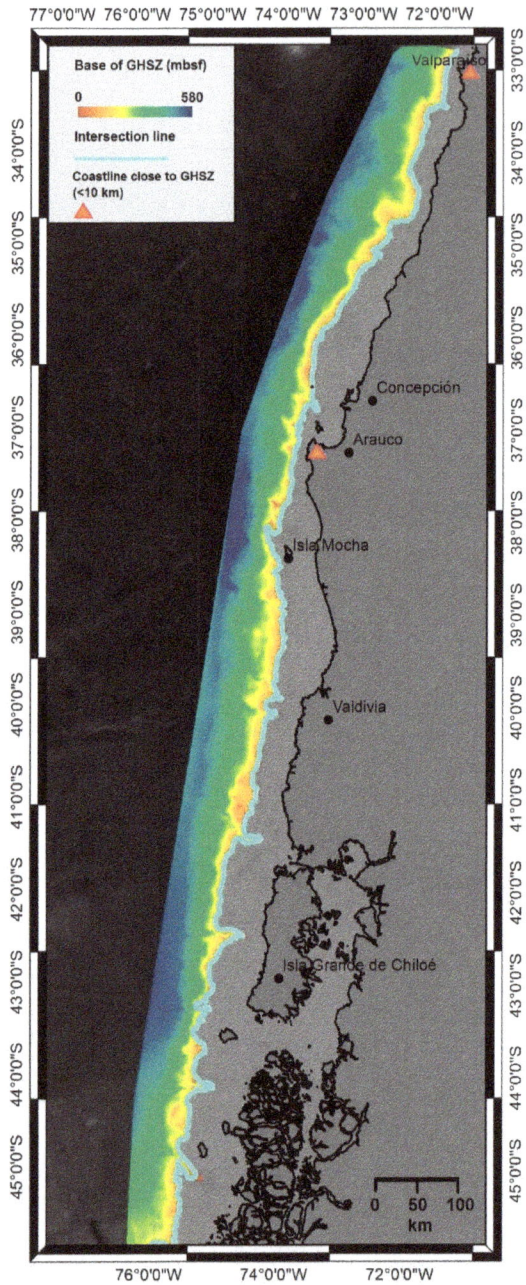

Figure 3. Depth of the base of the GHSZ below the seafloor for Scenario *S0* (present-day conditions). The base of the GHSZ is deeper for cool colors (lower slope) and shallower for warm colors (upper slope-shelf). The light-blue line marks the intersection between the base of the GHSZ and the seafloor. The red triangles highlight coastal locations very close to the intersection.

Figures 4 and 5 show the results of our modelling for present and future scenarios. For each of them, it was possible to model and calculate the thickness of the GHSZ. For future scenarios, these values were compared to the present Scenario *S0*, in order to observe possible variations in terms of thickness. Figure 4 shows for each scenario the average water depth at which the base of the GHSZ intersects the seafloor, considering the estimated error. Figure 5 reports for each scenario the average thickness of GHSZ sampled every 100 m.

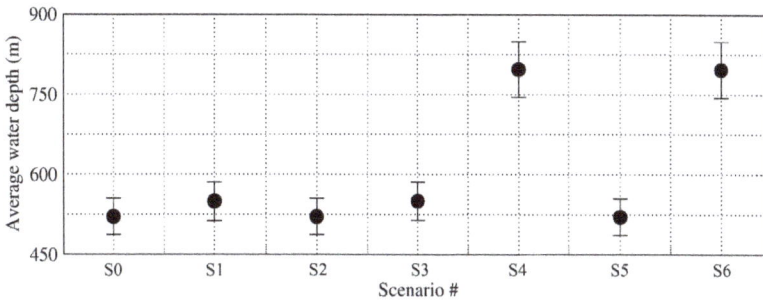

Figure 4. Average water depth at which the base of the GHSZ intersects the seafloor, considering the estimated error.

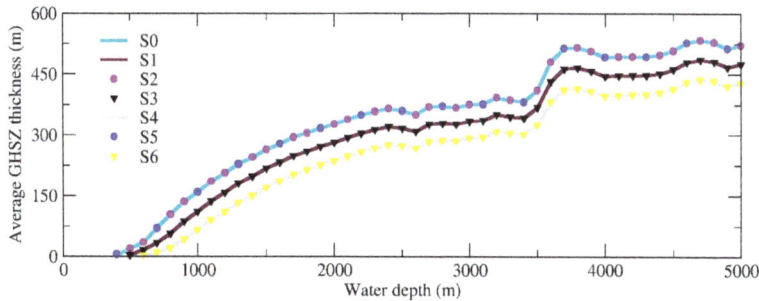

Figure 5. Average thickness of GHSZ. In every scenario, the thickness increases from the shelf-upper slope to the lower slope.

4. Discussion

The modelling results show that the predicted changes in climate could affect gas hydrate stability in the Chilean margin. Due to the small modelled increase in sea level, the *S2* and *S5* scenarios show a negligible variation compared to *S0* (Figures 4 and 5). In addition, *S3* and *S6* show a negligible variation compared to *S1* and *S4*, respectively (Figures 4 and 5). Considering different temperature increases (*S1* and *S4*), the modelled GHSZ would be reduced in terms of thickness and the average water depth for stability conditions would be greater with respect to *S0*.

Table 1 reports the two future scenarios in which seafloor temperature increase was considered. Global warming initially affects the stability of gas hydrate located in the proximity of the intersection between the base of the GHSZ and the seafloor [51] in the shallow upper slope-shelf. It is worth highlighting that the increase of 2 °C in seafloor temperature (*S1*) is almost the same observed for *S3*, where 2 °C temperature increase is combined with a pressure increase due to 1.6 m sea level rise. In both cases, in fact, there would be total gas hydrate dissociation in about 3% of the area in which gas hydrates are stable at present (*S0*). The effects of global warming in 100 years would be similar but amplified. In fact, in both *S4* and *S6*, there could be a potential release of the gas in 6.5% of the area in which the gas hydrate is stable at present (*S0*). It is possible to identify two zones (offshore Arauco

Peninsula and Valparaíso) in which the intersection between the seafloor and the base of the GHSZ is nearby to the shoreline (close to 10 km; Figure 3). Considering this, the potential dissociation could have dangerous consequences for coastal cities and infrastructure because of possible slope instability.

Table 1. Amount of gas released by thermal dissociation of gas hydrate, considering a porosity of 40% and a mean hydrate saturation of 3% according to the minimum value proposed by [21,30].

Scenario	Gas Hydrate Dissociation Area	Total Volume	Pore Volume	Hydrate Volume	Gas Volume
S1 ($\Delta T = 2$ °C)	3%	113 km^3	45 km^3	1.36 km^3	222 km^3
S4 ($\Delta T = 4$ °C)	6.5%	482 km^3	193 km^3	5.79 km^3	950 km^3

At standard pressure and temperature conditions, about 164 m^3 of methane are contained in 1 m^3 of gas hydrate. So, according to our modelling, in the next 50 years, there could be a potential release of 222 km^3 of methane from hydrate dissociation for *S1* (Table 1). On the other hand, in the next 100 years there could be a potential emission of 950 km^3 of methane for *S4* (Table 1). However, steady state models, which represent the hydrate system at the equilibrium after a warming or cooling period, could overestimate emission of gas from hydrate. In fact, the non-inclusion of thermodynamics processes, like self-preservation of gas hydrate [52] and the time of propagation of heat through the entire thickness of sediments could lead to a too rapid disappearance of the GHSZ during warming events [53].

To show clearly the possible effects of global warming, Figure 6 reports the intersections between the seafloor and the base of the GHSZ for all scenarios, focusing on an area characterized by strong slope gradients (35°–38° S). Note that the lines related to *S0*, *S2* and *S5* roughly overlap as well as *S1*–*S3* and *S4*–*S6*, as already discussed; arrows represent the dip direction and their size is directly proportional to the slope degree.

In the selected sector reported in Figure 6, due to the tectonic-sedimentary configuration, the area is characterized by high angle slopes and, for this reason, unstable in the long term. In fact, high basal frictions in convergent margins give rise to oversteepening and, therefore, more unstable accretionary prisms [40]. If combined with local high sedimentation rate, these two factors could create some critical areas with high risk for slope failure. Several authors have suggested that in the area reported in Figure 6 the above conditions are verified and several slope failures are documented (e.g., [54–56]). In fact, more than 60 submarine slopes were mapped in the area between 35° S to 38° S. Among these, Valdes Slide, Reloca Slide and the Northern, Central and Southern Embayments are the most noticeable lower-continental slides because of their size and volume [54–57]. In addition, most of the slope failures mapped in this area are related to submarine canyons, mainly on the upper-continental slope [56]. These critically-stable continental slopes are more susceptible areas to slide risk if gas hydrate dissociation is considered [16]. It is important to remind ourselves that the combination of critically-stable slopes and the proximity of the intersections to the coast contributes to define the potential instability of this area.

In Figure 6, average slope values along the modelled intersections are about 10°, up to 20° at submarine canyons (e.g., Bío Bío Canyon). It is clear that the effect of the temperature increase modelled in *S4* and *S6* could be potentially more critical in the proximity of the high degree slopes, located not far from the coast. Also, active and fossil fluid venting, such as authigenic carbonates, along the upper slope between 36.5° and 36.8° S seems to contribute to the potential weakening of sediment cohesion and help trigger submarine landslides [58]. Moreover, the critical submarine slope issue should be linked to the high seismicity characterizing the Chilean margin. In fact, areas characterized by high basal friction coefficients and steep slopes (Figure 6) could be particularly sensitive in case of an earthquake, promoting slides with potential gas hydrate dissociation. In particular, the selected area has been affected by the strongest earthquakes ever recorded, such as the Mw 9.5 (1960) and the Mw 8.8 (2010) events, both with very shallow hypocenters [59] (Figure 6).

Recent studies show that slope stability in active margins is higher due to seismic strengthening, hence suggesting an inverse relationship between seismicity and submarine landslides [60–62]. Despite that, different authors [61,63] show that high sedimentation rates in continental slopes seem to be able to counteract seismic strengthening, which is thought to be particularly high in this sector, as mentioned before.

Finally, it is worth mentioning that, as already remarked by several authors (e.g., [19]), an earthquake could trigger gas hydrate dissociation. So, better understanding of the link between gas hydrate, slope stability and earthquakes is required.

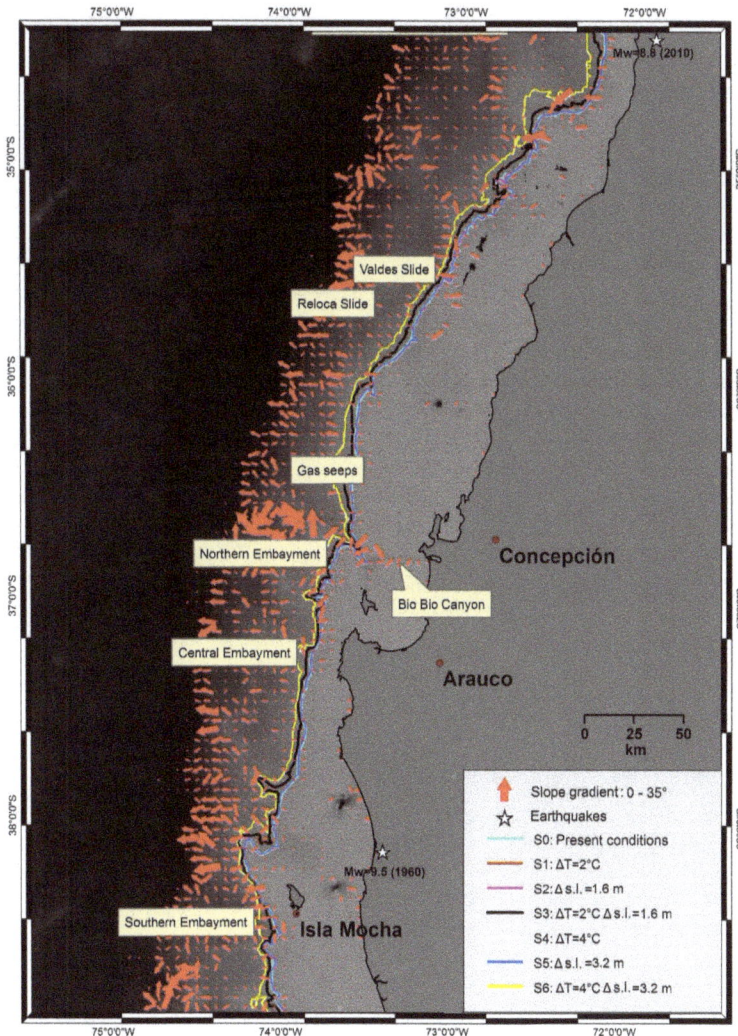

Figure 6. Study area between 35° S–38° S. The solid colored lines represent the intersections between the base of the GHSZ and the seafloor for each scenario. Here, the intersections are related to the slope gradient, marked by red arrows; the main slope failures are mapped, and major recent earthquakes are marked by stars.

5. Conclusions

We modelled the GHSZ using steady state modelling to verify where the gas hydrate could be stable along the Chilean margin and to evaluate in first approximation the possible effects of climate change on gas hydrate stability. Based on the model results, it was possible to estimate the thickness of marine sediments in which the conditions for the formation of gas hydrate are met.

Under present-day conditions (*S0*), depending on methane availability, gas hydrates can form down to 580 m b.s.f., along the lower slope, as confirmed by Coffin et al. [50]. These authors integrated data from seismic surveys, geochemical analysis of porewater samples from piston cores and heat flow probing.

Considering the IPCC [5] and NASA [49] forecasts for the future global warming over the next 50 and 100 years, we simulated the impact of climate change on the GHSZ, for the first time ever for the Chilean margin. The modelled future scenarios, considering an increase in temperature (*S1, S3, S4, S6*), would indicate total gas hydrate dissociation along the upper slope. This suggests that, despite higher pressure due to sea level rise, the effect of the seafloor temperature increase on gas hydrate stability is significant. Sea level rise seems to be insufficient to counteract the effect of temperature increase, which is primarily responsible for GHSZ thinning [64].

The potential methane release due to gas hydrate dissociation could cause slope instability. Due to the tectonic-sedimentary configuration, the 35° S–38° S sector has been identified as potentially critical for the long term. As a consequence, coastal cities could be seriously damaged if tsunamis were triggered due to gas hydrate dissociation. Furthermore, the high seismicity of this area could significantly affect slope failure and consequently gas hydrate stability. An integrated approach is needed to understand the link between these processes.

In conclusion, our scenarios suggested that climate change could affect hydrate stability in long term. For this reason, transient modelling is necessary to understand how the hydrate dissociation could happen, since it shows how the hydrate system changes during the warming or cooling period until the equilibrium state. Moreover, an integration with geophysical data, in particular seismic data, could contribute to calibrate future models and to make appropriate assumptions on the initial gas hydrate distribution and saturation, which are inhomogeneous from down-slope to up-slope [53,65]. Furthermore, our steady state modelling demonstrates that more effort should be devoted to gaining a better understanding of the relationship between the gas hydrate system and complex natural phenomena, such as climate change, slope stability and earthquakes.

Author Contributions: Conceptualization and methodology, G.A., U.T., M.G. and I.V.-C.; writing—original draft preparation, G.A.; writing—review and editing, G.A., U.T., M.G., I.V.-C. and S.C.; supervision, U.T., M.G., I.V.-C. and S.C.

Funding: This research was partially supported by the Italian Ministry of Education, Universities and Research (Decreto MIUR No. 631 dd. 8 August 2016) under the extraordinary contribution for Italian participation in activities related to the international infrastructure PRACE—The Partnership for Advanced Computing in Europe (www.prace-ri.eu).

Acknowledgments: We want to thank the COST-MIGRATE Action (reference code ES1405-050317-082155). We would like to thank the anonymous reviewers for valuable comments.

Conflicts of Interest: The authors declare no conflict of interest.

References

1. Makogon, Y.F. *Hydrate of Hydrocarbons*; PennWell Publishing Co.: Tulsa, Oklahoma, 1997.
2. Moridis, G.J.; Collett, T.S.; Boswell, R.; Kurihara, M.; Reagan, M.T.; Koh, C.; Sloan, E.D. Toward production from gas hydrates: Current status, assessment of resources, and simulation-based evaluation of technology and potential. *SPE Reserv. Eval. Eng.* **2009**, *12*, 745–771. [CrossRef]
3. Sloan, E.D., Jr. *Clathrate Hydrates of Natural Gases*, 2nd ed.; Revised and Expanded; Marcel Dekker, Inc.: New York, NY, USA, 1998; p. 705.

4. Marín-Moreno, H.; Giustiniani, M.; Tinivella, U. The potential response of the hydrate reservoir in the South Shetland Margin, Antarctic Peninsula, to ocean warming over the 21st century. *Polar Res.* **2015**, *34*, 27443. [CrossRef]
5. IPCC. Contribution of Working Groups I, II and III to the Fifth Assessment Report of the Intergovernmental Panel on Climate Change. In *Climate Change 2014: Synthesis Report*; Pachauri, R.K., Meyer, L.A., Eds.; IPCC: Geneva, Switzerland, 2014; 151p.
6. Wallman, K.; MIGRATE Consortium. Marine Gas Hydrate—An Indigenous Resource of Natural Gas for Europe (MIGRATE). 2015; Available online: https://www.migrate-cost.eu/ (accessed on 29 April 2019).
7. MacDonald, I.; Joye, S. Lair of the "Ice Worm". *Quarterdeck* **1997**, *5*, 5–7.
8. Kvenvolden, K.A. Methane hydrate in the global organic carbon cycle. *Terra Nova* **2002**, *14*, 302–306. [CrossRef]
9. Kennett, J.P.; Cannariato, K.G.; Hendy, I.L.; Behl, R.J. *Role of Methane Hydrates in Late Quaternary Climatic Change: The Clathrate Gun Hypothesis*; AGU: Washington, DC, USA, 2003; Volume 54, 216p.
10. Dickens, G.R. Modeling the global carbon cycle with a gas hydrate capacitor: Significance for the latest Paleocene thermal maximum. In *Natural Gas Hydrates: Occurrence, Distribution, and Detection*; Paull, C.K., Dillon, W.P., Eds.; Geophysical Monograph Series; American Geophysical Union: Washington, DC, USA, 2001; Volume 124, pp. 19–40.
11. Xu, W.; Lowell, R.P.; Peltzer, E.T. Effect of seafloor temperature and pressure variations on methane flux from a gas hydrate layer: Comparison between current and late Paleocene climate conditions. *J. Geophys. Res. Solid Earth* **2001**, *106*, 26413–26423. [CrossRef]
12. Milkov, A.V. Global estimates of hydrate-bound gas in marine sediments: How much is really out there? *Earth-Sci. Rev.* **2004**, *66*, 183–197. [CrossRef]
13. Vargas-Cordero, I.; Tinivella, U.; Villar-Muñoz, L.; Bento, J. High Gas Hydrate and Free Gas Concentrations: An Explanation for Seeps Offshore South Mocha Island. *Energies* **2018**, *11*, 3062. [CrossRef]
14. Kvenvolden, K.A. Potential effects of gas hydrate on human welfare. *Proc. Natl. Acad. Sci. USA* **1999**, *96*, 3420–3426. [CrossRef]
15. Sultan, N.; Cochonat, P.; Foucher, J.P.; Mienert, J. Effect of gas hydrates melting on seafloor slope instability. *Mar. Geol.* **2004**, *213*, 379–401. [CrossRef]
16. Sultan, N.; Cochonat, P.; Canals, M.; Cattaneo, A.; Dennielou, B.; Haflidason, H.; Laberg, J.S.; Long, D.; Mienert, J.; Trincardi, F.; et al. Triggering mechanisms of slope instability processes and sediment failures on continental margins: A geotechnical approach. *Mar. Geol.* **2004**, *213*, 291–321. [CrossRef]
17. Tinivella, U. The seismic response to over-pressure versus gas hydrate and free gas concentration. *J. Seism. Explor.* **2002**, *11*, 283–305.
18. Tinivella, U.; Giustiniani, M.; Accettella, D. 2011. BSR versus climate change and slides. *J. Geol. Res.* **2011**, *2011*, 390547.
19. Boobalan, A.J.; Ramanujam, N. Triggering mechanism of gas hydrate dissociation and subsequent sub marine landslide and ocean wide Tsunami after Great Sumatra—Andaman 2004 earthquake. *Arch. Appl. Sci. Res.* **2013**, *5*, 105–110.
20. Bangs, N.L.; Sawyer, D.S.; Golovchenko, X. Free gas at the base of the gas hydrate zone in the vicinity of the Chile triple junction. *Geology* **1993**, *21*, 905–908. [CrossRef]
21. Rodrigo, C.; González-Fernández, A.; Vera, E. Variability of the bottom-simulating reflector (BSR) and its association with tectonic structures in the Chilean margin between Arauco Gulf (37° S) and Valdivia (40° S). *Mar. Geophys. Res.* **2009**, *30*, 1–19. [CrossRef]
22. Vargas-Cordero, I.; Tinivella, U.; Accaino, F.; Loreto, M.F.; Fanucci, F.; Reichert, C. Analyses of bottom simulating reflections offshore Arauco and Coyhaique (Chile). *Geo-Mar. Lett.* **2010**, *30*, 271–281. [CrossRef]
23. Cande, S.C.; Leslie, R.B.; Parra, J.C.; Hobart, M. Interaction between the Chile Ridge and Chile Trench: Geophysical and geothermal evidence. *J. Geophys. Res: Solid Earth* **1987**, *92*, 495–520. [CrossRef]
24. Froelich, P.N.; Kvenvolden, K.A.; Torres, M.E.; Waseda, A.; Didyk, B.M.; Lorenson, T.D. Geochemical evidence for gas hydrate in sediment near the Chile Triple Junction. *Proc. ODP Sci. Results* **1995**, *141*, 279–287.
25. Sloan, E.D., Jr.; Koh, C. *Clathrate Hydrates of Natural Gases*, 3rd ed.; CRC Press: Boca Raton, FL, USA, 2007; p. 752.

26. Vargas-Cordero, I.; Tinivella, U.; Accaino, F.; Fanucci, F.; Loreto, M.F.; Lascano, M.E.; Reichert, C. Basal and frontal accretion processes versus BSR characteristics along the Chilean margin. *J. Geol. Res.* **2011**, *2011*, 846101. [CrossRef]

27. Vargas-Cordero, I.; Tinivella, U.; Villar-Muñoz, L.; Giustiniani, M. Gas hydrate and free gas estimation from seismic analysis offshore Chiloé island (Chile). *Andean Geol.* **2016**, *43*, 263–274. [CrossRef]

28. Vargas-Cordero, I.; Tinivella, U.; Villar-Muñoz, L. Gas Hydrate and Free Gas Concentrations in Two Sites inside the Chilean Margin (Itata and Valdivia Offshores). *Energies* **2017**, *10*, 2154.

29. Grevemeyer, I.; Villinger, H. Gas hydrate stability and the assessment of heat flow through continental margins. *Geophys. J. Int.* **2001**, *145*, 647–660. [CrossRef]

30. Vargas-Cordero, I.; Tinivella, U.; Accaino, F.; Loreto, M.F.; Fanucci, F. Thermal state and concentration of gas hydrate and free gas of Coyhaique, Chilean Margin (44 30′ S). *Mar. Pet. Geol.* **2010**, *27*, 1148–1156. [CrossRef]

31. Villar-Muñoz, L.; Bento, J.P.; Klaeschen, D.; Tinivella, U.; Vargas-Cordero, I.; Behrmann, J.H. A first estimation of gas hydrates offshore Patagonia (Chile). *Mar. Pet. Geol.* **2018**, *96*, 232–239. [CrossRef]

32. Bangs, N.L.; Cande, S.C. Episodic development of a convergent margin inferred from structures and processes along the southern Chile margin. *Tectonics* **1997**, *16*, 489–503. [CrossRef]

33. Ramos, V. Plate tectonic setting of the Andean Cordillera. *Episodes* **1999**, *22*, 183–190.

34. Grevemeyer, I.; Diaz-Naveas, J.L.; Ranero, C.R.; Villinger, H.W. Heat flow over the descending Nazca plate in central Chile, 32 S to 41 S: Observations from ODP Leg 202 and the occurrence of natural gas hydrates. *Earth Planet. Sci. Lett.* **2003**, *213*, 285–298. [CrossRef]

35. Melnick, D. Neogene Seismotectonics of the South-Central Chile Margin: Subduction-Related Processes over Various Temporal and Spatial Scales. Ph.D. Thesis, Universität Potsdam, Potsdam, Germany, 2007.

36. Cembrano, J.; Lara, L. The link between volcanism and tectonics in the southern volcanic zone of the Chilean Andes: A review. *Tectonophysics* **2009**, *471*, 96–113. [CrossRef]

37. Vargas-Cordero, I. Gas Hydrate Occurrence and Morpho-Structures along Chilean Margin. Ph.D. Dissertation, Fisiche e Naturali, Università di Trieste, Trieste, Italy, 2009.

38. Manea, V.C.; Pérez-Gussinyé, M.; Manea, M. Chilean flat slab subduction controlled by overriding plate thickness and trench rollback. *Geology* **2012**, *40*, 35–38. [CrossRef]

39. Melnick, D.; Echtler, H.P. Inversion of forearc basins in south-central Chile caused by rapid glacial age trench fill. *Geology* **2006**, *34*, 709–712. [CrossRef]

40. Maksymowicz, A. The geometry of the Chilean continental wedge: Tectonic segmentation of subduction processes off Chile. *Tectonophysics* **2015**, *659*, 183–196. [CrossRef]

41. GMRTMapTool. Available online: https://www.gmrt.org/GMRTMapTool/ (accessed on 29 April 2019).

42. National Oceanographic Data Center (NODC). Available online: https://www.nodc.noaa.gov/OC5/woa13/woa13data.html (accessed on 29 April 2019).

43. Bangs, N.L.; Brown, K.M. Regional heat flow in the vicinity of the Chile Triple Junction constrained by the depth of the bottom simulating reflection. *Proc. ODP Sci. Results* **1995**, *141*, 253–259.

44. Villar-Muñoz, L.; Behrmann, J.H.; Diaz-Naveas, J.; Klaeschen, D.; Karstens, J. Heat flow in the southern Chile forearc controlled by large-scale tectonic processes. *Geo-Mar. Lett.* **2014**, *34*, 185–198.

45. Mix, A.C.; Tiedemann, R.; Blum, P. *Shipboard Scientific Party Proceedings of the ODP*; Initial Reports; Ocean Drilling Program: College Station, TX, USA, 2003; pp. 1–145.

46. Tinivella, U.; Giustiniani, M. Variations in BSR depth due to gas hydrate stability versus pore pressure. *Glob. Planet. Chang.* **2013**, *100*, 119–128. [CrossRef]

47. Khan, M.J.; Ali, M. A Review of Research on Gas Hydrates in Makran. *Bahria Univ. Res. J. Earth Sci.* **2016**, *1*, 28–35.

48. Dickens, G.R.; Quinby-Hunt, M.S. Methane hydrate stability in pore water: A simple theoretical approach for geophysical applications. *J. Geophys. Res.* **1997**, *102*, 773–783. [CrossRef]

49. NASA Global Climate Change. Available online: https://climate.nasa.gov/scientific-consensus/ (accessed on 29 April 2019).

50. Coffin, R.; Pohlman, J.; Gardner, J.; Downer, R.; Wood, W.; Hamdan, L.; Walker, S.; Plummer, R.; Gettrust, J.; Diaz, J. Methane hydrate exploration on the mid Chilean coast: A geochemical and geophysical survey. *J. Pet. Sci. Eng.* **2007**, *56*, 32–41. [CrossRef]

51. Marín-Moreno, H.; Giustiniani, M.; Tinivella, U.; Piñero, E. The challenges of quantifying the carbon stored in Arctic marine gas hydrate. *Mar. Pet. Geol.* **2016**, *71*, 76–82. [CrossRef]

52. Thatcher, K.E.; Westbrook, G.K.; Sarkar, S.; Minshull, T.A. Methane release from warming-induced hydrate dissociation in the West Svalbard continental margin: Timing, rates, and geological controls. *J. Geophys. Res. Solid Earth* **2013**, *118*, 22–38. [CrossRef]

53. Ruppel, C.D.; Kessler, J.D. The interaction of climate change and methane hydrates. *Rev. Geophys.* **2017**, *55*, 126–168. [CrossRef]

54. Contreras-Reyes, E.; Völker, D.; Bialas, J.; Moscoso, E.; Grevemeyer, I. Reloca Slide: An ~24 km^3 submarine mass-wasting event in response to over-steepening and failure of the central Chilean continental slope. *Terra Nova* **2016**, *28*, 257–264. [CrossRef]

55. Geersen, J.; Völker, D.; Behrmann, J.H.; Reichert, C.; Krastel, S. Pleistocene giant slope failures offshore Arauco peninsula, southern Chile. *J. Geol. Soc.* **2011**, *168*, 1237–1248. [CrossRef]

56. Völker, D.; Geersen, J.; Behrmann, J.H.; Weinrebe, W.R. Submarine mass wasting off Southern Central Chile: Distribution and possible mechanisms of slope failure at an active continental margin. In *Submarine Mass Movements and their Consequences. Advances in Natural and Technological Hazards Research*; Yamada, Y., Kawamura, K., Ikehara, K., Ogawa, Y., Urgeles, R., Mosher, D., Chaytor, J., Strasser, M., et al., Eds.; Springer: Dordrecht, The Netherlands, 2012; pp. 379–389.

57. Geersen, J.; Völker, D.; Behrmann, J.H.; Kläschen, D.; Weinrebe, W.; Krastel, S.; Reichert, C. Seismic rupture during the 1960 Great Chile and the 2010 Maule earthquakes limited by a giant Pleistocene submarine slope failure. *Terra Nova* **2013**, *25*, 472–477. [CrossRef]

58. Klaucke, I.; Weinrebe, W.; Linke, P.; Kläschen, D.; Bialas, J. Sidescan sonar imagery of widespread fossil and active cold seeps along the central Chilean continental margin. *Geo-Mar. Lett.* **2012**, *32*, 489–499. [CrossRef]

59. USGS Science for a Changing World. Available online: https://earthquake.usgs.gov/earthquakes/ (accessed on 29 April 2019).

60. Sawyer, D.E.; DeVore, J.R. Elevated shear strength of sediments on active margins: Evidence for seismic strengthening. *Geophys. Res. Lett.* **2015**, *42*, 10–216. [CrossRef]

61. Brothers, D.S.; Andrews, B.D.; Walton, M.A.; Greene, H.G.; Barrie, J.V.; Miller, N.C.; Brink, U.T.; East, A.E.; Haeussler, P.J.; Kluesner, J.W.; et al. Slope failure and mass transport processes along the Queen Charlotte Fault, southeastern Alaska. In *Subaqueous Mass Movements*; Lintern, D.G., Mosher, D.C., Moscardelli, L.G., Bobrowsky, P.T., Campbell, C., Chaytor, J.D., Clague, J.J., Georgiopoulou, A., Lajeunesse, P., Normandeau, A., Eds.; Geological Society: London, UK, 2018; p. SP477-30.

62. Greene, H.G.; Barrie, J.V.; Brothers, D.S.; Conrad, J.E.; Conway, K.; East, A.E.; Enkin, R.; Maier, K.L.; Nishenko, S.P.; Walton, M.A.; et al. Slope failure and mass transport processes along the Queen Charlotte Fault Zone, western British Columbia. In *Subaqueous Mass Movements*; Lintern, D.G., Mosher, D.C., Moscardelli, L.G., Bobrowsky, P.T., Campbell, C., Chaytor, J.D., Clague, J.J., Georgiopoulou, A., Lajeunesse, P., Normandeau, A., Eds.; Geological Society: London, UK, 2018; p. SP477-30.

63. Sawyer, D.E.; Reece, R.S.; Gulick, S.P.; Lenz, B.L. Submarine landslide and tsunami hazards offshore southern Alaska: Seismic strengthening versus rapid sedimentation. *Geophys. Res. Lett.* **2017**, *44*, 8435–8442. [CrossRef]

64. Giustiniani, M.; Tinivella, U.; Jakobsson, M.; Rebesco, M. Arctic Ocean gas hydrate stability in a changing climate. *J. Geol. Res.* **2013**, *2013*, 783969. [CrossRef]

65. Gorman, A.R.; Senger, K. Defining the updip extent of the gas hydrate stability zone on continental margins with low geothermal gradients. *J. Geophys. Res. Solid Earth* **2010**, *115*, B07105. [CrossRef]

geosciences

MDPI

Article

Methane Hydrate Stability and Potential Resource in the Levant Basin, Southeastern Mediterranean Sea

Ziv Tayber [1], Aaron Meilijson [1], Zvi Ben-Avraham [1,2] and Yizhaq Makovsky [1,3,*]

[1] The Dr. Moses Strauss Department of Marine Geosciences, Leon H. Charney School of Marine Sciences, University of Haifa, Haifa 3498838, Israel
[2] The Department of Geophysics, Porter School of the Environment and Earth Sciences, Tel-Aviv University, Rammat-Aviv, Tel Aviv 6997801, Israel
[3] The Hatter Department of Marine Technologies, Leon H. Charney School of Marine Sciences, University of Haifa, Haifa 3498838, Israel
* Correspondence: yizhaq@univ.haifa.ac.il; Tel.: +972-52-302-0406

Received: 20 April 2019; Accepted: 3 July 2019; Published: 11 July 2019

check for updates

Abstract: To estimate the potential inventory of natural gas hydrates (NGH) in the Levant Basin, southeastern Mediterranean Sea, we correlated the gas hydrate stability zone (GHSZ), modeled with local thermodynamic parameters, with seismic indicators of gas. A compilation of the oceanographic measurements defines the >1 km deep water temperature and salinity to 13.8 °C and 38.8‰ respectively, predicting the top GHSZ at a water depth of ~1250 m. Assuming sub-seafloor hydrostatic pore-pressure, water-body salinity, and geothermal gradients ranging between 20 to 28.5 °C/km, yields a useful first-order GHSZ approximation. Our model predicts that the entire northwestern half of the Levant seafloor lies within the GHSZ, with a median sub-seafloor thickness of ~150 m. High amplitude seismic reflectivity (HASR), correlates with the active seafloor gas seepage and is distributed across the deep-sea fan of the Nile within the Levant Basin. Trends observed in the distribution of the HASR are suggested to represent: (1) Shallow gas and possibly hydrates within buried channel-lobe systems 25 to 100 mbsf; and (2) a regionally discontinuous bottom simulating reflection (BSR) broadly matching the modeled base of GHSZ. We therefore estimate the potential methane hydrates resources within the Levant Basin at ~100 trillion cubic feet (Tcf) and its carbon content at ~1.5 gigatonnes.

Keywords: gas hydrates; methane stability; seismic interpretation; Levant Basin; Eastern Mediterranean; climate change

1. Introduction

Gas hydrates are non-stoichiometric crystalline solids that are formed under a suitable thermodynamic pressure–temperature–salinity balance of water molecules, arranged in lattice-like crystal "cages" around gas molecules (e.g., [1,2]). Natural gas hydrates (NGH) deposits are widely distributed along the continental margins around the world, where gas fluxes are steadily available in the shallow sediment [2–7]. Their presence is bound in the marine environment between the top of the gas hydrate stability zone (GHSZ) in the water column, and its base within the sedimentary column below the seafloor. These are set by a thermodynamic balance between the impacts of pressure, temperature, and salinity on the hydrate stability [8]. In practice, the GHSZ is primarily controlled by the balance between the increase of water and sediment pressure with depth, the bottom water temperature and the increase in temperature with the depth beneath the seafloor. In many places, the presence of NGH is marked on seismic images by a bottom simulating reflection (BSR), suggested to represent the accumulation of free gas below the GHSZ [9–11]. However, NGH are also reported in areas without the presence of a distinct BSR (e.g., [12–14]), and a seismic BSR does not always indicate the presence of NGH (e.g., [14,15]).

NGH are estimated to contain a substantial portion of all organic carbon on Earth, and therefore play a crucial role in the global carbon cycle [2,16,17] and constitute a major potential energy resource [18]. Moreover, the precarious stability of NGH may lead to their dissociation in response to global warming. This may release vast amount of methane gas, which would induce a positive climate warming feedback [1,19–22]. Alternatively, dissociation may occur locally in the course of offshore activities, constituting a significant geohazard [23,24].

Isolated from the buffering effects of the global oceanic system, the Eastern Mediterranean Sea (EMS; Figure 1) is particularly sensitive to global climate changes [24–27]. With a major part of its seafloor within the GHSZ [28,29], this region would be expected to sustain a significant NGH potential [5,30,31]. Thus, global and regional changes would be expected to result in accentuated dissociation of hydrates in the EMS, which might result in a prominent climatic feedback response. In addition, the EMS can offer a relatively closed-system natural laboratory to study the linkage between environmental changes and NGH stability (e.g., [29]). Moreover, ongoing deep-sea discovery and development of prominent natural gas reservoirs across the southeastern Mediterranean Sea, and particularly in the Levant Basin (Figure 1) [32], invoke the need for addressing the possible presence of NGH in this area. However, across the entire EMS region, NGH were recovered to date only in a few mud volcanoes, and in spite of the pervasive exploration of the region no additional verified presence of hydrates was observed [33].

This study re-assesses the potential presence of NGH in the Levant Basin (Figure 1) through focused modeling of the GHSZ in the region and examining its correlation with a set of new observations. The results of this study are expected to serve as a base for future evaluation of the controls and possible impact of such presence.

1.1. Geological Setting of the Levant Basin

The Levant Basin was formed during the Early Mesozoic by the breakup of the northern edge of Gondwanaland and subsequent collision with Eurasia but attained its present appearance during the Neogene [34,35]. Seafloor spreading of the Herodotus and Cyprus basins was followed by the development of subduction along the Cypriot arc, and a forearc accretionary system along the Florence Rise-Latakia Ridge [36] (Figure 1). Coincident continental collision of Cyprus with the Eratosthenes Seamount (Figure 1) probably resulted in ~1 km subsidence of the latter and its surrounding since Late Miocene (e.g., [35]). Restriction of the connectivity with the Atlantic during the Messinian Salinity Crisis resulted in the deposition of a thick evaporitic sequence [37–39]. This sequence reaches thicknesses of ~2 km in the central part of the Levant Basin and pinches out upslope toward the basin margins [40,41]. The top of this sequence is generally imaged as a pronounced high amplitude seismic reflection, the M reflection [42]. Minor (generally <100 m) folding of the M reflector and seafloor, and a network of faults affecting the post-Messinian sedimentary section accommodate the flow of the Messinian evaporites away from the Nile delta and the eastern margin of the basin [43]. No diapirism of the Messinian salt is observed across the Levant Basin. An outpour of clastic sediments since the Oligocene and the formation of the present-day Nile formed an extensive sedimentary cone, which extends into the Herodotus and Levant Basins (Figure 1) and reaches thicknesses of >8 km [44,45]. The eastern deep-sea fan of the Nile (Figure 1), stretching across a major part of the basin, is riddled throughout with deep-sea channel and lobe systems accommodating direct transport of Nilotic sediments toward the Cypriot Deep (Figure 1) [43,46,47]. Concurrently, a sedimentary bypass of Nilotic sediments, carried northeastward by currents and then transported down slope, constructs the northeastward prograding southeastern continental margin sedimentary wedge [48–50]. Both the Nile deep sea fan and margin sedimentary wedge prograded over the evaporites layer since the Pliocene, reaching at present thicknesses of ~0.5 and ~1.5 km respectively (e.g., [43]). Estimated quaternary sedimentation rates in the Levant Basin range between ~2.2 cm/ka on its northeastern margin, southeast of Cyprus, to ~6 to ~20 cm/ka in its southeastern part [51–53]. Organic-rich sapropel units deposited recurrently in the EMS since the Miocene [54,55]. Their deposition co-occurred with periods of insolation maxima and increased monsoonal activity, which caused increased Nilotic discharge into the EMS ([55,56] and reference therein). This leads to the breakdown in deep-water formation and production of anoxic conditions at the seafloor in the deeper parts of the basin [57]. Furthermore, increased primary production augmented the organic matter flux to the deep-water [58,59], and its preservation was enabled due to the anoxic conditions at the seafloor [56].

This study focuses on the evidence for shallow gas accumulation and potential NGH formation within the widely distributed deep-sea channels of the Nile fan. The channels in the Levant Basin are probably similar in their sedimentary content to the channels observed on the western Nile fan, which transported mixed marine and terrigenous siliciclastic material [60]. Such sediments are characterized by relatively large pore-space and grain-size, and therefore constitute a favorable media for hosting free gas or NGH [61].

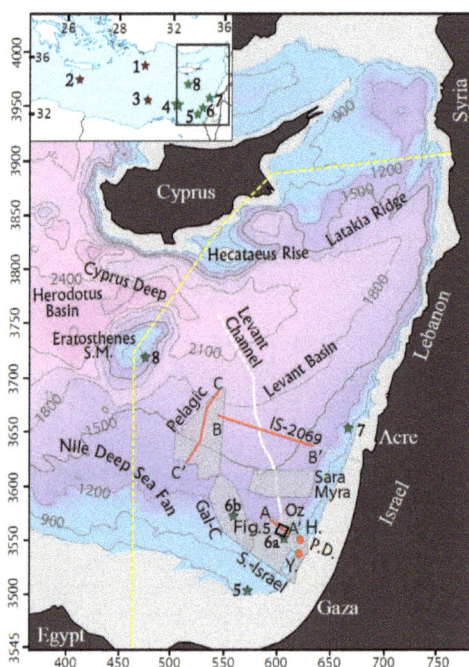

Figure 1. A color-coded and contoured bathymetric map of the Levant Basin [62], overlaid with the outlines of the 3D seismic blocks (gray areas) and the TGS IS-2069 seismic profile (gray line) analyzed in this work. The outlines of the seismic traverses displayed in Figures 6–8 are marked (red lines) and labeled. Also marked are the locations of Hanna (H.) and Yam (Y.) wells (red dots), the outline of Figure 5 (black rectangle), and the northern and western border of the study area (yellow dashed line). The inset (upper left corner) displays a map of the Eastern Mediterranean Sea (EMS), marked with natural gas hydrates (NGH) (brown stars) and methane seepage (green stars) observations locations: 1. The Thessaloniki mud volcano and Anaximander Seamount region (e.g., [33,63,64]); 2. the Olimpi mud volcanos field (e.g., [33], and references therein); 3. observations of hydrates formation during sampling [65]; 4. the Nile Delta and deep sea fan seepage domain (e.g., [66,67]); 5. seafloor seepage offshore the Sinai Peninsula [68]; 6. (a) the Palmahim disturbance (P.D.) and Levant Channel (white line), and (b) Gal-C, seepage sites [69–71]; 7. methane sampling offshore acre [70]; 8. Eratosthenes Seamount seepage sites [72,73]. The main map is plotted in UTM coordinates (zone 36N, datum WGS84) and labeled in km, while the inset is plotted in geographical coordinates.

1.2. Natural Gas and Hydrates in the EMS

To date, NGH deposits were sampled or inferred to exist only on several mud volcanoes along the accretionary complex traversing the northern part of the EMS ([33] and references therein; Figure 1). Most NGH sampled there were predominantly found within relatively fine muddy sediments. On the Thessaloniki mud volcano, in the Anaximander Seamount region, the predominantly methane bearing NGH are present at a water depth of ~1260 m, just below the calculated top of the GHSZ [63,64]. A single set of direct indications of NGH stability in the Nile deep sea fan, in the southern part of the EMS, was described by [65]. They observed formation of hydrates within a sampling funnel during collection of gas emitted from the seafloor, and hydrate coating that formed on ascending bubbles, dissolving near the top of GHSZ. Their analysis of the sampled gas composition demonstrated a predominance of methane with minor portions of ethane and propane, for which the top of the GHSZ was estimated at the water depth of ~1350 m. In accordance, echosounder imaging observed ascending gas bubbles flares that dissipated just below a water depth of ~1350 m, presumably due to dissolution of the bubbles hydrate skins at the top of the GHSZ [65].

In spite of extensive exploration activity across the EMS, including a broad coverage by 2D and 3D commercial and academic seismic data and multiple drill wells, no additional observation of hydrates or a seismic BSR was ever documented in peer-reviewed publications. Albeit, several meeting abstracts reported observations of BSR in the Nile cone (e.g., [74–76]).

A multitude of pockmarks and other seepage edifice have been identified over the last two decades across the Nile deep-sea fan and adjacent Levant Basin and Eratosthenes Seamount, with their scope continuously expanding as new data becomes available (e.g., [68–73,77,78]). Similar intra- to post-Messinian buried features are also abundant in the geophysical record [79–83]. Together these provide potential sources for hydrates formation in the EMS, at present or over climatic changes. In particular, this study is motivated by the recent discovery of active methane seeps within large scale (hundreds of meters) pockmarks at a water depth of 1100–1250 m. These were identified on the crests of compressional folds in the toe of the Palmahim disturbance [69–71,84] (Figure 1). The latter is a large-scale (15 × 50 km) rotational slide detached on the Messinian evaporites offshore southern Israel [85]. Additional seepage was discovered within the Levant Channel, and ~40 km west of there on a seafloor fold in the Gal-C exploration block [71] (Figure 1). The Levant Channel is a major deep-sea channel marking the eastern flank of the deep-sea fan of the Nile and bounding the Palmahim disturbance on its west [43]. The prevalence of methane and scarcity of heavier hydrocarbons imply that gas emitted from these surface features originates predominantly from microbial methanogenesis (e.g., [65,70]). We note that also the commercial gas reservoirs, discovered recently at sediment depths reaching ~5 km in the Levant Basin and below the deep-sea fan of the Nile, contain predominantly microbial methane (e.g., [86,87]). However, no clear link has been delineated to date between these reservoirs and seafloor seepage. Seafloor authigenic carbonates composition in the central deep-sea fan of the Nile reveal spatio-temporal variations of Holocene seepage ages, suggested to be related with sediment transport variations [88]. However, such variations may have alternatively, or additionally, been associated with glacial-interglacial eustatic sea level cycles, affecting the stability of NGH (e.g., [29]).

2. Materials and Methods

This study combines two stages: Modeling the GHSZ in the Levant Basin based on locally constrained thermodynamic parameters; and analysis of seismic data for the distribution of high amplitude seismic reflectivity (HASR) and additional indicators for the presence of gas.

2.1. Hydrate Modeling Methodology

To model the GHSZ in the Levant Basin, we first constrained the thermodynamic parameters (temperature, salinity, and pressure) in the relevant water depths (>800 m) of the basin (Figure 1) by compiling data that was acquired in oceanographic surveys (Figure 2) and making assumptions with respect to the sub-seafloor conditions. Modeling used the CSMHyd software (1998, Colorado School of Mines Center for Hydrate Research, Golden, CO, USA) [89] to calculate the top of the GHSZ, and the depth below the seafloor of its base for pure methane hydrates as a function of the Levant Basin seafloor depth (Figure 3). Finally, using Matlab we mapped the base hydrate stability thickness throughout the Levant Basin based on the seafloor bathymetry (Figure 4). In general, our analysis is bound by the shallowing of the Levant Basin flanks. The deeper western limit was arbitrarily set approximately at the crest of Eratosthenes Seamount, connecting it with the African coastline approximately along the 40° E latitude, with the Cyprus margins along the line connecting to the crest of Hecataeus Rise and eastern Cyprus, and east to the Syrian margin (Figure 1). The details of our modeling procedures are detailed below.

2.2. Seismic Data and Analysis Methodology

The seismic component of this work is based on the analysis of five commercially acquired and processed 3D seismic blocks, and one 2D seismic profile (Table 1). The 3D blocks cover together (with some minor overlaps) a significant portion of the southern to central parts of the Levant Basin (Figure 1) between water depths of 900 to 1900 m, while the 2D profile connects the deepest 3D coverage with the eastern margin of the basin. The different data products that were available for our analysis vary in their exact processing and amplitude levels (Table 1), but were all processed through standard commercial workflows yielding zero phase amplitude preserved data. Thus, although not rigorously accurate the relative amplitudes are meaningful, at least in the region of interest within the first hundreds of meters beneath the seafloor. This assumption was asserted by us through manual visualization of the data, as well as through the calculation of amplitude histograms for each dataset.

All data were loaded to a Paradigm multi-survey project desktop for analysis. The time-migrated data two-way-times (TWT) were scaled to depth using constant seismic velocities of 1520 and 2000 m/s for the water and post-Messinian sedimentary column respectively. These velocity approximations were established through a comparison of overlapping regions of available depth migrated volumes and time migrated volumes and 2D profiles transecting them. We estimate the resulting depth errors to be ±5 and ±10 m for the seafloor and top Messinian (M) horizon respectively. These errors are approximately identical to the nominal resolution of the seismic reflections from the same depths. Paradigm's propagator module was utilized for supervised automatic picking of the seafloor horizon, yielding a detailed bathymetric map at the resolution of each of the 3D seismic blocks. 3D shaded relief views of these bathymetric maps (e.g., Figure 5) were used to manually map the distribution of seafloor pockmarks and additional seafloor features.

Table 1. The seismic datasets used.

Survey/Block	Type	Acquisition	Imaging	Area (km²) /Length (km)	Spacing (m)
Southern Israel	3D	Gebco 2000	Time Migration	1900	12.5 × 12.5
Gal-C	3D	Gebco 2000	Time Migration	1400	12.5 × 12.5
Oz	3D	Ion-GTX	Depth Migration	400	25 × 12.5
Sara-Mira	3D	CGG-Veritas 2011	Depth Migration	1350	25 × 12.5
Pelagic	3D	CGG-Veritas 2009	Depth Migration	2350	25 × 12.5
TGS-IS209	2D	TGS-Nopec 2000	Time Migration	140	12.5

This study examines the correlation of high amplitude reflectivity identified in the seismic datasets across the basin (the HASR; Figures 6–8) with the GHSZ modeling results. The distribution of the HASR within the post-Messinian sedimentary stack was evaluated by two independent methods. Initially the HASR was manually picked on every 100th inline section and then on every 100th crossline section throughout each of the 3D blocks. The seafloor and HASR picks were then jointly outputted and their distributions were plotted using Matlab (Figure S1 in Supplementary Materials). To verify the correlation revealed, we repeated the process through a more rigorous automatic picking procedure. A sub-volume was extracted from each of the seismic blocks stretching 5 to 300 mbsf, eliminating the seafloor reflection above and the Messinian reflection below. The amplitude histograms of the sub-volumes extracted from each block were calculated, and a scaling factor to normalize the histograms of the different blocks was determined. Each of the sub-volumes was then loaded to Paradigm Voxel utility, where the HASR was picked by threshold detection. Following testing we established the threshold at the top 0.65% of the normalized histogram negative amplitudes tail. The picks were then converted to multi-valued horizons and outputted to Matlab distribution plots (Figure 9) and Paradigm spatial plots (Figure 10).

3. Results

3.1. Establishing the Local Environmental Conditions in the Deep Levant Basin

3.1.1. Bottom Water Temperature and Salinity

Bottom water temperatures and salinities of the Levant Basin were acquired from two independent data sets. The first consists of four vessel-based conductivity and temperature depth (CTD) casts surveys, collected between the years 2009 to 2012 to water depths >1500 m, and extracted from the EU PERSEUS consortium on-line repository [90]. The second data set consists of underwater remotely operated vehicle (ROV) based CTD measurements collected in the course of E/V Nautilus 2011 survey offshore Israel [69].

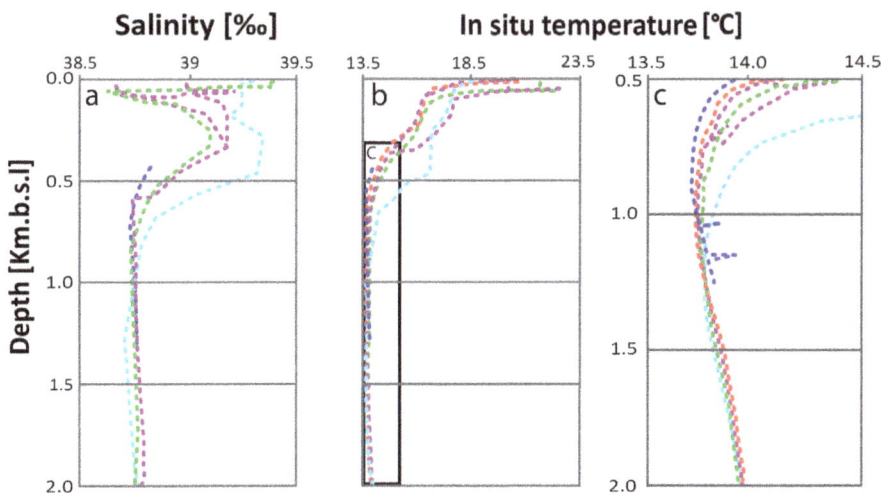

Figure 2. Seawater salinity (**a**) and temperature (**b**) profiles, measured in various locations in the EMS by CTD casts in the course of four different cruises (6901084-06/2012-red, 6901043-09/2012-green, 6900850-10/2011-purple, 6900794-01/2009-azure; from [90]) and during EV Nautilus 11/2011 ROV survey ([69]; blue). (**c**) A zoom of the in situ temperature profiles (**b**) in the water depth range of 0.5 to 2 km.

These datasets combined constrain water body salinities in the range of 38.74 to 38.83‰ between water depths of 800 to 2000 m (Figure 2). We therefore used an average salinity of 38.80‰ for the Levant Basin GHSZ model. The temperatures measured at the sea surface show sub-annual variability in the range of 16 °C to 28 °C (Figure 2). However, at a water depth of 800 m, constituting the top of the EMS deep water mass [91], water temperatures converge to a constant value of ~13.69 °C. The water temperature then increases at a rate of 0.015 to 0.02 °C per 100 m of water depth to in situ water temperatures of ~13.94 °C at a water depth of 2000 m. We therefore used a mean water temperature value of 13.80 °C for the deep water mass. This is the actual permanent temperature at the water depth interval of 900 to 1400 m. The values constrained here are consistent with other published values for the Levant region (e.g., [27,92,93]).

3.1.2. Sediment Salinity and Geothermal Gradient

As no data is currently published on pore water salinities in the Levant Basin seafloor, we used the same bottom water value of 38.8‰ also for the sub-seafloor salinity. This is probably a reasonable approximation considering a relatively high seawater content within the bottom sediments. Moreover, sensitivity tests (discussed below) demonstrate that the possible effects of salinity mis-estimations on our modeling results are minor.

The sediment temperature was calculated based on a constant seafloor temperature of 13.8 °C and a linear increase with depth amounting to the geothermal gradient. Published estimations of the geothermal gradient in the Levant Basin range between 20 to 37 °C/km, constituting the lower and upper bound estimates respectively. [94] used seafloor measurements to estimate geothermal gradients of ~37 °C/km at two stations within the Levant Basin. In contrast, [95] estimated based on the bottom-hole temperatures from several onshore wells as average geothermal gradient for northern and central Israel in the range of 22 to 25 °C/km. [96] estimated a vertical geothermal gradient in the range of 20 to 26 °C/km for the northern inland and offshore areas of the Nile fan. Their study is based on temperature data from 48 wells located adjacent to our study area. [97] estimated an average vertical geothermal gradient in the range of 20 to 30 °C/km based on temperature logs from wells in southern Israel. Most recently [86] suggested the average geothermal gradient of 28.5 °C/km, measured in the Yam-1 and Yam-2 wells in the southeastern margin of the basin (Figure 1), as an estimate for the Levant Basin geothermal gradient. However, [98] modeled the thermal history of the Levant Basin based on an interpreted chronostratigraphic framework of the basin and the measurements in four wells along its flanks (including apparently the Yam wells). In particular, this framework includes the presence of the ~2 km thick Messinian salt unit within the Basin and its absence in the flaks. Their modeling predicts geothermal gradients in the ranges between 20 and 28 °C/km and 13 and 20 °C/km in the Basin flanks and center, respectively. However, in the lack of published measurements from wells within the Levant Basin, the validity of these results to our GHSZ modeling is uncertain. It is notable that, in contrast to the significant impact of salt diapirs on NGH distribution in salt basins, such as the Gulf of Mexico (e.g., [99–101]), the relatively minor deformation of the Messinian salt in the Levant Basin is expected to inflict only limited variability on the GHSZ. Considering the high sensitivity of the GHSZ model to the geothermal gradient, and the uncertainty of its value, we created three versions of the GHSZ model using geothermal gradients of 20, 28.5, and 37 °C/km.

3.1.3. Pressure

Hydrates stability within the sediments depends on the interstitial pore pressure [1], generally bound between the hydrostatic and lithostatic pressure profiles [102,103]. At relatively shallow sediment depths of the GHSZ (normally <500 m [104]) in normally compacting basins sediments, porosity and permeability are generally high and the pore pressure is approximately hydrostatic or slightly above (e.g., [102,105]). We therefore assume a hydrostatic pore pressure profile in our GHSZ modeling. This assumption is supported by the pore pressure profile measured in Hanna-1 well, located at a water depth of 972 m in the eastern boundary of the study area (Figure 1), showing only a slight deviation from hydrostatic pressure within <500 m below the seafloor [106].

The hydrostatic pressure in the Levant Basin was calculated as a function of the depth below the surface using the equations of [107] in the range of 0.1–35.4 MPa. These equations estimate the pressure within maximal error bounds of 2 kPa, which are equivalent to depth errors <0.2 m. This is a negligible value in comparison with the water and sediment column depth-range of the GHSZ (>1000 m). These calculations were evaluated for each of the bathymetric grid cells as described below.

3.1.4. Bathymetry

To model the seafloor bathymetry of the Levant Basin we used a 250 m digital elevation model based on the bathymetric map of [62]. Across the exclusive economic zone of Israel, the bathymetry was updated based on a 250 m resolution bathymetric digital elevation model (DEM) released by the State of Israel, Ministry of Energy [108].

3.2. Electing a GHSZ Modeling Approach for the Levant Basin

To evaluate the adequate thermodynamic conditions for NGH formation, three different models of the GHSZ were tested [8,109,110]. These models use a phase diagram of solid methane hydrate versus liquid water and free gas phases (Figure 3). The models presented by [8,109] are empirical models, using a narrow range of temperature–salinity (T–S) conditions. In contrast, the model presented by [110] relies on statistical thermodynamics of the pressure–temperature (P–T) equilibrium conditions for methane hydrate stability and uses a wider range of T–S values.

Figure 3a presents the chemical equilibrium points predicted for the Levant Basin by these three models as a function of depth (i.e., pressure) and temperature, for a variety of constant salinity concentrations permitted by each of the methods and noted. The NGH stability zone is represented by the area below the curve predicted by each model; while above the curve water and free gas are predicted. The GHSZ is determined by cross-referencing the methane hydrate stability in the phase diagram with the seafloor bathymetric depth and the geothermal gradient. The results of the three different models diverge substantially from the Levant Basin conditions (Figure 3a). The results obtained by the models of [8,109] represent end-member solutions, while the results obtained by the [110] model fall between them.

Selection of the appropriate modeling scheme for this study is based on the following considerations: (1) The models by [8,109] are based on experimental data using pressure conditions that are below those predicted for the relevant depths in the Levant Basin; (2) the model suggested by [110] considers a salinity of 35.0‰, which is 3.8‰ lower than the Levantine average deep water salinity value; and (3) the use of the CSMHyd software implementation of [110] has become a standard for predicting GHSZ in related studies (e.g., [111,112]). Particularly, modeling of GHSZ at overlapping areas in the EMS using CSMHyd was recently performed by [29,65]. The Levant Basin GHSZ is therefore calculated in the present study based on the algorithms of [110], as implemented in the CSMHyd software [89].

In practice, CSMHyd was used to calculate the thermodynamic equilibrium pressure lookup table for temperatures values at increments of 0.02, 0.0285, and 0.037 °C, corresponding to the tested geothermal gradient, for pure methane hydrates and the EMS salinity. Then a Matlab code was used to search for the base GHSZ depth, by recursively calculating the pressure and temperature below the bathymetric depth. These were calculated based on the hydrostatic and geothermal gradients at a depth increment of 1 m, until the calculated equilibrium conditions were reached. The process was repeated for every point in the bathymetric DEM of the Levant with water depth >1250 m and the different geothermal gradients considered, yielding modeled maps of the base GHSZ (Figure 4). Finally, the modeled maps were loaded to the Paradigm Epos software and plotted over the seismic data to be compared with the trend of the observed seismic anomalies.

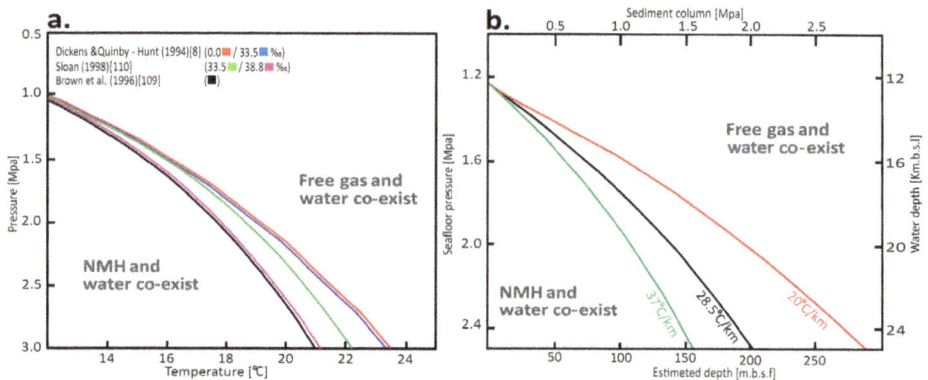

Figure 3. A comparison of modeled methane hydrate stability curves. (**a**) The equilibrium curves predicted by the different modeling schemes considered in this work [8,109,110] under different constant salinity conditions (as color-coded at the top). (**b**) The effect of different temperature gradients (color-coded as noted) on the depth below the seafloor of the base gas hydrate stability zone (GHSZ) predicted by CSMhyd [89] with a constant salinity of 38.8‰, as a function of the seafloor water depth.

3.3. Model Sensitivities

Potential uncertainties in the modeling performed in this study might be introduced by the modeling methodology. Based on comparison with a large set of published experimental data of salt-inhibited hydrates stabilities, [110] evaluated CSMHyd modeled pressure (and therefore depth) predictions absolute errors at ~15%, which is unacceptable. However, these uncertainties cannot be attributed to CSMHyd alone, as they necessarily incorporate the experimental data measurement errors. These measurement errors are symmetrically distributed around CSMHyd's predictions, and therefore may be averaged out [112]. Moreover, a divergence of the predicted and modeled values appears to be associated with larger temperatures and pressures than considered in this study (e.g., Figure 4 in [113]). In addition, the reliability of CSMHyd predictions was verified in a variety of studies, by their correlation with geophysical, well-log, and experimental results (e.g., [111,112,114–118]). Particularly, [119] employed a Monte-Carlo style simulation of their modeling uncertainties, which are similar to the uncertainties in this study (as discussed below), including explicitly CSMHyd 15% pressure prediction errors. They obtained a combined 1σ (one standard deviation) of most likely error estimate of ~±20 m at a water depth of 1240 m (Figure S2 in [119]) and decreasing values of the most likely error at increasing depths. Such uncertainties are acceptable in the context of the first-order evaluation performed in this study.

Additional errors in our modeling might be introduced by the average estimates and assumed values of the environmental parameters. Biased estimates of the bottom water temperature and salinity will affect the estimated water depth to the top of the GHSZ. This is also the seafloor depth at which the GHSZ pinches out laterally, and therefore defines the spatial extent of the GHSZ. Thus, with the ~1° slope gradient of the seafloor in the Levant, any bias on the top of the GHSZ will bias by a factor of c. ×50 the estimated potential lateral extent of hydrates across the basin seafloor. Similarly, biases in the interstitial salinity and geothermal gradient would bias the estimated depth below the seafloor of the GHSZ base. To evaluate the sensitivity of the model to our estimated salinity and temperature values we varied each of these parameters while keeping the other parameters fixed.

3.3.1. Water Temperature Effects

Temperature is the main factor affecting methane hydrate formation (e.g., [120,121]). The effects of water temperature uncertainties on the GHSZ model was evaluated by varying the modeled temperatures in steps of 0.1 °C within the range of 10–15 °C, keeping a constant salinity of 38.8‰. The results show a deepening of the GHSZ top by 4.2 m for every increase of 0.1 °C. The in situ temperature range of the Levant Basin bottom waters, as measured in the CTD surveys and the Nautilus expedition, is 13.69 to 13.94 °C (Figure 2). Thus, the expected temperature variation from the average temperature value is up to 0.13 °C, corresponding to a maximum shift of 5.9 m in the depth of the GHSZ base and a decrease of the estimated potential lateral extent of hydrates by ~600 m. This deviation represents error bounds of 0.05% in the modeled water depth on the top of the GHSZ. These error estimates are negligible in the context of this work.

3.3.2. Salinity Effects

An increase in salt content acts as an inhibitor to the formation of methane hydrate [110], and therefore would increase the water depth on the top of the GHSZ and decrease the depth of the GHSZ base beneath the seafloor. The sensitivity of the model results in an error in the estimated water salinity, which was examined by varying the salinity in steps of 0.25‰ in the range of 36.5 to 39.5‰, and by modeling the GHSZ with a constant water temperature. The results show a weak sensitivity of the variance in the GHSZ top boundary to the change in salinity. The slope of the salinity versus depth curve predicts a variance of 1.66 m in the depth of the GHSZ top for every shift of 0.25‰ in salinity. Consequently, a maximum error of 0.4 m, representing a deviation of 0.3‰, is introduced to the model by averaging the in situ salt concentrations of the Levant Basin bottom water values in the range of 38.74 to 38.83‰ (Figure 2).

3.3.3. Geothermal Effects

In order to determine the model sensitivity to different geothermal gradients, we modeled the two endmember models with the 20 and 37 °C/km geothermal gradients, and the in-between model with the 28.5 °C/km geothermal gradient (Figure 3). Changing the geothermal gradient does not change the top of the GHSZ, which is within the water body and therefore independent of the geothermal gradient (e.g., [8]). However, the different geothermal gradients result in significant deviations of the base of the GHSZ. The geothermal effect is best reflected in the differences between the different modeled base of GHSZ curves in Figure 3b. The depth of the base GHSZ at a water depth of 1800 m ranges between ~150 m under the low geothermal gradient of 20 °C/km, ~110 m calculated under the medium geothermal gradient of 28.5 °C/km, and ~85 m calculated under the high geothermal gradient of 37 °C/km. This paper discusses therefore the implications of these three alternative geothermal gradient models.

3.4. The Modelled Distribution of the GHSZ in the Levant Basin

Integrating the Levant pressure–temperature–salinity parameters into the model of [110] reveals that the GHSZ stretches widely across the Levant Basin (Figure 4). An initial estimate for the potential gas hydrates occurrence zone (GHOZ) (e.g., [1]) is provided by the sub-seafloor thickness of the GHSZ. Figure 4a,b presents the potential GHOZ thickness calculated and mapped for two of the alternative geothermal gradient estimates discussed above, namely the 20 °C/km and 28.5 °C/km geothermal gradients, respectively. A comparison between these maps (Figure 4a,b) provides an insight to the possible uncertainty in the implications of our modeling. The top of the GHSZ is located in both cases at the water depths of 1250 m, which limits the extent of the GHSZ to the northwestern two thirds of the basin. The thickest potential GHOZ is observed within the Cyprian trench at water depths of up to 2700 m, and it gradually thins to the south and east. The constrained depth to the base of the GHSZ (and therefore the potential thickness below the seafloor of the potential GHOZ) in the

Levant is highly dependent on the geothermal gradient assumed, as discussed above. The base of the potential GHOZ at the deepest part of the study area is 431 and 260 m below the seafloor for the two alternative geothermal gradient estimates of 20 °C/km and 28.5 °C/km, respectively (Figure 4). The median modeled thicknesses of the potential GHOZ within the Levant Basin are ~200 and ~150 m respectively, when applying the same geothermal gradients.

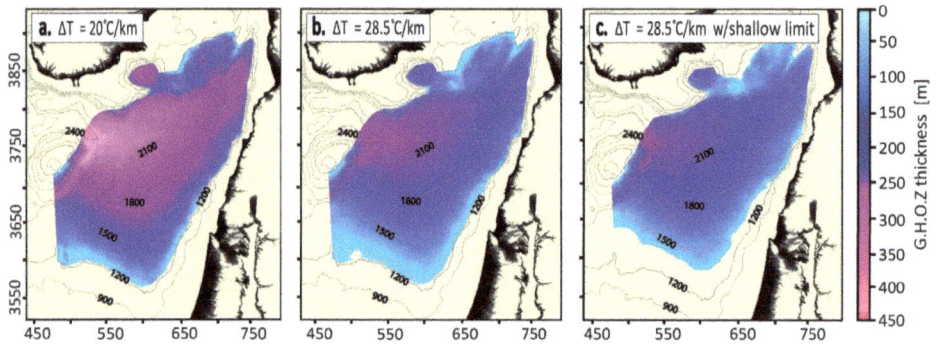

Figure 4. The model predicted the extent and thickness (color coded on the right) of the potential gas hydrates occurrence zone (GHOZ) below the Levant Basin seafloor (plotted with UTM coordinates in km). (**a**) The GHOZ estimated by the sub-seafloor part of the GHSZ under a geothermal gradient of 20 °C/km. (**b**) The GHOZ estimated by the sub-seafloor part of the GHSZ under a geothermal gradient of 28.5 °C/km. (**c**) The GHOZ, estimated from the GHSZ under a geothermal gradient of 28.5 °C/km and top bound 25 m below the seafloor, as inferred from the presence of the shallow high amplitude seismic reflectivity (HASR) (see Section 4.5 text). This map constitutes our conservative estimate for the potential GHOZ in the Levant Basin.

3.5. Seismic Evidence for the Presence of Gas and Hydrates in the Southwestern Levant Basin

Considering the wide distribution of the GHSZ within the Levant Basin, the presence of hydrates is mainly conditioned on the presence of methane within the sediments. As pervasive sampling of the Levant sub-seafloor sediments is lacking, we seek preliminary evidence for the presence of methane through the analysis of an extensive set of available 3D seismic data.

3.5.1. Pockmarks in the Southeastern Levant Basin

Precise picking of the seafloor reflection in the 3D seismic volumes allows a detailed search for bathymetric features commonly associated with the presence of gas. Analysis of the high-resolution bathymetry available across the 3D seismic blocks identified the Palmahim disturbance fold crests group consisting of seven oddly shaped pockmark clusters measuring hundreds of meters across (Type-A; Figure 5); and a total of 160 smaller pockmark structures ranging in diameter between 50 to 150 m (Type-B; Figure 5). While a few of the identified pockmarks may represent errors in the seafloor maps, the majority of the features are robustly mapped and identified. The pockmarks are identified only in the Southern Israel, Gal-C, and Oz seismic surveys covering the base of the continental slope of southern Israel and the adjacent eastern part of the deep sea fan of the Nile region (Figures 1 and 10). The pockmarks were detected in water depths of 1000 to 1300 m, and 90% of them are concentrated in four large clusters. Three of the clusters, which include 70% of the identified pockmarks, are located around a water depth of 1200 m, closely corresponding to the top of the GHSZ at 1250 m. Two of the clusters are located in the Nile deep sea fan region, and the other two (including the Palmahim Disturbance group) are located at the base of the continental slope of southern Israel.

3.5.2. Characteristics of Seismic Reflectivity in the Southern Levant Basin

A variety of reflectivity patterns characterize the top (post-Messinian) sedimentary section below the seafloor of the southeastern part of the Levant Basin, as imaged by the pervasive seismic dataset investigated in this work. The continental margin of Israel, in the eastern part of the investigated area, is characterized by relatively continuous and generally moderate amplitude reflections interleaved with chaotic intervals (Figures 6 and 7). The latter representing mass transport complexes (MTC). This westward thinning, regionally up to ~6° dipping sedimentary section, was described in detail by [122], was based primarily on the Southern Israel 3D seismic volume. The top sedimentary section of the Nile submarine fan, to the west of the Israeli margin, was described by [46,47] based on detailed investigations of the Gal-C 3D seismic volume. The seismic reflectivity in this area (Figures 6–8) is characterized primarily of locally segmented and dipping sets of reflections, representing channel levee complexes, embedded between coherent packages of locally continuous sub-parallel reflections, representing layered hemi-pelagic sediments. Intervals of chaotic reflectivity correspond to MTCs. This entire sedimentary package gently thins northward, and is folded and truncated by numerous thin-skinned sin-depositional halokinetic faults and ~100 m high folds. The bottom ca. third of this sedimentary stack, above the prominent M reflection, is generally characterized by relatively continuous high-amplitude reflections. Above this unit and up to ~100 m below the seafloor the section is generally characterized by moderate to low amplitude reflections.

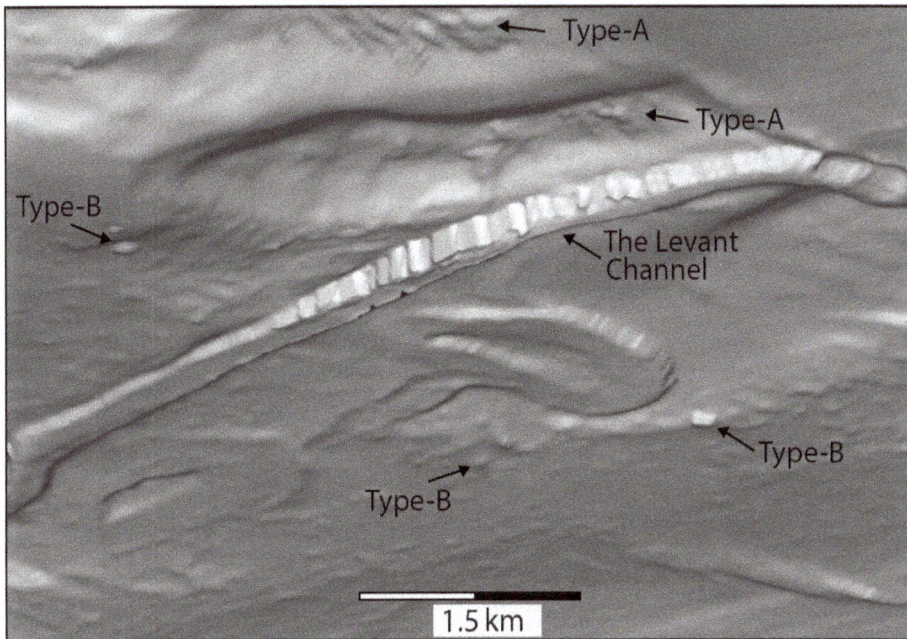

Figure 5. A 3D shaded relief view of the high-resolution bathymetry, extracted from the Southern Israel 3D seismic data, across the southern part of the Levant channel within the Southern Israel block (see Figure 1 for position), viewed from the northwest. The pockmarks observed are classified into two types based on their sizes and morphologies. Type-A are oddly shaped large (hundreds of meters) scale pockmarks, while Type-B are generally rounded pockmarks ranging in diameter between 50 to 150 m.

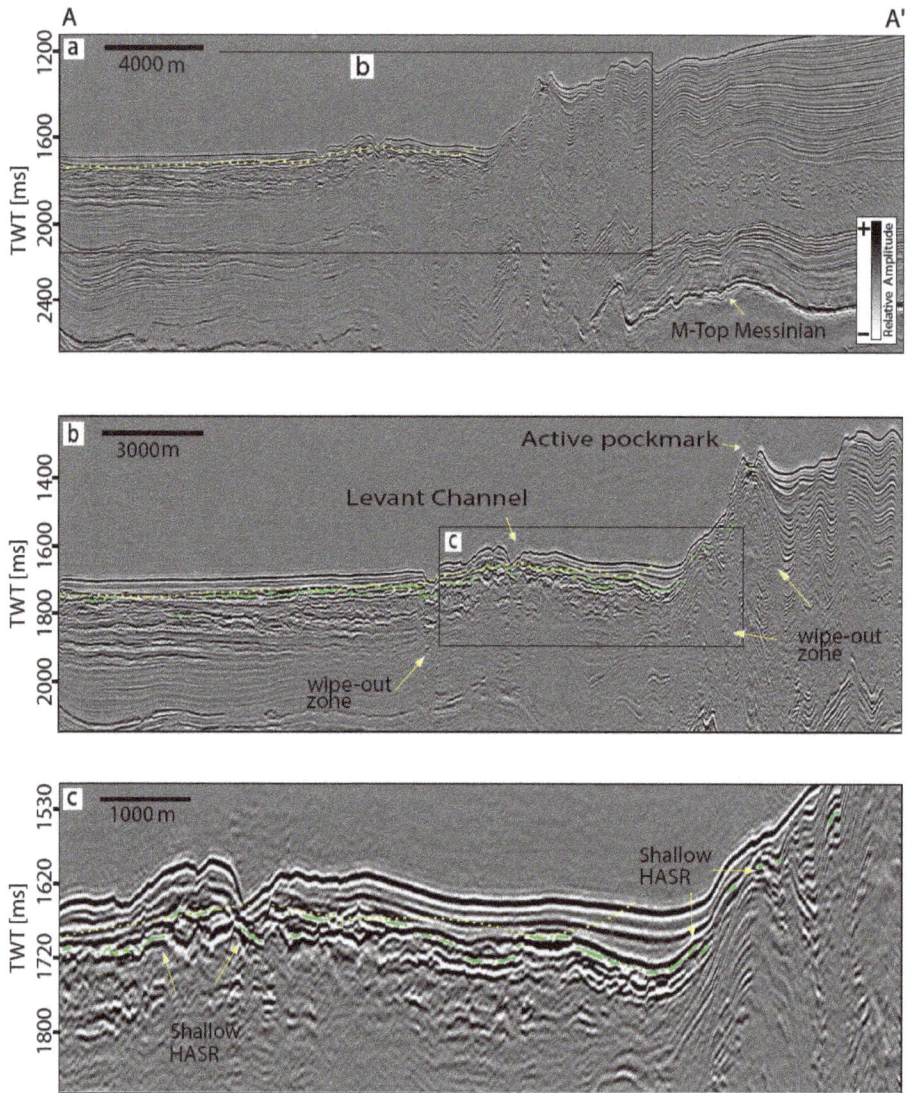

Figure 6. Seismic time migrated section A-A' (Figure 1) of the Southern Israel 3D data. The yellow dashed line represents the base GHSZ modeled with a thermal gradient of 28.5 °C/km on the regional bathymetric digital elevation model (DEM), while green dots represent the automatic HASR picks. (**a**) The full extent of the section down to the top of the Messinian evaporites (M). (**b**) A zoom on the rectangle in (**a**), showing the positions of ROV-surveyed active methane seepage site at the top pockmark of Palmahim disturbance and in the Levant Channel [69–71,84] and the Shallow HASR. (**c**) A zoom on the Levant channel area (the rectangle in (**b**)) showing the segmented appearance and mostly negative polarity of the HASR.

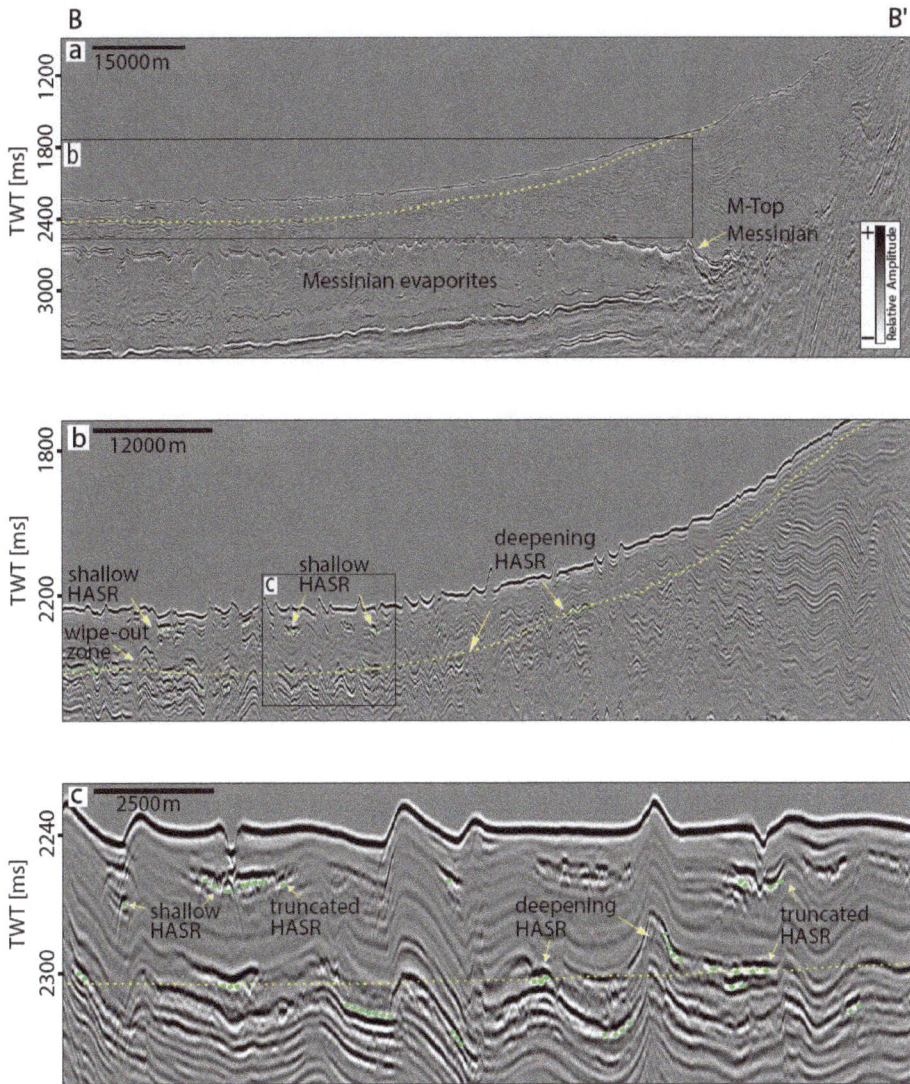

Figure 7. Seismic time migrated section B-B′ (Figure 1) of the TGS IS-2069 2D profile. The yellow dashed line represents the base GHSZ modeled with a thermal gradient of 28.5 °C/km, while green dots represent the automatic HASR picks. (**a**) The full extent of the section down through the Messinian evaporites. (**b**) A zoom between 1780 to 2500 ms TWT in (**a**) showing the Shallow and Deepening HASR within the post-Messinian section. (**c**) A zoom on the rectangle in (**b**) depicting the truncated appearance and changing polarity of both the Shallow and Deepening HASR.

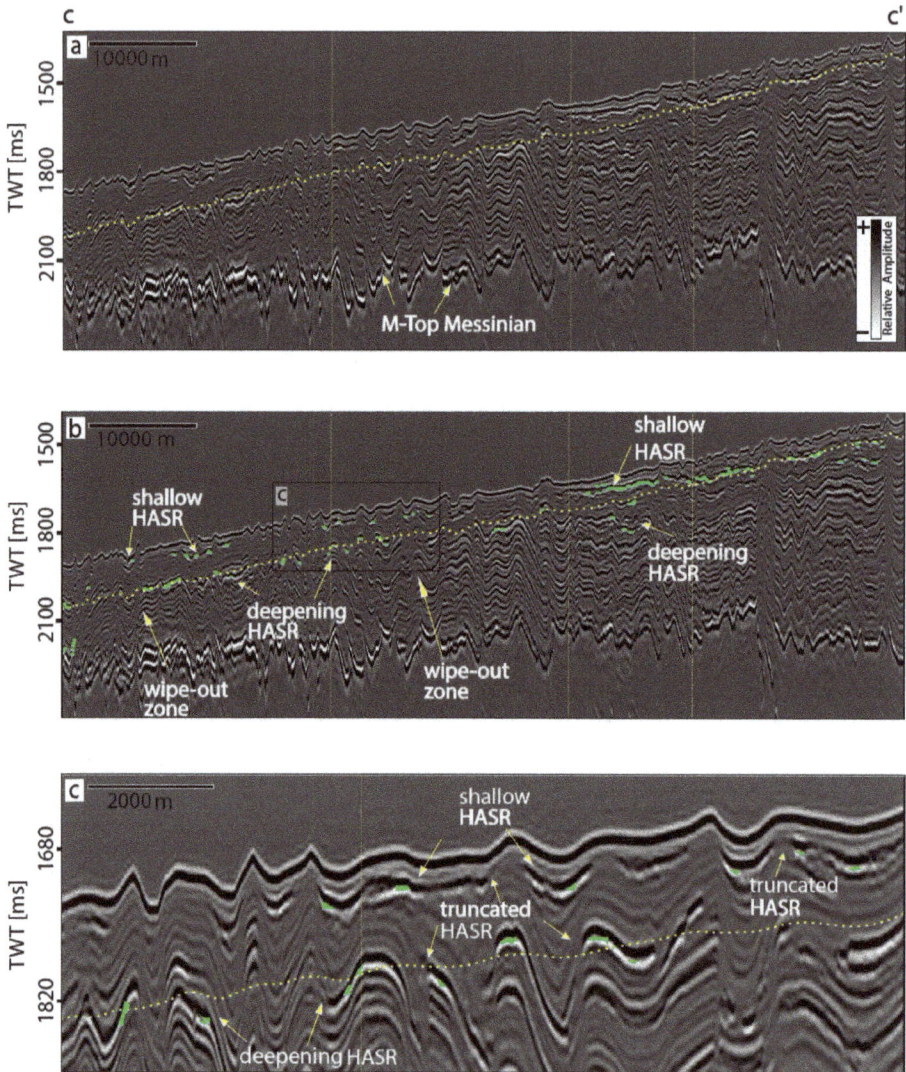

Figure 8. Seismic depth migrated section C-C′ (Figure 1) of the Pelagic 3D data. The yellow dashed line represents the base GHSZ modeled with a thermal gradient of 28.5 °C/km, while green dots represent the automatic HASR picks. (**a**) The full extent of the section down to the top of the Messinian evaporites (M). (**b**) The same as in (**a**) with an overlay of the automatic HASR picks (green dots). (**c**) A zoom on the rectangle in (**b**) depicting the truncated appearance and mostly negative polarity of both the Shallow and Deepening HASR.

A discontinuous band of segmented and scattered anomalously high-amplitude reflectivity (the HASR) is consistently observed within the top of the sedimentary stack, down to >100 m beneath the seafloor, throughout the Nile deep sea fan domain of the southeastern Levant Basin (Figures 6–8). Much of the HASR display clear reversed polarity with respect to the seafloor reflections (e.g., white primary phase in Figures 6c, 7c and 8c), which usually implies a reduction in the seismic impedance across the sub-surface reflector. However, in many cases the HASR polarity is normal (the same as the seafloor, e.g., Figure 8c), or indistinguishable and laterally changing (e.g., Figure 6c). These changes in the polarity of the HASR do not seem to consistently represent certain datasets, or distinct ranges of water or sediment depths. High-amplitude reversed polarity reflectivity, similar to the reverse polarity HASR, is commonly considered as a direct indication for the presence of free gas within the sediments, while hydrates are commonly expected to be associated with normal phase high amplitude seismic reflectivity (e.g., [24]). However, similar reflectivity characteristics may alternatively represent abrupt changes between lithologies, physical integrities (e.g., porosities), or the impact of thin layers tuning (e.g., [24]). The HASR appears as anomalous segments within more continuous moderate amplitude reflections, or as separate reflectivity phases (Figures 6–8). Segments of the HASR extend ~0.1 to several kilometers horizontally, and predominantly one, but sometimes up to several, seismic reflectivity phases (i.e., tens of meters) vertically. Neighboring segments are frequently vertically offset by up to several tens of meters with respect to each other; and in many cases, segments extend parallel to each other. Zones of amplitude wipe-outs (seismic amplitude blanking) extend in places hundreds of meters below the HASR concentrations, presumably the effects of seismic scattering and pronounced attenuation at the HASR levels. In general, no HASR is observed below the seafloor of the Israeli continental margin, with the exception of the folds at the western end of the Palmahim disturbance and their vicinity (Figures 1 and 6). The active methane seeps discovered within large-scale pockmarks at the crests of these folds [69–71] are underlain by pronounced high amplitude and reverse polarity reflections, just (<10 m) beneath the seafloor (Figure 6). These reflections and additional reflectivity segments observed beneath the slopes of these folds to the west of the pockmarks, appear to represent a prolongation of the Nile deep sea fan HASR. Below water depths of 1400 to 1500 m the HASR band becomes less distinct and appears to extend deeper (down to ~200 m) into the sedimentary section (Figures 7 and 8). Below water depths of >1500 m the HASR appears to separate into two branches: the first remains relatively coherent and limited to <100 m below the seafloor, while the other is characterized by localized enhancement of reflections of the moderate and higher amplitude sequences discussed above. This deeper branch of the HASR is sometimes harder to distinguish from the deeper higher amplitude reflections. However, in places it is distinctly apparent, mainly where amplitudes increase along a reflection and then abruptly diminish to a moderate amplitude reflection (Figures 7 and 8).

3.5.3. Vertical Distribution of the HASR

To assess the relation of the HASR with the possible presence of NGH in the southeastern Levant Basin we evaluate the spatial distribution of the HASR and compare it to the GHSZ modeling predictions. This evaluation was done first through manual picking and then repeated through automatic threshold detection of high amplitude negative phase reflections, representing the HASR, throughout our combined seismic dataset (as detailed above). Figure 9 displays the distribution of the automatically detected HASR in depth within the sediments (depth below the seafloor) as a function of the water depth, and compares it to the modeled depths below the seafloor of the base of GHSZ. This plot lumps together the different surveys representing a wide range of water depths and geological settings across the continental slope of Israel and the deep sea fan of the Nile. Figure 9 also compares the statistics of the HASR picks to the statistics of the entire dataset, inspecting for possible biases of the picking results by the depth distribution of the available data. This plot shows substantial data coverage through the water depths range of 900 to 1900 m, albeit with most of the data covering the water depth range between 900 to 1600 m (primarily around 1200 m). Thus, distribution of the picks is probably over emphasizing the distribution of the HASR around the water depth of 1200 m and under-emphasizing the distribution of the HASR at water depths >1500 m. Considering these reservations we suggest that the trends observed in Figure 9 are significant. We note that the HASR picking represents only the highest amplitudes of the HASR, above the threshold selected for the automatic picking. In practice, the HASR phenomenon is more widely spread in the seismic data, as observed by us during the manual inspection and picking.

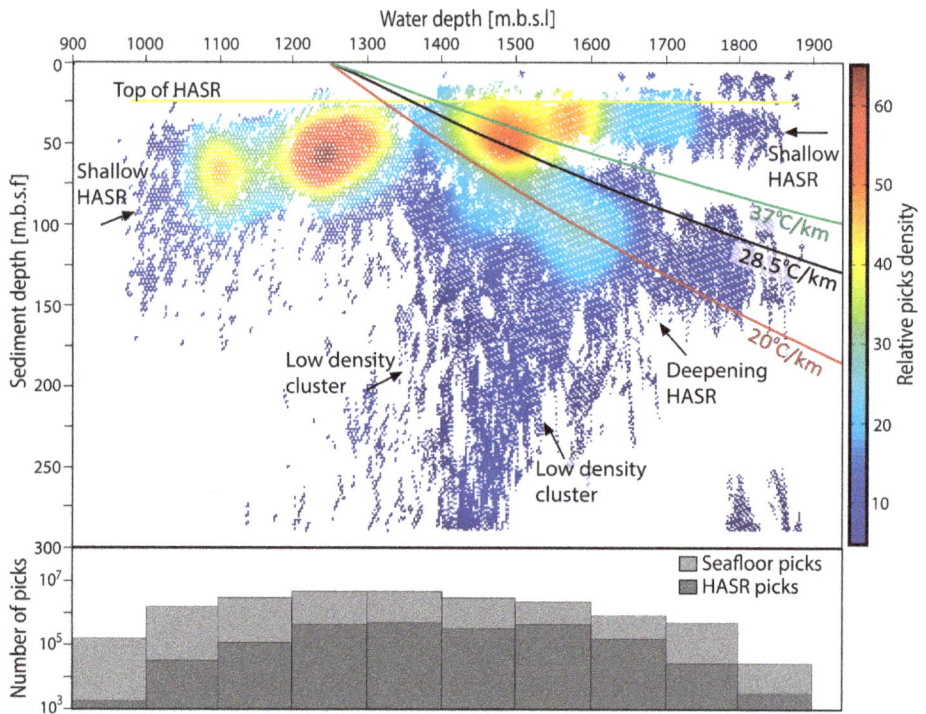

Figure 9. The distribution of automatic HASR picks with respect to the seafloor water depth and the sediments depth below the seafloor. The color scale (right) represents the relative density of picks, i.e., the number of picks found in every 1 × 1 m bins. Overlain curves are the base GHSZ models for geothermal gradients of 20 °C/km (red), 28.5 °C/km (black), and 37 °C/km (green). The Shallow and Deepening HASR clusters are evident as distinct trends of high picks distributions. The histogram at the bottom shows the total number of positions (traces) in the seismic data (light gray) and the number of automatic HASR picks in 100 m intervals of the seafloor depth. This histogram demonstrates the validity of our picks distribution.

The HASR picks plotted in Figure 9 combine three clusters in accordance with the general observations of the HASR described above. The first cluster trends sub-parallel to the seafloor throughout the water depths >1000 m, while essentially no HASR picks were detected at water depths <1000 m. Between water depths of ~1000 to 1350 m this cluster is distributed primarily between 25 to ~120 m below the seafloor. The HASR in this population trends slightly to shallower sediments depths (beneath the seafloor) with increasing water depth, which is expressed in three ways: (1) The top boundary of the picks starts at a sediments depth of 40 m at the water depth of 1000 m, and ascends to a sediments depth of 25 m at a water depth of 1350 m; (2) high density patches of HASR picks are centered around the sediments depth of 70 m at the water depth range of 1050 to 1150 m, and around a sediments depth of 55 m at the water depths of 1200 to 1300 m (Figure 9); (3) the bottom boundary of the HASR ascends from a sediments depth of 90 to 150 m as the water depths increase between 1000 to 1350 m respectively, although this trend is characterized by a high scatter (Figure 9). In the water depth range of 1350 to 1900 m the first cluster of picks continues trending at a generally constant depth below the seafloor (Figure 9). The top boundary of the picks is generally 25 m below the seafloor, except for the several limited patches of picks appearing at shallower depths (Figure 9). The bottom boundary of this part of the first cluster is between 50 and 60 m below the seafloor. This first cluster of picks is referred to as "Shallow HASR" hereafter.

A second cluster of picks branches down from the Shallow HASR around the water depth of 1400 m (between water depths of 1300 and 1500 m) and a sediments depth of ~60 m, and trends to higher sediments depths with increasing water depths (Figure 9). The upper boundary of this cluster deepens to a sediment depth of 100 m at a water depth of 1900 m. A general lack of picks clearly distinguishes this cluster at water depths >1500 m from the Shallow HASR above. The majority of HASR picks in this cluster are concentrated between water depths of 1350 to 1670 m, with a more sporadic distribution continuing along the same trend to deeper water. This reduction in the density of the picks is probably at least partly reflecting the significant reduction in the distribution of available data covering this water depth range. Overall this cluster trends from 60 to 125 m below the seafloor (Figure 9). The bottom boundary of this population is not distinct as that of the upper HASR and is determined mostly by the decline in picks density with depth (Figure 9). This second cluster of picks is referred to as "Deepening HASR" hereafter. An additional low-density cluster of HASR picks appears at water depths of 1200–1650 m, extending to a sediment depth of 300 m. This low density cluster does not display a consistent orientation and is detached from the main populations described above. This third cluster of picks represents to a great extent the lower unit of the sedimentary stack described above (e.g., Figure 6a), which is characterized by generally relatively continuous high-amplitude reflections. Owing to the deepening of the seafloor, this unit crosses from 200 to 300 m sediments depth at a water depth of ~1300 m, to a depth <100 at water depths >1600 m. Note however, that the Deepening HASR is represented by a substantial increase in picks density, and its trend clearly cuts through the lower reflective sedimentary unit, represented by the third low density cluster of picks (Figure 9).

Overlaying the HASR distribution with the modeled base-GHSZ depth curves (Figure 9) reveals that the trend and projected intercept of the Deepening HASR cluster is generally matched by the curve modeled with the 20 °C/km geothermal gradients, while the curve modeled with the 28.5 °C/km geothermal gradient appears to bound the upper extent of the Deepening HASR. This is verified also by overlaying the depths, predicted by the 28.5 °C/km geothermal gradient curve and bathymetric DEM, on the seismic data (Figures 6–8). The resulting depth curve generally matches the upward truncations of the imaged HASR segments. Thus, there is a viable match between the HASR trend and a reasonable estimate of the base of GHSZ in the Levant Basin.

3.5.4. Spatial Distribution of the HASR

Map plots (Figure 10) reveal that the spatial distribution of HASR picks is uneven. The Shallow HASR is found in the deep sea fan of the Nile region, and is absent from the continental slope of Israel except for the tow region of the Palmahim disturbance. The Shallow HASR picks are mostly concentrated in patch sets reminiscent of foliage. Overlaying the distribution of the Shallow HASR with the bathymetry (Figure 10) reveals that the foliage-like patterns are aligned along the turbidite channels etched into the current seafloor and are branching downstream from them. Within each patch, the HASR picks depths below the seafloor are highly scattered, reflecting the laterally discontinuous nature of the HASR described above.

The distribution patterns of the Deepening HASR are reminiscent of the Shallow HASR patterns, but the spatial distributions across the basin of the two are considerably different. The Deepening HASR is mostly found in the western part of Sara-Mira and Pelagic 3D seismic surveys, as well as in the western part of the TGS 2D line (Figures 1 and 6, Figures 7 and 8). The deeper Deepening HASR are distributed at the top of the relatively high amplitude unit discussed above (Figures 6–8), where they appear as discontinuous patches mostly limited to the elevated anticlinal parts of folding structures. A small portion of the picks is also sparsely distributed within the high amplitude unit (Figures 6–8). Yet, these picks do not show any continuity and cannot be attributed to a specific reflecting horizon of this unit. Minor occurrences of the Deepening HASR are recognized also in the deeper parts of Gal-C, Sothern Israel, and Oz surveys, where the rare picks mainly appear along the surface of meandering channel paths around the paleo-channel canyons and in close proximity to faulted or folded structures (Figure 10).

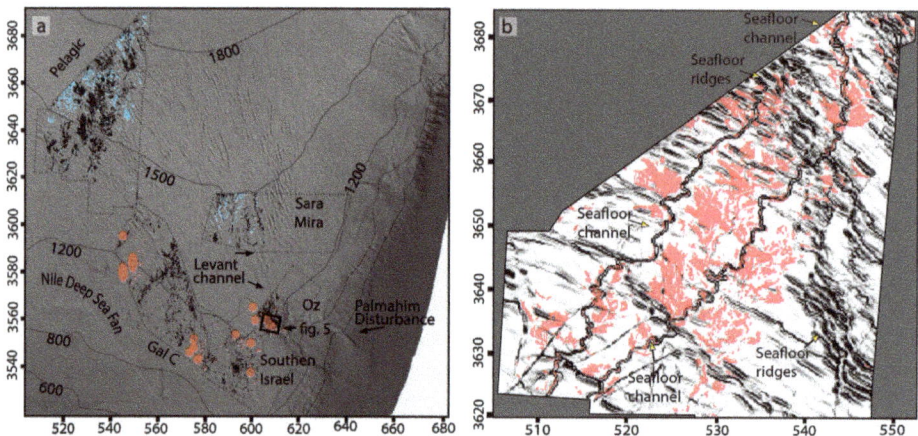

Figure 10. (**a**) A shaded relief map of the southeastern part of the Levant Basin bathymetric DEM [62,108] overlaid with the spatial distribution of the HASR automatic picks and pockmark clusters within the different 3D seismic blocks analyzed. Black dots mark the Shallow HASR, while blue dots mark the Deepening HASR. Red ellipses mark the location of pockmark clusters. This figure demonstrates the wide distribution of Shallow HASR picks across the Levant Basin; the concentration of pockmark clusters between water depths of 900 to 1300 m across the deep fan of the Nile; and the distribution of the Deepening HASR at water depths >1300 m. (**b**) A zoom on the Pelagic 3D seismic block with the pronounced shading of the bathymetry (enhanced gray scale) overlaid with the automatic picks of the Shallow HASR (red). This figure demonstrates the relation between the Shallow HASR and seafloor channels of the Nile fan.

4. Discussion

4.1. The GHSZ in the Levant Basin

The main purpose of this study is to review possible evidence and constrain the potential for the presence of NGH in the Levant Basin, southeastern Mediterranean, notwithstanding the current lack of direct evidence. For this purpose, we modeled the GHSZ in the Levant Basin based on the local thermodynamic conditions and salinity of the bottom water and interstitial pores in the sub-seafloor sediments. The bottom seawater conditions are relatively well constrained by data and therefore the definition of the top of the GHSZ to 1250 ± 5 m water depth is robust. This modeled top of the GHSZ is consistent with the shallowest water depth in which NGH occur in the Anaximander region, at a water depth of 1264 m at the summit of the Thessaloniki mud volcano [63,64]. [63] argued that this occurs at the top boundary of their modeled GHSZ, considering very similar local thermodynamic conditions to those used in this study. [65] observed the disappearance of the hydrates coating at a water depth of 1350 m, and estimated the top of the GHSZ in the western submarine fan of the Nile as 110 m deeper than our estimate for the Levant. This inconsistency may be explained by possible slightly elevated water temperature, and by the different gas composition in the cold seep sampled by [65].

The intersection of the modeled top of GHSZ with the bathymetry outlines the spatial extent of the potential GHOZ (Figure 4). The GHSZ stretches over more than half of the Levant Basin seafloor, constituting the vast majority of its central and northern parts. With the shortage of reliable data, approximations had to be made for the geothermal, pressure, and salinity profiles within the seafloor sediments. Yet, supported by indirect evidence and sensitivity analyses we argue that the modeled GHSZ presented here provide useful first-order approximated bounds of the possible distribution of

NGH in the Levant Basin. The differences between the alternative maps of Figure 4a,b probably offer an over estimation of the uncertainty bounds in determining the depth below the seafloor to the base of the GHSZ. We suggest that the model calculated with a geothermal gradient of 28.5 °C/km (Figure 4) provides a conservative estimate of the depth to base of the GHSZ in the Levant Basin.

4.2. Methane and Hydrates in the Levant Seafloor Sediments

The actual in-place occurrence of NGH within the GHSZ is dependent on the availability and steady flow of methane gas within the sediment (e.g., [123]), and on the storage capacity of the sedimentary medium (e.g., [124]). The presence of an active gas system within the Levant Basin seafloor sediments is indicated by two main lines of evidence: direct evidence for seafloor gas seepage and geophysical indicators for the potential presence of gas.

4.2.1. Direct Evidence for Gas Seepage at the Seafloor

The multiple direct observations of active seafloor gas seeps within the perimeters of the Levant Basin [68–71,84] were all documented within a water depth range of 1000–1200 m, while the upper bound of the modeled GHSZ is at a water depth of 1250 m. Consequently, our model of the GHSZ in the Levant does not directly link the observed gas seeps and the possible upward percolation of methane from NGH within the GHSZ. Such a connection would require significant upslope lateral flow of methane. A possible indirect evidence for shallow gas emission is presented by the occurrence of pockmarks across the seafloor, as mapped based on the analyzed 3D seismic surveys (e.g., Figures 5 and 10). These pockmarks appear to be limited to a water depth range of 1000–1300 m (Figure 10), correlating well with the spatial distribution of the edge of the modeled GHSZ and suggesting a possible tunneling of free methane gas along the base of the GHSZ toward its edge. Tunneling of methane to the edges of the GHSZ is attributed to the base of the GHSZ functioning as a seal that prevents gas escapes and tunneling the gas toward the GHOZ pinch out (e.g., [2,23,125]). Additionally, the pockmarks depth distribution might represent a record of the shift of the top of the GHSZ, and therefore the migration of the pinch out of the GHOZ to deeper water. The warming event at the end of the last glacial period, ~14.5 ka, was estimated to have raised the bottom water temperature in the Western Mediterranean by ~4 °C [126]. Assuming accordingly ~4°C cooler bottom waters in the Levant Basin (being ~10 °C) at ~14.5 ka, the modeled top of the GHSZ would occur at a water depth of ~890 m. This estimate is also consistent with the results of [29]. Phase delays between faster sea level changes [127] and slower warming of the Levant bottom water, such as suggested for the Arctic Ocean by [111], may have further facilitated the formation of NGH deposits and their subsequent destabilization. These processes may have resulted in enhanced gas release and seepage along the retreat path, and formed the observed pockmarks and authigenic carbonates. This, however, remains a hypothesis until the ages of seepage and carbonate precipitation are constrained.

4.2.2. The HASR–Evidence for Free Gas and Possibly Hydrates

The second line of evidence for the presence of gas in the seafloor sediments of the Levant Basin relies on the interpretation of the distinct HASR imaged off the deep-water shallow sediments of the studied area. We suggest that the HASR is associated with the presence of free gas, and possibly in at least some of the region with hydrate accumulations, within the upper sedimentary section. The correlation of high amplitude seismic reflectivity and underlying seismic amplitude blanking with subsurface gas bearing intervals is a commonly observed result of the strong response of seismic waves to the presence of even minor free gas content (e.g., [103,128]). High amplitude reflections may also be associated with hydrate accumulations within the GHSZ (e.g., [24,101,129]). Notably, high amplitude seismic reflections may alternatively represent other sub seafloor features in our study area, such as the prominent lithological and porosity contrast between shale layers and porous sand bodies lacking any presence of gas or hydrates. The reverse polarity of much of the HASR indicates low impedance intervals within the generally clay rich seafloor of the Levant basin, reinforcing the suggested presence of free gas (e.g., [117]). The normal or indistinguishable polarity observed in many of the cases may represent hydrate bearing intervals within the GHSZ, but seems puzzling outside and below the GHSZ. However, reflection polarities may be elusive due to thin layers tuning and structural complexities, particularly when the reflectors are segmented or discontinuous. Such polarity complexities are commonly observed where free gas and hydrates occur within discordant sedimentary intervals (e.g., [101,129]). In these cases, unraveling the polarity information requires focused analysis [130], which is outside the scope of this study. A central base for our interpretation is the unambiguous correlation of seafloor methane gas seepages, which were verified by seafloor surveying and sampled in the study area [68–71,84], with HASR just below the seafloor (e.g., Figure 5). Taking together the seismic characteristics with these direct verifications of methane gas seepage, we argue that at least part of the HASR represents the presence of free methane gas and possibly also the hydrates within the seafloor sediments. We note that the generation of reflectivity by the presence of free gas bubbles within the sediments requires methane saturation within the interstitial pore water (e.g., [131]). Thus, the observation of gas related HASR below the seafloor implies a significantly larger availability of dissolved methane for the formation of hydrates within the sediments, at least in parts of the Levant Basin.

4.2.3. The Shallow HASR–Evidence for Shallow Gas beneath the Levant Basin Seafloor

The vertical distribution of the observed Shallow HASR picks cluster is sub-parallel to the seafloor (Figure 9) and does not appear to depend strongly on the water depth or pressure. Neither does it appear to match the trend of any of the possible hydrate stability curves. Moreover, it clearly extends laterally significantly outside of the GHSZ, which is limited by the 1250 m water depth contour (Figure 10). We therefore suggest that at least outside the GHSZ the HASR represents primarily free methane gas accumulations in the shallow sub-seafloor sediments. Within the GHSZ region, bounded by the 1250 m water depth contour (Figure 10), the Shallow HASR may similarly represent hydrate concentrations. If the HASR represents hydrates then the upper cutoff of the HASR probably represents the top of the GHOZ, which is usually located tens of meters below the seafloor (e.g., [1,132,133]). Alternatively, the HASR may also represent the presence of free gas within the upper part of the GHSZ. Such occurrences of free gas within the GHSZ are indicated by seafloor bubbles emanations in cold seeps above hydrate bearing intervals [134]. The presence of free gas within the GHSZ have been suggested based on seismic reflectivity and velocity variations [135,136] as well as cone penetration test results [128]. In addition, high salinity, measured in boreholes drilled through hydrate bearing regions, is argued to be the product of free gas supply and hydrates formation within the GHSZ (e.g., [116]). Some mechanisms suggested for the maintenance of free gas within the GHSZ, which may apply for the Levant Basin, are: Transient focused flow through structural or lithological pathways (e.g., [116,135]); local increase of pore water salinities as a result of proximate NGH formation, allowing the stability of free gas within the GHSZ [114,116]; and low sediment permeability resulting in insufficient water supply for NGH formation to the zone containing free gas [128].

Whether the Shallow HASR represents free gas or NGH, both alternatives imply that the saturation of dissolved methane is exceeded in the interstitial pore water at their level [110]. The shallow cut-off of methane saturation in the marine environment, and therefore presumably also the top of the HASR distribution, is generally constrained by anaerobic oxidation of methane (AOM). The AOM is maintained below the sulfate methane transition zone (SMTZ) by balanced diffusion fluxes of methane from the saturated zone and sulfate from the seafloor (e.g., [137]). Reference [138] showed that the depth below the seafloor to the top of the free gas zone can be used to estimate the upward flux of dissolved methane, given in situ methane solubility (saturation). To examine the possibility of our interpretation of the HASR as related to methane free gas or hydrates, we followed the approach of [138] in estimating the upward methane flux based on the upper cutoff of the Shallow HASR distribution 25 m below the seafloor (see detailed derivation in the Supplementary Information). Based on [139,140] and the same thermodynamic conditions used in our GHSZ modeling we estimate that the methane solubility, related with the top of the Shallow HASR, is between 0.98 and 1.03 mmol, depending on the water depth. To estimate the balance of sulfate and methane fluxes at the SMTZ we assumed negligible methane production between the HASR and seafloor and linear concentration gradients. The resulting estimated methane fluxes are 9 to 20 mmol·a^{-1}·m^{-2}, corresponding to water depths rage of 1200 to 2000 m. These results are in agreement with the SMTZ distribution models and methane fluxes reported from NGH provinces across the globe, ranging between 20 and 250 mmol·a^{-1}·m^{-2} [141–147]. Thus, these results support our interpretation of the seismic HASR as representing free gas or in-place NGH concentrations, both of which support the possible occurrence of NGH deposits within the modeled GHSZ.

The essentially exclusive occurrence of the Shallow HASR in the Nile deep sea fan (Figure 10) and their clear correlation with seafloor channels suggests a genetic relation between these phenomena. [46,47] demonstrated that the upper sedimentary section within the eastern deep sea fan of the Nile is composed of densely spaced relatively sand rich channel-levee complexes, encased within hemipelagic sediments. We suggest that these relatively sand rich bodies constitute localized reservoirs of free gas, and possibly also NGH, represented by the Shallow HASR. The foliage like pattern of the Shallow HASR, diverging from the current seafloor channels, suggest that the primary reservoirs for the Shallow HASR are paleo-lobes, associated with these channel systems, buried tens of meters below the seafloor. The scattered nature of the HASR within the observed patches probably represent the complexity of the channel-levee-lobe systems (e.g., as discussed by [148]) and their truncation by recent salt tectonics (as discussed by [46]). Similar associations of scattered free gas and hydrate accumulations with buried relatively sandy channel systems, and the observations of a diffuse BSR, have been reported from various large river deep sea fans. Examples are the Congo River [149], Godavari River [150,151], and the Pearl River [129].

4.3. The Deepening HASR–a Distributed BSR in the Levant Basin (?)

The locus and trend of the Deepening HASR picks cluster, mapped in the study area, appear to be approximately bounded by the modeled base-GHSZ with the 20 and 28.5 °C/km thermal gradients (Figure 9). The identification of the Deepening HASR on the seismic sections may appear somewhat arbitrary (Figures 7 and 8), as the picked phases appear is some cases to be conformal with the generally relatively high amplitude reflections of the lower sedimentary unit in the study area. Indeed, picks scattered below the Deepening HASR and toward shallower water-depths do detect reflections within that layer. However, the Deepening HASR cluster represents an order of magnitude larger community of picks that are concentrated along this trend, demonstrating that it is subjectively anomalous with respect to the rest of the unit. A more careful inspection of the sections (Figures 7 and 8) reveals the clear intensification of reflection amplitudes as they approach the Deepening HASR alignment (approximately demarcated in the figures by the overlaid base GHSZ modeled curves). Moreover, the pronounced amplitude reflections associated with the Deepening HASR commonly appear truncated in the upward direction, but in many cases can actually be traced farther with diminished amplitudes. Taking together the first-order match between the GHSZ model and the Deepening HASR, and their observed characters, we suggest that the Deepening HASR constitutes a segmented and distributed BSR, one of the primary indicators for the presence of NGH. Thus, the Deepening HASR is suggested to represent the presence of free gas trapped below the base of the GHSZ and possibly also hydrates deposited at the lower part of the GHOZ. The distributed nature of the Deepening HASR probably results from the distribution of free gas and hydrates in localized relatively sandy channel lobe systems (as discussed above for the Shallow HASR), the truncation of the lower unit sandy bodies by salt tectonics and possible variations of methane supply.

4.4. The Possible Sources of Methane

Two possible mechanisms may supply methane to the pervasive gas and possibly hydrate systems proposed here to exist in the Levant Basin. Methane may be supplied by in situ methanogenesis through bacterial decomposition of organic material (e.g., [152,153]). The methane generated by the bacteria within the GHSZ is incorporated in the NGH inside the GHSZ, while methane generated outside the GHSZ may form free gas bubbles. [154] modeled the case of methane diffusion to fine sandy layers from methanogenesis occurring in the surrounding fine grain intervals containing modest amounts (<0.5% of dry weight) of organic matter. This mechanism is concluded to be sufficient to supply the required methane for the formation of the hydrates observed in the Cascadia margin at IODP site U1325. Alternatively, methane transport through upward fluid flow from lower sedimentary levels may be important (e.g., [133,153]). In this case, the ascending gas is incorporated into the NGH structure or remains trapped below the layer of the NGH-containing sediments [1,155,156]. In either case, a possible source for hydrocarbons production is the abundant organic matter (up to 7%) buried in sapropel deposits throughout much of the post-Messinian sediments of the deep-water part of the Levant Basin [157–159]. The evident association of the HASR to the meandering channels and associated buried lobes may suggest an additional direct contribution of transported organic matter from the Nile River. Taken together the proximity of organic-rich sapropel units and coarse grain sizes (high porosity sediments) characterizing the infill of submarine channel-Levee-fan complexes [47,59] provide both possible sources and reservoirs for gas and hydrate formation and accumulation. The possible sources for organic matter discussed above, support the possible in situ formation of methane outside, below, and within the GHSZ. The deep giant gas reservoirs recently found in the Levant Basin [86,87], or related systems, are also possible gas sources for methane that may have been transported via migration paths such as faults and folds to the GHSZ. The existence of such migration paths within the southeastern Levant Basin was suggested by [80,82], but thorough connectivity has not been demonstrated as of yet.

4.5. Volume Assessment of Potential NGH Deposits in the Levant

Lacking conclusive indications, this study offers only suggestive evidences, raising the possibility for the presence of hydrates in the Levant Basin. Yet, our combined modeling and observations offer a first-order assessment of the NGH potential in the Levant Basin, with direct implications on potential energy resources, geohazard risk estimation, and models related to carbon cycles and its possible relation with present and future climate-change.

For this first-order assessment of potential NGH resources in the Levant Basin, we take the simplistic steady state assumption, while noting [123] discussion of the deficiencies of this assumption. In that case, NGH may form throughout the sedimentary column within the GHOZ, constrained by the isopach between the seafloor bathymetry and our modeled base GHSZ (Figure 4). Assuming that the intercept and trend of the Deepening HASR (Figure 9) represent a segmented and distributed BSR, the fit of the modeled base of the GHSZ curves with this trend constrains the relevant geothermal gradient between 20 to 28.5 °C/km within the post-Messinian sediments of the Levant Basin. Larger geothermal gradients, toward the 37 °C/km end member geothermal gradient, are not in agreement with the Deepening HASR being aligned with the base of the GHSZ. Thus, a maximum estimate of NGH potential is constrained by the map that was modeled using the lower geothermal gradient end member of 20 °C/km (Figure 4), while a more modest estimate is constrained by the map calculated with the higher 28.5 °C/km geothermal gradient (Figure 4). Consumption of methane at the SMTZ reduces its saturation, inhibiting the possible formation of NGH [141]. Thus, discarding the top of the sedimentary section, above the top cutoff of the Shallow HASR cluster, defines presumably a more reliable estimate of the sedimentary column available for hydrate accumulation i.e., the GHOZ. This estimate of the GHOZ is represented by the thickness map of Figure 4c with a total volume of ~3250 km^3. The geothermal gradient of the Levant Basin (20 to 28.5 °C/km [86,94–97] is expected to be lower than in the adjacent Herodotus Basin and the Cypriot arc regions [160–162]. Thus, the GHOZ thickness map, which is based on the Levant Basin parameters, is expected to be less accurate toward these areas.

The discussion above suggests that hydrates in the Levant are primarily concentrated within relatively sandy channel-levee-lobe complexes. Based on our partial mapping of these silty to sandy bodies (Figure 10), we estimate them to occupy only ~3% of the sedimentary volume defined by Figure 4c. Considering a lower bound porosity of ~35% and hydrate saturation of ~50% for silt and sand rich host sediments [2,133] we obtain that NGH occupy on average ~0.5% of the sedimentary volume within the potential GHOZ (Figure 4c). Thus, considering a conservative NGH to gas yield factor of ~160 [4], the current study provides a first-order conservative estimation of ~100 Tcf (~2750 km^3) for the potential volume (at standard temperature and pressure) of locked methane, and ~1.5 gigatonnes of carbon in NGH in the Levant Basin. These estimations constitute between ~0.1 to 3‰ of the estimated global NGH methane volumes and carbon content in marine sediments [2,4,5,163].

5. Conclusions

Seafloor gas seepages discovered in recent years in and around the Levant Basin, and sparse observations of hydrates in the broader EMS context are suggestive of the presence of NGH in the basin. Motivated by these findings this study combines thermodynamic modeling and analysis of a pervasive seismic dataset to examine the potential for such a presence.

- Thermodynamic modeling, using the CSMHyd software, robustly constrained by locally measured southeastern Mediterranean water temperature and salinity profiles reveals that the top of the GHSZ is at a water depth of 1250 ± 5 m. Intersecting this depth with the bathymetry reveals that more than half of the Levant Basin seafloor, namely its central to northern part, lies within the GHSZ.
- Modeling the base of the GHSZ is constrained by the lack of measured sub-seafloor thermodynamic parameters, and associated uncertainties. The primary modeling uncertainty is related with the availability and confidence of published sub-seafloor geothermal gradients. Yet, using simplistic approximations for the thermodynamic parameters yields a useful first-order approximation of the GHSZ within the Levant. The base of the GHSZ lies at a depth of 150 to 200 mbsf at a water depth of 1750 m, and may reach a depth of 430 mbsf at the northwestern edge of the studied area.
- Seafloor pockmark clusters observed in our seismic data are concentrated at a water depth of ~1200 m, just upslope from the modeled top bound of the GHSZ and mostly above the deep sea fan of the Nile. These pockmarks suggest the prominence of seafloor gas in the region, and may represent a partly ongoing gas seepage episode associated with the presumed downslope retreat of NGH since the last glacial time.
- Scattered high amplitude seismic reflectivity (HASR) is pervasively distributed beneath the seafloor across the deep sea fan of the Nile, correlating in several sites with observed active gas seepages. The HASR is therefore suggested to represent the wide spread presence of free gas, and possibly NGH, captured within buried distributed channel–levee related sandy/silty units.
- The distribution of the HASR depth beneath the seafloor vs. the water depth bifurcates into two main clusters. Most of the HASR across the study area cluster between 25 to 100 mbsf, and sub-parallel to the seafloor. This cluster is suggested to represent shallow free gas, and possibly hydrates, whose top is bounded by the SMTZ. An additional major HASR cluster trends to greater sub-seafloor depth with increasing water depth. The trend of this cluster broadly matches the modeled base of the GHSZ with a thermal gradient between 20 to 28.5 °C/km, and it is therefore suggested to represent a regionally discontinuous BSR beneath the Levant Basin. The discontinuity is attributed to the distributed nature of channel-lobes systems, into which the NGH and presumably underlying free gas are accumulated.

- Taken together, the modeling results and seismic analysis suggest the probable presence of NGH in the Levant Basin, within confined lithological bodies across the deep sea fan of the Nile. The presence of NGH is conservatively bounded between the base GHSZ modeled using the thermal gradient of 28.5 °C/km, and the SMTZ, which is estimated to occur ~25 m below the seafloor. Thus, the potential methane resource within the Levant Basin is estimated at ~100 Tcf, which would contain ~1.5 gigatonnes of carbon.

Supplementary Materials: The following are available online at http://www.mdpi.com/2076-3263/9/7/306/s1, Section S1: Methane flux and the sulfate methane transition zone (SMTZ); Section S2: Manual picking of the HASR; Figure S1: The distribution of manually picked HASR.

Author Contributions: Conceptualization, Y.M.; formal analysis, Z.T. and Y.M.; supervision, Y.M.; writing—original draft, Z.T., A.M., and Y.M.; writing—review and editing, Z.T., A.M., Z.B.-A., and Y.M.

Funding: This research was funded by the Moses Strauss Department of Marine Geosciences and Leon H. Charney School of Marine Sciences, University Haifa.

Acknowledgments: We thank the Oil Commissioner's Office in the Ministry of Energy of the State of Israel, Modiin Energy, Genesis Energy (Israel) Ltd. for data sharing and permitting. We thank Paradigm for sponsoring the use of their software. We also thank O.M. Bialik, S. Pinkert, and three anonymous reviewers for their helpful comments, which significantly improved this paper.

Conflicts of Interest: The authors declare no conflict of interest. The funders had no role in the design of the study; in the collection, analyses, or interpretation of data; in the writing of the manuscript, or in the decision to publish the results.

References

1. Sloan, E.D.; Koh, C. *Clathrate Hydrates of Natural Gases*, 3rd ed.; CRC Press: Boca Raton, FL, USA, 2008; p. 730.
2. Boswell, R.; Waite, W.F.; Kvenvolden, K.; Koh, C.A.; Klauda, J.B.; Buffett, B.A.; Frye, M.; Maslin, M. What are gas hydrates. In *Frozen Heat: A UNEP Global Outlook on Methane Gas Hydrates?* Beaudoin, Y.C., Waite, W., Boswell, R., Dallimore, S.R., Eds.; United Nations Environment Programme; GRID-Arendal; Birkland Trykkeri A/S: Birkeland, Norway, 2014; Volume 1, pp. 11–30.
3. Kvenvolden, K.A. Methane hydrate—A major reservoir of carbon in the shallow geosphere? *Chem. Geol.* **1988**, *71*, 41–51. [CrossRef]
4. Milkov, A.V. Global estimates of hydrate-bound gas in marine sediments: How much is really out there? *Earth Sci. Rev.* **2004**, *66*, 183–197. [CrossRef]
5. Klauda, J.B.; Sandler, S.I. Global distribution of methane hydrate in ocean sediment. *Energy Fuels* **2005**, *19*, 459–470. [CrossRef]
6. Tréhu, A.M.; Ruppel, C.; Holland, M.; Dickens, G.R.; Torres, M.E.; Collett, T.S.; Schultheiss, P. Gas hydrates in marine sediments: Lessons from scientific ocean drilling. *Oceanography* **2006**, *19*, 124–142. [CrossRef]
7. Piñero, E.; Marquardt, M.; Hensen, C.; Haeckel, M.; Wallmann, K. Estimation of the global inventory of methane hydrates in marine sediments using transfer functions. *Biogeosciences* **2013**, *10*, 959–975. [CrossRef]
8. Dickens, G.R.; Quinby-Hunt, M.S. Methane hydrate stability in seawater. *Geophys. Res. Lett.* **1994**, *21*, 2115–2118. [CrossRef]
9. Shipley, T.H.; Houston, M.H.; Buffler, R.T.; Shaub, F.J.; McMillen, K.J.; Ladd, J.W.; Worzel, J.L. Seismic evidence for widespread possible gas hydrate horizons on continental slopes and rises. *AAPG Bull.* **1979**, *63*, 2204–2213.
10. Hyndman, R.D.; Spence, G.D. A seismic study of methane hydrate marine bottom simulating reflectors. *J. Geophys. Res.* **1992**, *97*, 6683–6698. [CrossRef]
11. MacKay, M.E.; Jarrard, R.D.; Westbrook, G.K.; Hyndman, R.D. Origin of bottom-simulating reflectors: Geophysical evidence from the Cascadia accretionary prism. *Geology* **1994**, *22*, 459–462. [CrossRef]
12. Berndt, C.; Bünz, S.; Clayton, T.; Mienert, J.; Saunders, M. Seismic character of bottom simulating reflectors: Examples from the mid-Norwegian margin. *Mar. Pet. Geol.* **2004**, *21*, 723–733. [CrossRef]
13. Shedd, W.; Boswell, R.; Frye, M.; Godfriaux, P.; Kramer, K. Occurrence and nature of "bottom simulating reflectors" in the northern Gulf of Mexico. *Mar. Pet. Geol.* **2012**, *34*, 31–40. [CrossRef]

14. Majumdar, U.; Cook, A.E.; Shedd, W.; Frye, M. The connection between natural gas hydrate and bottom-simulating reflectors. *Geophys. Res. Lett.* **2016**, *43*, 7044–7051. [CrossRef]

15. Tsuji, Y.; Fujii, T.; Hayashi, M.; Kitamura, R.; Nakamizu, M.; Ohbi, K.; Saeki, T.; Yamamoto, K.; Namikawa, T.; Inamori, T.; et al. Methane-hydrate occurrence and distribution in the eastern Nankai Trough, Japan: Findings of the Tokai-oki to Kumano-nada methane-hydrate drilling program. In *Natural Gas Hydrates—Energy Resource Potential and Associated Geologic Hazards*; Collett, T., Johnson, A., Knapp, C., Boswell, R., Eds.; AAPG Memoir; American Association of Petroleum Geologists: Tulsa, OK, USA, 2009; Volume 89, pp. 228–246.

16. Kvenvolden, K.A. Methane hydrate in the global organic carbon cycle. *Terra Nova* **2002**, *14*, 302–306. [CrossRef]

17. Dickens, G.R. Rethinking the global carbon cycle with a large, dynamic and microbially mediated gas hydrate capacitor. *Earth Planet. Sci. Lett.* **2003**, *213*, 169–183. [CrossRef]

18. Beaudoin, Y.C.; Waite, W.; Boswell, R.; Dallimore, S.R. (Eds.) *Frozen Heat: A UNEP Global Outlook on Methane Gas Hydrates*; United Nations Environment Programme, GRID-Arendal; Birkland Trykkeri A/S: Birkeland, Norway, 2014; Volume 2, p. 96.

19. Dickens, G.R. Carbon cycle: The blast in the past. *Nature* **1999**, *401*, 752–755. [CrossRef]

20. Kennett, J.P.; Cannariato, K.G.; Hendy, I.L.; Behl, R.J. Carbon isotopic evidence for methane hydrate instability during Quaternary interstadials. *Science* **2000**, *288*, 128–133. [CrossRef] [PubMed]

21. Buffett, B.; Archer, D. Global inventory of methane clathrate: Sensitivity to changes in the deep ocean. *Earth Planet. Sci. Lett.* **2004**, *227*, 185–199. [CrossRef]

22. Wallmann, K.; Dallimore, S.; Biastoch, A.; Westrook, G.; Shakova, N.; Severinghaus, J.; Dickens, G.; Mienert, J. Assessment of the sensitivity and response of Methane hydrate to global climatic change. In *Frozen Heat: A UNEP Global Outlook on Methane Gas Hydrates*; Beaudoin, Y.C., Waite, W., Boswell, R., Dallimore, S.R., Eds.; United Nations Environment Programme, GRID-Arendal; Birkland Trykkeri A/S: Birkeland, Norway, 2014; Volume 1, pp. 50–75.

23. Sultan, N.; Cochonat, P.; Foucher, J.P.; Mienert, J. Effect of gas hydrates melting on seafloor slope instability. *Mar. Geol.* **2004**, *213*, 379–401. [CrossRef]

24. McConnell, D.R.; Zhang, Z.; Boswell, R. Review of progress in evaluating gas hydrate drilling hazards. *Mar. Pet. Geol.* **2012**, *34*, 209–223. [CrossRef]

25. Hecht, A. Abrupt changes in the characteristics of Atlantic and Levantine intermediate waters in the Southeastern Levantine Basin. *Oceanol. Acta* **1992**, *15*, 25–42.

26. Lejeusne, C.; Chevaldonné, P.; Pergent-Martini, C.; Boudouresque, C.F.; Pérez, T. Climate change effects on a miniature ocean: The highly diverse, highly impacted Mediterranean Sea. *Trends Ecol. Evol.* **2010**, *25*, 250–260. [CrossRef] [PubMed]

27. Sisma-ventura, G.; Yam, R.; Kress, N.; Shemesh, A. Water column distribution of stable isotopes and carbonate properties in the South-eastern Levantine basin (Eastern Mediterranean): Vertical and temporal change. *J. Mar. Syst.* **2016**, *158*, 13–25. [CrossRef]

28. Miles, P.R. Potential distribution of methane hydrate beneath the European continental margins. *Geophys. Res. Lett.* **1995**, *22*, 3179–3182. [CrossRef]

29. Praeg, D.; Geletti, R.; Wardell, N.; Unnithan, V.; Mascle, J.; Migeon, S.; Camerlenghi, A. The Mediterranean Sea: A natural laboratory to study gas hydrate dynamics. In Proceedings of the 7th International Conference on Gas Hydrates (ICGH 2011), Edinburgh, UK, 17–21 July 2011; pp. 17–21.

30. Klauda, J.B.; Sandler, S.I. Phase behavior of clathrate hydrates: A model for single and multiple gas component hydrates. *Chem. Eng. Sci.* **2003**, *58*, 27–42. [CrossRef]

31. Merey, S.; Longinos, N. Does the Mediterranean Sea have potential for producing gas hydrates? *J. Nat. Gas Sci. Eng.* **2018**, *55*, 113–134. [CrossRef]

32. Fylaktos, N.; Papanicolas, C.N. New technologies for Eastern Mediterranean offshore gas exploration. In *European Parliamentary Research Service*; Scientific Foresight Unit (STOA): Brussels, Belgium, 2019; PE 634.419; p. 68.

33. Minshull, A.M.; Marín-Moreno, H.; Betlem, P.; Bialas, J.; Buenz, S.; Burwicz, E.; Cameselle, A.L.; Cifci, G.; Giustiniani, M.; Hillman, J.I.T.; et al. Hydrate occurrence in Europe: A review of available evidence. *Mar. Pet. Geol.*, in review.

34. Garfunkel, Z. Constrains on the origin and history of the Eastern Mediterranean basin. *Tectonophysics* **1998**, *298*, 5–35. [CrossRef]

35. Robertson, A.H.F. Mesozoic-Tertiary tectonic evolution of the easternmost Mediterranean area: Integration of marine and land evidence. In *Proceedings of the Ocean Drilling Program, Scientific Results*; Robertson, A.H.F., Emeis, K.C., Richter, C., Camerlenghi, A., Eds.; ODP: College Station, TX, USA, 1998; Volume 160, Chapter 54; pp. 723–782.

36. Aksu, A.E.; Hall, J.; Yaltırak, C. Miocene–recent evolution of Anaximander Mountains and Finike Basin at the junction of Hellenic and Cyprus arcs, eastern Mediterranean. *Mar. Geol.* **2009**, *258*, 24–47. [CrossRef]

37. Briand, F. (Ed.) The Messinian Salinity Crisis from mega-deposits to microbiology—A consensus report. In *N° 33 in CIESM Workshop Monographs*; CIESM: Monaco, 2008; p. 168.

38. Meilijson, A.; Hilgen, F.; Sepúlveda, J.; Steinberg, J.; Fairbank, V.; Flecker, R.; Waldmann, N.D.; Spaulding, S.A.; Bialik, O.M.; Boudinot, F.G.; et al. Chronology with a pinch of salt: Integrated stratigraphy of Messinian evaporites in the deep Eastern Mediterranean reveals long-lasting halite deposition during Atlantic connectivity. *Earth Sci. Rev.* **2019**, *194*, 374–398. [CrossRef]

39. Madof, A.S.; Bertoni, C.; Lofi, J. Discovery of vast fluvial deposits provides evidence for drawdown during the late Miocene Messinian salinity crisis. *Geology* **2019**, *47*, 171–174. [CrossRef]

40. Gardosh, M.; Druckman, Y.; Buchbinder, B.; Rybakov, M. *The Levant Basin Offshore Israel: Stratigraphy, Structure, Tectonic Evolution and Implications for Hydrocarbon Exploration*; Geophysical Institute of Israel Report 429/218/06; Geophysical Institute of Israel: Lod, Israel, 2006; p. 118.

41. Steinberg, J.; Gvirtzman, Z.; Folkman, Y.; Garfunkel, Z. Origin and nature of the rapid late Tertiary filling of the Levant Basin. *Geology* **2011**, *39*, 355–358. [CrossRef]

42. Ryan, W.B. Messinian badlands on the southeastern margin of the Mediterranean Sea. *Mar. Geol.* **1978**, *27*, 349–363. [CrossRef]

43. Gvirtzman, Z.; Reshef, M.; Buch-Leviatan, O.; Groves-Gidney, G.; Karcz, Z.; Makovsky, Y.; Ben-Avraham, Z. Bathymetry of the Levant basin: Interaction of salt-tectonics and surficial mass movements. *Mar. Geol.* **2015**, *360*, 25–39. [CrossRef]

44. Macgregor, D.S. The development of the Nile drainage system: Integration of onshore and offshore evidence. *Pet. Geosci.* **2012**, *18*, 417–431. [CrossRef]

45. Sestini, G. Nile Delta: A review of depositional environments and geological history. In *Deltas; Sites and Traps for Fossil Fuels*; Whateley, M.K.G., Pickering, K.T., Eds.; Geological Society of London: London, UK, 1989; pp. 99–127.

46. Folkman, Y.; Mart, Y. Newly recognized eastern extension of the Nile deep-sea fan. *Geology* **2008**, *36*, 939–942. [CrossRef]

47. Clark, I.R.; Cartwright, J.A. Interactions between submarine channel systems and deformation in deepwater fold belts: Examples from the Levant Basin, Eastern Mediterranean Sea. *Mar. Pet. Geol.* **2009**, *26*, 1465–1482. [CrossRef]

48. Tibor, G.; Ben-Avraham, Z.; Steckler, M.; Fligelman, H. Late Tertiary subsidence history of the Southern Levant Margin, Eastern Mediterranean Sea, and its implications to the understanding of the Messinian event. *J. Geophys. Res.* **1992**, *97*, 17593–17614. [CrossRef]

49. Buchbinder, B.; Zilberman, E. Sequence stratigraphy of Miocene-Pliocene carbonate-siliciclastic shelf deposits in the eastern Mediterranean margin (Israel): Effects of eustasy and tectonics. *Sediment. Geol.* **1997**, *112*, 7–32. [CrossRef]

50. Schattner, U.; Lazar, M. Hierarchy of source-to-sink systems—Example from the Nile distribution across the eastern Mediterranean. *Sediment. Geol.* **2016**, *343*, 119–131. [CrossRef]

51. Almogi-Labin, A.; Bar-Matthews, M.; Shriki, D.; Kolosovsky, E.; Paterne, M.; Schilman, B.; Matthews, A. Climatic variability during the last 90ka of the southern and northern Levantine Basin as evident from marine records and speleothems. *Quat. Sci. Rev.* **2009**, *28*, 2882–2896. [CrossRef]

52. Castañeda, I.S.; Schefuß, E.; Pätzold, J.; Damsté, J.S.S.; Weldeab, S.; Schouten, S. Millennial-scale sea surface temperature changes in the eastern Mediterranean (Nile River Delta region) over the last 27,000 years. *Paleoceanography* **2010**, *25*, PA1208. [CrossRef]

53. Ehrmann, W.; Schmiedl, G.; Seidel, M.; Krüger, S.; Schulz, H. A distal 140 kyr sediment record of Nile discharge and East African monsoon variability. *Clim. Past* **2016**, *12*, 713–727. [CrossRef]

54. Emeis, K.C.; Sakamoto, T.; Wehausen, R.; Brumsack, H.J. The sapropel record of the eastern Mediterranean Sea—Results of Ocean Drilling Program Leg 160. *Palaeogeogr. Palaeoclimatol. Palaeoecol.* **2000**, *158*, 371–395. [CrossRef]

55. Rohling, E.J.; Marino, G.; Grant, K.M. Mediterranean climate and oceanography, and the periodic development of anoxic events (sapropels). *Earth Sci. Rev.* **2015**, *143*, 62–97. [CrossRef]

56. Calvert, S.E.; Fontugne, M.R. On the late Pleistocene-Holocene sapropel record of climatic and oceanographic variability in the eastern Mediterranean. *Paleoceanogr. Paleoclimatol.* **2001**, *16*, 78–94. [CrossRef]

57. Tachikawa, K.; Vidal, L.A.; Cornuault, M.; Garcia, M.; Pothin, A.; Sonzogni, C.; Bard, E.; Menot, G.; Revel, M. Eastern Mediterranean Sea circulation inferred from the conditions of S1 sapropel deposition. *Clim. Past* **2015**, *11*, 855–867. [CrossRef]

58. Calvert, S.E.; Nielsen, B.; Fontugne, M.R. Evidence from nitrogen isotope ratios for enhanced productivity during formation of eastern Mediterranean sapropels. *Nature* **1992**, *359*, 223–225. [CrossRef]

59. De Lange, G.J.; Brumsack, H. Pore-water indications for the occurrence of gas hydrates in Eastern Mediterranean mud dome structures. In *Proceedings of the Ocean Drilling Program, Scientific Results*; Robertson, A.H.F., Emeis, K.C., Richter, C., Camerlenghi, A., Eds.; ODP: College Station, TX, USA, 1998; Volume 160, pp. 569–574.

60. Ducassou, E.; Migeon, S.; Mulder, T.; Murat, A.; Capotondi, L.; Bernasconi, S.M.; Mascle, J. Evolution of the Nile deep-sea turbidite system during the Late Quaternary: Influence of climate change on fan sedimentation. *Sedimentology* **2009**, *56*, 2061–2090. [CrossRef]

61. Long, D.; Lovell, M.A.; Rees, J.G.; Rochelle, C.A. (Eds.) Sediment-Hosted Gas Hydrates: New Insights on Natural and Synthetic Systems. *Geol. Soc. Lond. Spec. Publ.* **2009**, *319*, 1–9. [CrossRef]

62. Hall, J.K.; Udintsev, G.B.; Odinikov, Y.Y. *Geologic Structure of the Northeastern MEDITERRANEAN (Cruise 5 of the Research Vessel Akademik Nikolay Strakhov)*; Historical Productions-Hall Ltd.: Jerusalem, Israel, 1994; pp. 5–32.

63. Lykousis, V.; Alexandri, S.; Woodside, J.; De Lange, G.; Dählmann, A.; Perissoratis, C.; Rousakis, G. Mud volcanoes and gas hydrates in the Anaximander mountains (Eastern Mediterranean Sea). *Mar. Pet. Geol.* **2009**, *26*, 854–872. [CrossRef]

64. Perissoratis, C.; Ioakim, C.; Alexandri, S.; Woodside, J.; Nomikou, P.; Dählmann, A.; Lykousis, V. Thessaloniki mud volcano, the shallowest gas hydrate-bearing mud volcano in the Anaximander Mountains, Eastern Mediterranean. *J. Geol. Res.* **2011**, *2011*, 247983. [CrossRef]

65. Römer, M.; Sahling, H.; Pape, T.; dos Santos Ferreira, C.; Wenzhöfer, F.; Boetius, A.; Bohrmann, G. Methane fluxes and carbonate deposits at a cold seep area of the Central Nile Deep Sea Fan, Eastern Mediterranean Sea. *Mar. Geol.* **2014**, *347*, 27–42. [CrossRef]

66. Loncke, L.; Mascle, J. Mud volcanoes, gas chimneys, pockmarks and mounds in the Nile deep-sea fan (Eastern Mediterranean): Geophysical evidences. *Mar. Pet. Geol.* **2004**, *21*, 669–689. [CrossRef]

67. Dupré, S.; Woodside, J.; Foucher, J.P.; De Lange, G.; Mascle, J.; Boetius, A.; Harmégnies, F. Seafloor geological studies above active gas chimneys off Egypt (Central Nile Deep Sea Fan). *Deep Sea Res. Part I Oceanogr. Res. Pap.* **2007**, *54*, 1146–1172. [CrossRef]

68. Coleman, D.F.; Ballard, R.D. A highly concentrated region of cold hydrocarbon seeps in the Southeastern Mediterranean Sea. *Geo-Mar. Lett.* **2001**, *21*, 162–167. [CrossRef]

69. Coleman, D.F.; Austin, J.A., Jr.; Ben-Avraham, Z.; Makovsky, Y.; Tchernov, D. Seafloor pockmarks, deepwater corals, and cold seeps along the continental margin of Israel. *Oceanography* **2012**, *25*, 41–44.

70. Rubin-Blum, M.; Antler, G.; Turchyn, A.V.; Tsadok, R.; Goodman-Tchernov, B.N.; Shemesh, E.; Tchernov, D. Hydrocarbon-related microbial processes in the deep sediments of the Eastern Mediterranean Levantine Basin. *FEMS Microbiol. Ecol.* **2014**, *87*, 780–796. [CrossRef] [PubMed]

71. Makovsky, Y.; Rüggeberg, A.; Bialik, O.; Foubert, A.; Almogi-Labin, A.; Alter, Y.; Bampas, V.; Basso, D.; Feenstra, E.; Fentimen, R.; et al. *R/V AEGAEO Cruise EUROFLEETS2 SEMSEEP 20.09.–01.10.2016, Piraeus (Greece)–Piraeus (Greece)*; EUROFLEETS2 Cruise Summary Report; Université de Fribourg: Fribourg, Switzerland, 2016; p. 62.

72. Mayer, L.; Bell, K.L.C.; Ballard, R.; Nicolaides, S.; Konnaris, K.; Hall, J.; Shank, T.M. Discovery of sinkholes and seeps on Eratosthenes Seamount. *Oceanography* **2011**, *24*, 28–29.

73. Mitchell, G.; Mayer, L.; Bell, L.C.; Ballard, R.D.; Raineault, N.A.; Roman, C.; Ballard, W.B.A.; Cornwell, K.; Hine, A.; Shinn, E.; et al. Exploration of Eratosthenes Seamount—A continental fragment being forced down an oceanic trench. *Oceanography* **2013**, *26*, 36–41.

74. Praeg, D.; Geletti, R.; Mascle, J.; Unnithan, V.; Harmegnies, F. Geophysical exploration for gas hydrates in the Mediterranean Sea and a bottom-simulating reflection on the Nile Fan. In Proceedings of the 2008 GNGTS, Trieste, Italy, 6–8 October 2008; pp. 467–469.

75. Sharaf El Din, S.; Nassar, M. Gas hydrates over the Egyptian Med. Coastal waters. In Proceedings of the 2010 EGU General Assembly Conference Abstracts, Vienna, Austria, 2–7 May 2010; Volume 12, p. 78.

76. Praeg, D.; Migeon, S.; Mascle, J.; Unnithan, V.; Wardell, N.; Geletti, R.; Ketzer, J.M. Geophysical evidence of gas hydrates associated with widespread gas venting on the central Nile Deep-Sea Fan, offshore Egypt. In Proceedings of the 9th International Conference on Gas Hydrates, Denver, CO, USA, 25–30 June 2017.

77. Dimitrov, L.; Woodside, J. Deep sea pockmark environments in the eastern Mediterranean. *Mar. Geol.* **2003**, *195*, 263–276. [CrossRef]

78. Bertoni, C.; Kirkham, C.; Cartwright, J.; Hodgson, N.; Rodriguez, K. Seismic indicators of focused fluid flow and cross-evaporitic seepage in the Eastern Mediterranean. *Mar. Pet. Geol.* **2017**, *88*, 472–488. [CrossRef]

79. Bertoni, C.; Cartwright, J.A. 3D seismic analysis of circular evaporite dissolution structures, Eastern Mediterranean. *J. Geol. Soc.* **2005**, *162*, 909–926. [CrossRef]

80. Lazar, M.; Schattner, U.; Reshef, M. The great escape: An intra-Messinian gas system in the eastern Mediterranean. *Geophys. Res. Lett.* **2012**, *39*, L20309. [CrossRef]

81. Bertoni, C.; Cartwright, J.; Hermanrud, C. Evidence for large-scale methane venting due to rapid drawdown of sea level during the Messinian Salinity Crisis. *Geology* **2013**, *41*, 371–374. [CrossRef]

82. Eruteya, O.E.; Waldmann, N.; Schalev, D.; Makovsky, Y.; Ben-Avraham, Z. Intra-to post-Messinian deep-water gas piping in the Levant Basin, SE Mediterranean. *Mar. Pet. Geol.* **2015**, *66*, 246–261. [CrossRef]

83. Cartwright, J.; Kirkham, C.; Bertoni, C.; Hodgson, N.; Rodriguez, K. Direct calibration of salt sheet kinematics during gravity-driven deformation. *Geology* **2018**, *46*, 623–626. [CrossRef]

84. Ezra, O. Topology and Formation Settings of Deep Water Carbonates at the Boundaries of the Palmahim Disturbance, Offshore Israel. Master's Thesis, The Moses Strauss Department of Marine Geosciences, University of Haifa, Haifa, Israel, 2017.

85. Garfunkel, Z. Large-scale submarine rotational slumps and growth faults in the eastern Mediterranean. *Mar. Geol.* **1984**, *55*, 305–324. [CrossRef]

86. Gardosh, M.A.; Tannenbaum, E. The petroleum systems of Israel. In *Petroleum Systems of the Tethyan Region*; Marlow, L., Kendall, C., Yose, L., Eds.; AAPG Memoir; American Association of Petroleum Geologists: Tulsa, OK, USA, 2014; Volume 106, pp. 179–216.

87. Esestime, P.; Hewitt, A.; Hodgson, N. Zohr—A newborn carbonate play in the Levantine Basin, East-Mediterranean. *First Break* **2016**, *34*, 87–93.

88. Bayon, G.; Loncke, L.; Dupré, S.; Caprais, J.C.; Ducassou, E.; Duperron, S.; Etoubleau, J.; Foucher, J.P.; Fouquet, Y.; Gontharet, S.; et al. Multi-disciplinary investigation of fluid seepage on an unstable margin: The case of the Central Nile deep sea fan. *Mar. Geol.* **2009**, *261*, 92–104. [CrossRef]

89. Colorado School of Mines Center for Hydrate Research Web Site. Available online: http://hydrates.mines.edu/CHR/Software.html (accessed on 2 December 2012).

90. PERSEUS Web Site. Available online: http://isramar.ocean.org.il/Perseus_data/CastMap.aspx (accessed on 17 March 2013).

91. Robinson, A.R.; Malanotte-Rizzoli, P.; Hecht, A.; Michelato, A.; Roether, W.; Theocharis, A.; Bishop, J.; POEM Group. General circulation of the Eastern Mediterranean. *Earth Sci. Rev.* **1992**, *32*, 285–309. [CrossRef]

92. Zavatarielli, M.; Mellor, G.L. A numerical study of the Mediterranean Sea circulation. *J. Phys. Oceanogr.* **1995**, *25*, 1384–1414. [CrossRef]

93. Kress, N.; Manca, B.B.; Klein, B.; Deponte, D. Continuing influence of the changed thermohaline circulation in the eastern Mediterranean on the distribution of dissolved oxygen and nutrients: Physical and chemical characterization of the water masses. *J. Geophys. Res. Oceans* **2003**, *108*, 8109. [CrossRef]

94. Erickson, A.J. The Measurements and Interpretation of Heat Flow in the Mediterranean and Black Sea. Ph.D. Thesis, Massachusetts Institute of Technology, Cambridge, MA, USA, 1970.

95. Levitte, D.; Greitzer, Y. Geothermal update report from Israel 2005. In *World Geothermal Congress 2005*; International Geothermal Association: Auckland, New Zealand, 2005; pp. 1–5.

96. Riad, S.; Abdelrahman, E.M.; Refai, E.; El-Ghalban, H.M. Geothermal studies in the Nile Delta, Egypt. *J. Afr. Earth Sci. (Middle East)* **1989**, *9*, 637–649. [CrossRef]

97. Eppelbaum, L.; Modelevsky Jr, M.; Pilchin, A. Geothermal investigations in the Dead Sea Rift zone, Israel: Implications for petroleum geology. *J. Pet. Geol.* **1996**, *19*, 425–444. [CrossRef]

98. Marlow, L.; Kornpihl, K.; Kendall, C.G. 2-D basin modeling study of petroleum systems in the Levantine Basin, Eastern Mediterranean. *GeoArabia* **2011**, *16*, 17–42.

99. Ruppel, C.; Dickens, G.R.; Castellini, D.G.; Gilhooly, W.; Lizarralde, D. Heat and salt inhibition of gas hydrate formation in the northern Gulf of Mexico. *Geophys. Res. Lett.* **2005**, *32*. [CrossRef]

100. Portnov, A.; Cook, A.E.; Heidari, M.; Sawyer, D.E.; Santra, M.; Nikolinakou, M. Salt-driven evolution of a gas hydrate reservoir in Green Canyon, Gulf of Mexico. *AAPG Bull.* **2019**. [CrossRef]

101. Boswell, R.; Collett, T.S.; Frye, M.; Shedd, W.; McConnell, D.R.; Shelander, D. Subsurface gas hydrates in the northern Gulf of Mexico. *Mar. Pet. Geol.* **2012**, *34*, 4–30. [CrossRef]

102. Mann, D.M.; Mackenzie, A.S. Prediction of pore fluid pressures in sedimentary basins. *Mar. Pet. Geol.* **1990**, *7*, 55–65. [CrossRef]

103. Judd, A.; Hovland, M. *Seabed Fluid Flow: The Impact on Geology, Biology and the Marine Environment*; Cambridge University Press: New York, NY, USA, 2009; p. 475.

104. Kvenvolden, K.A. Natural gas hydrate: Background and history of discovery. In *Natural Gas Hydrate in Oceanic and Permafrost Environments*; Max, M.D., Ed.; Springer: Dordrecht, The Netherlands, 2000; pp. 9–16.

105. Bjørlykke, K.; Høeg, K.; Mondol, N.H. Introduction to Geomechanics: Stress and strain in sedimentary basins. In *Petroleum Geoscience from Sedimentary Environments to Rock Physics*; Bjørlykke, K., Ed.; Springer: Berlin/Heidelberg, Germany, 2015; pp. 301–318.

106. Ridout-Jamieson, R.H. *Geological Well Summary Hannah-1, Samedan Mediterranean Sea*; Cambrian Consultants Ltd.: New York, NY, USA, 2003; p. 30.

107. Leroy, C.C.; Parthiot, F. Depth-pressure relationships in the oceans and seas. *J. Acoust. Soc. Am.* **1998**, *103*, 1346–1352. [CrossRef]

108. Hall, J.K.; Lippman, S.; Gardosh, M.; Tibor, G.; Sade, A.R.; Sade, H.; Golan, A.; Amit, G.; Gur-Arie, L.; Nissim, I. *A New Bathymetric Map for the Israeli EEZ: Preliminary Results*; Ministry of National Infrastructures, Energy and Water Resources and the Survey of Israel: Jerusalem, Israel, 2015; p. 11.

109. Brown, K.M.; Bangs, N.L.; Froelich, P.N.; Kvenvolden, K.A. The nature, distribution, and origin of gas hydrate in the Chile Triple Junction region. *Earth Planet. Sci. Lett.* **1996**, *139*, 471–483. [CrossRef]

110. Sloan, E.D. *Clathrate Hydrates of Natural Gases*, 2nd ed.; CRC Press: New York, NY, USA, 1998.

111. Mienert, J.; Vanneste, M.; Bünz, S.; Andreassen, K.; Haflidason, H.; Sejrup, H.P. Ocean warming and gas hydrate stability on the mid-Norwegian margin at the Storegga Slide. *Mar. Pet. Geol.* **2005**, *22*, 233–244. [CrossRef]

112. Camps, A.P.; Long, D.; Rochelle, C.A.; Lovell, M.A. Mapping hydrate stability zones offshore Scotland. *Spec. Publ. Geol. Soc. Lond.* **2009**, *1*, 81–91. [CrossRef]

113. Duan, Z.; Li, D.; Chen, Y.; Sun, R. The influence of temperature, pressure, salinity and capillary force on the formation of methane hydrate. *Geosci. Front.* **2011**, *2*, 125–135. [CrossRef]

114. Milkov, A.V.; Dickens, G.R.; Claypool, G.E.; Lee, Y.J.; Borowski, W.S.; Torres, M.E.; Schultheiss, P. Co-existence of gas hydrate, free gas, and brine within the regional gas hydrate stability zone at Hydrate Ridge (Oregon margin): Evidence from prolonged degassing of a pressurized core. *Earth Planet. Sci. Lett.* **2004**, *222*, 829–843. [CrossRef]

115. Wang, S.; Wen, Y.; Song, H. Mapping the thickness of the gas hydrate stability zone in the South China Sea. *TAO Terr. Atmos. Ocean. Sci.* **2006**, *17*, 815–828. [CrossRef]

116. Liu, X.; Flemings, P.B. Passing gas through the hydrate stability zone at southern Hydrate Ridge, offshore Oregon. *Earth Planet. Sci. Lett.* **2006**, *241*, 211–226. [CrossRef]

117. Macelloni, L.; Simonetti, A.; Knapp, J.H.; Knapp, C.C.; Lutken, C.B.; Lapham, L.L. Multiple resolution seismic imaging of a shallow hydrocarbon plumbing system, Woolsey Mound, Northern Gulf of Mexico. *Mar. Pet. Geol.* **2012**, *38*, 128–142. [CrossRef]

118. Somoza, L.; León, R.; Medialdea, T.; Pérez, L.F.; González, F.J.; Maldonado, A. Seafloor mounds, craters and depressions linked to seismic chimneys breaching fossilized diagenetic bottom simulating reflectors in the central and southern Scotia Sea, Antarctica. *Glob. Planet. Chang.* **2014**, *123*, 359–373. [CrossRef]
119. Plaza-Faverola, A.; Vadakkepuliyambatta, S.; Hong, W.L.; Mienert, J.; Bünz, S.; Chand, S.; Greinert, J. Bottom-simulating reflector dynamics at Arctic thermogenic gas provinces: An example from Vestnesa Ridge, offshore west Svalbard. *J. Geophys. Res. Solid Earth* **2017**, *122*, 4089–4105. [CrossRef]
120. Sun, R.; Duan, Z. An accurate model to predict the thermodynamic stability of methane hydrate and methane solubility in marine environments. *Chem. Geol.* **2007**, *244*, 248–262. [CrossRef]
121. Yang, D.; Xu, W. Effects of salinity on methane gas hydrate system. *Sci. China Ser. D Earth Sci.* **2007**, *50*, 1733–1745. [CrossRef]
122. Frey-Martinez, J.F.; Cartwright, J.; Hall, B. 3D seismic interpretation of slump complexes: Examples from the continental margin of Israel. *Basin Res.* **2005**, *17*, 83–108. [CrossRef]
123. Xu, W.; Ruppel, C. Predicting the occurrence, distribution, and evolution of methane gas hydrate in porous marine sediments. *J. Geophys. Res. Solid Earth* **1999**, *104*, 5081–5095. [CrossRef]
124. Henry, P.; Thomas, M.; Clennell, M.B. Formation of natural gas hydrates in marine sediments: 2. Thermodynamic calculations of stability conditions in porous sediments. *J. Geophys. Res.* **1999**, *104*, 23005–230022. [CrossRef]
125. Dillon, W.P.; Lee, M.W. Gas Hydrates on the Atlantic Continental Margin of the United States—Controls on Concentration. *US Geol. Surv. Prof. Pap.* **1993**, *1570*, 313–330.
126. Vergnaud-Grazzini, C.; Devaux, M.; Znaidi, J. Stable isotope "anomalies" in Mediterranean Pleistocene records. *Mar. Micropaleontol.* **1986**, *10*, 35–69. [CrossRef]
127. Fleming, K.; Johnston, P.; Zwartz, D.; Yokoyama, Y.; Lambeck, K.; Chappell, J. Refining the eustatic sea-level curve since the Last Glacial Maximum using far-and intermediate-field sites. *Earth Planet. Sci. Lett.* **1998**, *163*, 327–342. [CrossRef]
128. Sultan, N.; Voisset, M.; Marsset, T.; Vernant, A.M.; Cauquil, E.; Colliat, J.L.; Curinier, V. Detection of free gas and gas hydrate based on 3D seismic data and cone penetration testing: An example from the Nigerian Continental Slope. *Mar. Geol.* **2007**, *240*, 235–255. [CrossRef]
129. Sha, Z.; Liang, J.; Zhang, G.; Yang, S.; Lu, J.; Zhang, Z.; Humphrey, G. A seepage gas hydrate system in northern South China Sea: Seismic and well log interpretations. *Mar. Geol.* **2015**, *366*, 69–78. [CrossRef]
130. Zhang, Z.; McConnell, D.R.; Han, D.H. Rock physics-based seismic trace analysis of unconsolidated sediments containing gas hydrate and free gas in Green Canyon 955, Northern Gulf of Mexico. *Mar. Pet. Geol.* **2012**, *34*, 119–133. [CrossRef]
131. Wever, T.F.; Abegg, F.; Fiedler, H.M.; Fechner, G.; Stender, I.H. Shallow gas in the muddy sediments of Eckernförde Bay, Germany. *Cont. Shelf Res.* **1998**, *18*, 1715–1739. [CrossRef]
132. Tréhu, A.M.; Long, P.E.; Torres, M.E.; Bohrmann, G.R.R.F.; Rack, F.R.; Collett, T.S.; Goldberg, D.S.; Milkov, A.V.; Riedel, M.; Schultheiss, P.; et al. Three-dimensional distribution of gas hydrate beneath southern Hydrate Ridge: Constraints from ODP Leg 204. *Earth Planet. Sci. Lett.* **2004**, *222*, 845–862. [CrossRef]
133. Collett, T.S.; Johnson, A.H.; Knapp, C.C.; Boswell, R. (Eds.) Natural Gas Hydrates: A Review. In *Natural gas Hydrates-Energy Resource Potential and Associated Geologic Hazards*; AAPG Memoir; American Association of Petroleum Geologists: Tulsa, OK, USA, 2009; Volume 89, pp. 146–219.
134. Peltzer, E.T.; Brewer, P.G. Practical physical chemistry and empirical predictions of methane hydrate stability. In *Natural Gas Hydrate in Oceanic and Permafrost Environments*; Max, M.D., Ed.; Springer: Dordrecht, The Netherlands, 2000; pp. 17–28.
135. Crutchley, G.J.; Pecher, I.A.; Gorman, A.R.; Henrys, S.A.; Greinert, J. Seismic imaging of gas conduits beneath seafloor seep sites in a shallow marine gas hydrate province, Hikurangi Margin, New Zealand. *Mar. Geol.* **2010**, *272*, 114–126. [CrossRef]
136. Gorman, A.R.; Holbrook, W.S.; Hornbach, M.J.; Hackwith, K.L.; Lizarralde, D.; Pecher, I. Migration of methane gas through the hydrate stability zone in a low-flux hydrate province. *Geology* **2002**, *30*, 327–330. [CrossRef]
137. Regnier, P.; Dale, A.W.; Arndt, S.; LaRowe, D.E.; Mogollón, J.; Van Cappellen, P. Quantitative analysis of anaerobic oxidation of methane (AOM) in marine sediments: A modeling perspective. *Earth Sci. Rev.* **2011**, *106*, 105–130. [CrossRef]

138. Dale, A.W.; Regnier, P.; Van Cappellen, P.; Fossing, H.; Jensen, J.B.; Jørgensen, B.B. Remote quantification of methane fluxes in gassy marine sediments through seismic survey. *Geology* **2009**, *37*, 235–238. [CrossRef]

139. Tishchenko, P.; Hensen, C.; Wallmann, K.; Wong, C.S. Calculation of the stability and solubility of methane hydrate in seawater. *Chem. Geol.* **2005**, *219*, 37–52. [CrossRef]

140. Duan, Z.; Mao, S.A. thermodynamic model for calculating methane solubility, density and gas phase composition of methane-bearing aqueous fluids from 273 to 523 K and from 1 to 2000 bar. *Geochim. Cosmochim. Acta* **2006**, *70*, 3369–3386. [CrossRef]

141. Borowski, W.S.; Paull, C.K.; Ussler, W. Marine pore-water sulfate profiles indicate in situ methane flux from underlying gas hydrate. *Geology* **1996**, *24*, 655–658. [CrossRef]

142. Niewöhner, C.; Hensen, C.; Kasten, S.; Zabel, M.; Schulz, H.D. Deep sulfate reduction completely mediated by anaerobic methane oxidation in sediments of the upwelling area off Namibia. *Geochim. Cosmochim. Acta* **1998**, *62*, 455–464. [CrossRef]

143. Borowski, W.S.; Paull, C.K.; Ussler, W. Global and local variations of interstitial sulfate gradients in deep-water, continental margin sediments: Sensitivity to underlying methane and gas hydrates. *Mar. Geol.* **1999**, *159*, 131–154. [CrossRef]

144. Dickens, G.R. Sulfate profiles and barium fronts in sediment on the Blake Ridge: Present and past methane fluxes through a large as hydrate reservoir. *Geochim. Cosmochim. Acta* **2001**, *65*, 529–543. [CrossRef]

145. Treude, T. Anaerobic Oxidation of Methane in Marine Sediments. Ph.D. Thesis, University of Bremen, Bremen, Germany, 2003.

146. Coffin, R.; Pohlman, J.; Gardner, J.; Downer, R.; Wood, W.; Hamdan, L.; Diaz, J. Methane hydrate exploration on the mid Chilean coast: A geochemical and geophysical survey. *J. Pet. Sci. Eng.* **2007**, *56*, 32–41. [CrossRef]

147. Coffin, R.; Hamdan, L.; Plummer, R.; Smith, J.; Gardner, J.; Hagen, R.; Wood, W. Analysis of methane and sulfate flux in methane-charged sediments from the Mississippi Canyon, Gulf of Mexico. *Mar. Pet. Geol.* **2008**, *25*, 977–987. [CrossRef]

148. Abreu, V.; Sullivan, M.; Pirmez, C.; Mohrig, D. Lateral accretion packages (LAPs): An important reservoir element in deep water sinuous channels. *Mar. Pet. Geol.* **2003**, *20*, 631–648. [CrossRef]

149. Gay, A.; Lopez, M.; Cochonat, P.; Séranne, M.; Levaché, D.; Sermondadaz, G. Isolated seafloor pockmarks linked to BSRs, fluid chimneys, polygonal faults and stacked Oligocene–Miocene turbiditic palaeochannels in the Lower Congo Basin. *Mar. Geol.* **2006**, *226*, 25–40. [CrossRef]

150. Ramana, M.V.; Ramprasad, T.; Paropkari, A.L.; Borole, D.V.; Rao, B.R.; Karisiddaiah, S.M.; Desa, M.; Kocherla, M.; Joao, H.M.; Lokabharati, P.; et al. Multidisciplinary investigations exploring indicators of gas hydrate occurrence in the Krishna–Godavari Basin offshore, east coast of India. *Geo-Mar. Lett.* **2009**, *29*, 25–38. [CrossRef]

151. Riedel, M.; Collett, T.S.; Shankar, U. Documenting channel features associated with gas hydrates in the Krishna–Godavari Basin, offshore India. *Mar. Geol.* **2011**, *279*, 1–11. [CrossRef]

152. Claypool, G.E.; Kaplan, I.R. The origin and distribution of methane in marine sediments. In *Natural Gases in Marine Sediments*; Kaplan, I.R., Ed.; Plenum Press: New York, NY, USA, 1974; pp. 99–139.

153. Paull, C.K.; Ussler, W.; Borowski, W.S. Sources of biogenic methane to form marine gas hydrates. In situ production or upward migration? *Ann. N. Y. Acad. Sci.* **1994**, *715*, 392–409. [CrossRef]

154. Malinverno, A. Marine gas hydrates in thin sand layers that soak up microbial methane. *Earth Planet. Sci. Lett.* **2010**, *292*, 399–408. [CrossRef]

155. Davie, M.K.; Buffett, B.A. Sources of methane for marine gas hydrate: Inferences from a comparison of observations and numerical models. *Earth Planet. Sci. Lett.* **2003**, *206*, 51–63. [CrossRef]

156. Haacke, R.R.; Westbrook, G.K.; Hyndman, R.D. Gas hydrate, fluid flow and free gas: Formation of the bottom-simulating reflector. *Earth Planet. Sci. Lett.* **2007**, *261*, 407–420. [CrossRef]

157. Kroon, D.; Alexander, I.; Little, M.; Lourens, L.J.; Matthewson, A.; Robertson, A.H.; Sakamoto, T. Oxygen isotope and sapropel stratigraphy in the eastern Mediterranean during the last 3.2 million years. In *Proceedings of the Ocean Drilling Program, Scientific Results*; Robertson, A.H.F., Emeis, K.-C., Richter, C., Camerlenghi, A., Eds.; ODP: College Station, TX, USA, 1998; Volume 160, Chapter 14; pp. 181–189.

158. Cramp, A.; O'Sullivan, G. Neogene sapropels in the Mediterranean: A review. *Mar. Geol.* **1999**, *153*, 11–28. [CrossRef]

159. Bar-Matthews, M.; Ayalon, A.; Kaufman, A. Timing and hydrological conditions of Sapropel events in the Eastern Mediterranean, as evident from speleothems, Soreq cave, Israel. *Chem. Geol.* **2000**, *169*, 145–156. [CrossRef]
160. Eckstein, Y. Review of heat flow data from the eastern Mediterranean region. *Pure Appl. Geophys.* **1978**, *117*, 150–159. [CrossRef]
161. Makris, J.; Stobbe, C. Physical properties and state of the crust and upper mantle of the Eastern Mediterranean Sea deduced from geophysical data. *Mar. Geol.* **1984**, *55*, 347–363. [CrossRef]
162. Camerlenghi, A.; Cita, M.B.; Vedova, B.D.; Fusi, N.; Mirabile, L.; Pellis, G. Geophysical evidence of mud diapirism on the Mediterranean Ridge accretionary complex. *Mar. Geophys. Res.* **1995**, *17*, 115–141. [CrossRef]
163. Wallmann, K.; Pinero, E.; Burwicz, E.; Haeckel, M.; Hensen, C.; Dale, A.; Ruepke, L. The global inventory of methane hydrate in marine sediments: A theoretical approach. *Energies* **2012**, *5*, 2449–2498. [CrossRef]

geosciences

MDPI

Article

Evidences for Paleo-Gas Hydrate Occurrence: What We Can Infer for the Miocene of the Northern Apennines (Italy)

Claudio Argentino [ID], Stefano Conti, Chiara Fioroni [ID] and Daniela Fontana *[ID]

Department of Chemical and Geological Sciences, University of Modena and Reggio Emilia, 41125 Modena, Italy; claudio.argentino@unimore.it (C.A.); stefano.conti@unimore.it (S.C.); chiara.fioroni@unimore.it (C.F.)
* Correspondence: daniela.fontana@unimore.it

Received: 25 February 2019; Accepted: 15 March 2019; Published: 20 March 2019

check for updates

Abstract: The occurrence of seep-carbonates associated with shallow gas hydrates is increasingly documented in modern continental margins but in fossil sediments the recognition of gas hydrates is still challenging for the lack of unequivocal proxies. Here, we combined multiple field and geochemical indicators for paleo-gas hydrate occurrence based on present-day analogues to investigate fossil seeps located in the northern Apennines. We recognized clathrite-like structures such as thin-layered, spongy and vuggy textures and microbreccias. Non-gravitational cementation fabrics and pinch-out terminations in cavities within the seep-carbonate deposits are ascribed to irregularly oriented dissociation of gas hydrates. Additional evidences for paleo-gas hydrates are provided by the large dimensions of seep-carbonate masses and by the association with sedimentary instability in the host sediments. We report heavy oxygen isotopic values in the examined seep-carbonates up to +6‰ that are indicative of a contribution of isotopically heavier fluids released by gas hydrate decomposition. The calculation of the stability field of methane hydrates for the northern Apennine wedge-foredeep system during the Miocene indicated the potential occurrence of shallow gas hydrates in the upper few tens of meters of sedimentary column.

Keywords: gas hydrates; seep-carbonates; clathrites; Miocene; northern Apennines

---

1. Introduction

The sensitivity of gas hydrates to climate changes and tectonic activity is still poorly constrained and more efforts are needed to understand how they respond to these forcing processes [1–4]. In the last decades, much work has been done in terms of monitoring and modelling the gas hydrate dynamics. The investigation of seepage activities and bottom simulating reflectors (BSR) has provided quantitative results that improved the accuracy of existing models [5].

Gas hydrates are widely distributed in present-day continental margins and their stability depends on temperature, pressure and availability of gas and water [6–8]. Gas hydrates continuously dissociate and recrystallize in order to maintain their position in the sedimentary column within the stability field. When gas hydrates cannot keep pace with sedimentation and burial, they no longer fall in the stability zone and their destabilization causes the rapid release of huge amounts of fluids that induces mud diapirism, soft-sediment deformation and seafloor collapses or may trigger large-scale continental slope instability (i.e., slumps, slides) [9–13]. Shallow gas hydrates are generally associated with authigenic ^{13}C-depleted carbonates [14,15]. Gas hydrate-associated carbonates, called clathrites [16], have been sampled from present-day seafloor or a few meters below [17–23]. These carbonates may form bodies of remarkable dimension and show peculiar structures such as vacuolar or vuggy-like fabrics, association of pure aragonitic and gas hydrate layers (zebra-like structures) and breccias produced by the destabilization of gas hydrates [18,24–26]. Geochemical indicators in porewaters (i.e., chlorine anomalies, oxygen isotopes) are also used to identify and quantify gas hydrate distribution and formation/destabilization processes [27–30]. Gas hydrate decomposition releases methane and fresh water into porewaters (low chlorine content) generating a ^{18}O-enriched signature [18,27,31–33]. Formation of gas hydrates produces enrichments in dissolved chloride content and depleted $\delta^{18}O$ values [27]. Peculiar minerals (greigite, pyrrhotite) [34,35] are also reported as possible markers of gas hydrate dissociation and their identification in the marine sedimentary record may allow the recognition of paleo-gas hydrate occurrence.

Compared to the abundant literature on present-day gas hydrates, only few studies deal with their past occurrence [36–38] or with fossil seep-carbonates recording the dissociation of gas hydrates [39–46]. In fossil sediments, the paleo-occurrence of gas hydrates is particularly challenging to assess, due to the lack of well-established proxies and to the uncertainties on the reconstruction of paleoenvironmental conditions (pressure, temperature, depth) controlling the hydrate stability field. Clathrite-like structures have been reported in fossil deposits and can be used as an indication of past gas hydrate destabilization [47,48]. Additional evidences can be yielded by geochemical signatures, the large dimensions of seep-carbonate deposits (several hundred meters in lateral extent and tens of meters in thickness) and the association with sedimentary instability (soft-sediment deformations) in hosting sediments [49].

Several Miocene seep-carbonate outcrops in different geological settings of the northern Apennines [50,51] (Figure 1) show characters suggesting paleo-gas hydrate occurrence. In this paper, we report new data and the results of twenty years of studies on seep-carbonates, obtained from field-work, facies analysis, geochemistry and biostratigraphy in order to support this hypothesis. The concentration of gas hydrate-associated carbonates in specific interval of the Miocene and their relationships with soft-sediment deformations may contribute to the understanding of factors that lead to their destabilization. Moreover, the investigation of paleo-gas hydrate in the sedimentary record may shed light into their long-term evolution and the interplay with sea level changes and tectonics [45,51].

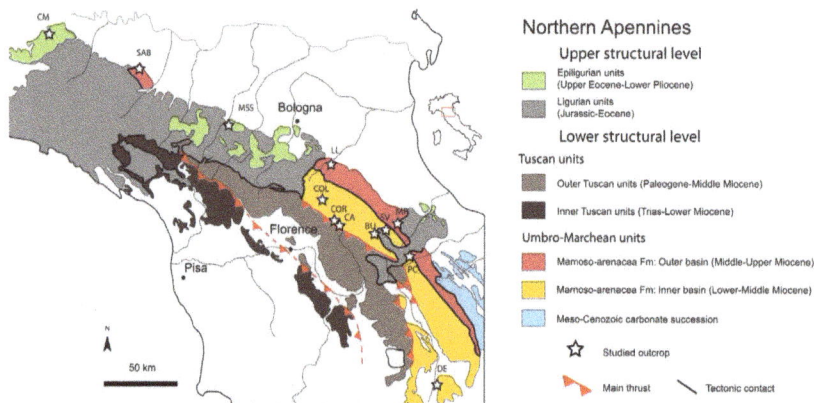

Figure 1. Simplified geological map of the northern Apennines (Italy) showing the main structural units and the location of the studied outcrops: CM = Cappella Moma [44°56'19.7" N; 9°04'54.7" E], SAB = Salsomaggiore [44°44'22.8" N; 10°04'56.5" E], MSS = Montardone, Sasso Streghe [44°28'40.3" N; 10°47'40.7" E], LL = Lame, Rontana [44°14'08.3" N; 11°42'57.0" E], COL = Colline, Mondera [44°06'41.6" N; 11°34'22.8" E], COR = Corella, Casellino [43°56'59" N, 1134'59.8" E], CA = Castagno d'Andrea [43°53'15.1" N; 11°40'32.6" E], MP = Montepetra [43°55'50.3" N; 12°11'38.2" E], SV = San Vernicio [43°54'58.4" N; 11°57'58.3" E], BU = Case Buscarelle [43°54'50.6" N; 11°55'30.4" E], PC = Poggio Campane [44°43'52.0" N; 12°14'18.9" E], DE = Deruta [42°58'33.3" N; 12°26'21.1" E].

2. Geological Setting

2.1. The Northern Apennines

The northern Apennine chain is an orogenic, NE-verging wedge, characterized by the stacking of several structural units of oceanic and continental origin (Figure 1). The complex structure of the chain is the result of the convergence and collision between the European and Africa plates, with the interposition of Adria and Corsica-Sardinia microplates. The collisional stage is accomplished by the subduction of the Adria under the Corsica-Sardinia lithosphere, coupled with the flexuring of the foreland and the formation of foredeep basins, progressively migrating towards NE [52]. The oceanic units (Ligurian units, Jurassic to Eocene), deposited within the Piedmont-Ligurian Ocean, a portion of the Tethys, comprehend a complex assemblage of highly deformed deep-marine sediments. The Epiligurian units represent the filling of wedge-top basins and are characterized by basinal to shelfal deposits (Eocene-Lower Pliocene) [52]. Ligurian and Epiligurian units represent the uppermost portion of the chain. They overlay the deformed Tuscan and Umbro-Marchean units (Figure 1) deposited on the Adria microplate and mainly consist of Mesozoic to Paleogene carbonate successions folded and segmented by thrusts (Figure 1). Foredeep basins are filled by sheet-like turbidites (Marnoso arenacea Fm); the depocenter of the basin migrated in response to the advancing Apenninic accretionary wedge incorporating the previously formed foredeep deposits. During the outward migration, the development of intrabasinal highs (Figure 2), related to blind faults and thrusts, caused the fragmentation of the foredeep into an inner and outer part [53,54]. Sedimentation on top of intrabasinal highs was mainly represented by hemipelagites and diluted turbidites (drape mudstones) forming some hundred metre thick fine-grained intervals. These structural highs represent favourable conditions to gas hydrate accumulation, seepage phenomena and sediment instability along their flanks; their deactivation heralds the successive involvement of the inner foredeep in the accretionary wedge linked to the uplift and the closure of the foredeep. During this phase, the foredeep turbidites are covered by slope marls representing the closure facies [55]: Vicchio Marls, Verghereto Marls, Ghioli di letto Fm (Figure 2) (in Figure 1, slope marls are included in Tuscan and Umbro-Marchean units). Slope marls are characterized by abundant slumps and extra formational slides, sourced by previously accreted units.

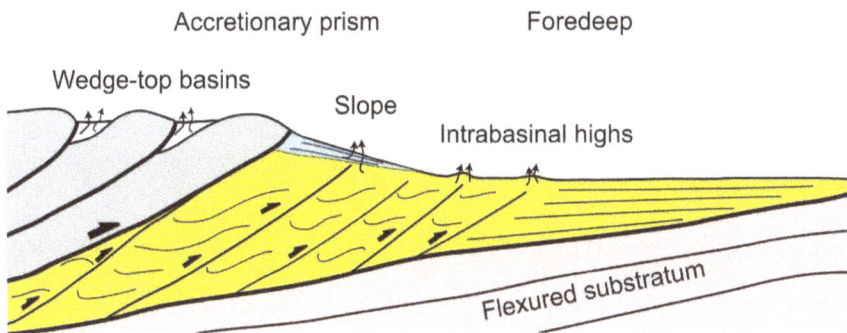

Figure 2. Schematic profile showing the main structural elements of the northern Apennine wedge-foredeep system hosting hydrate-related seep-carbonates during the Miocene: wedge-top basins (Epiligurian units, white colour) developed on top of allochthonous units (Ligurian units, grey colour), slope sediments (closure facies, light blue colour) and foredeep basin (Umbro-Marchean units, yellow colour). Intrabasinal highs developed in the inner sector of the foredeep. The position of methane seeps on the seafloor is here indicated by black arrows.

2.2. Miocene Seep-Carbonates

Large seep-carbonate bodies occur in three different positions of the Apennine wedge-top-foredeep system, from the inner to the outer setting (Figures 1 and 2):

(1) In wedge-top basins within the Epiligurian succession. The largest number of seep-carbonate bodies is within a 50 meters-thick interval in the basal portion of slope marls of the Termina Fm. The widest carbonates bodies are distributed in the peripheral portion of chaotic deposits (Montardone melange) interpreted as mud diapir [49].

(2) Along the outer slope of the accretionary prism, close to the front of the orogenic wedge. Seep-carbonates are hosted in fine-grained sediments draping thrust-bounded folds and buried ridges constituted by the older accreted turbiditic units [51]. Slope sediments including seep-carbonates have a wide extent up to 100 km parallel to the structural trend of the chain and mark the closure stage of the foredeep before the overriding of Ligurian units.

(3) At the leading edge of the deformational front in the inner foredeep, in fault-related anticlines, standing above the adjacent deep seafloor forming intrabasinal highs. Seep-carbonates are hosted in fine-grained intervals sedimented above these structures surrounded by basinal turbidites [51]. The ridges extend laterally for 10–15 km. Thrust faults are connected to the basal detachment through growing splay faults.

3. Methods

We applied different proxies to identify the occurrence of paleo-gas hydrates in the northern Apennines. It is worth mentioning that none of them is conclusively decisive when considered alone but integrated with others allow more accurate interpretations [56]. Based on a detailed field-work (Table 1), we determined number and dimension of seep-carbonate bodies in primary position (not reworked) and considered only larger outcrops, up to 1000 m of lateral extension, made up of several carbonate bodies, laterally and vertically repeated. In each outcrop we analysed larger bodies with lateral extension wider than 10 m and thickness higher than 5 m (Table 1). A detailed facies analysis allowed us to identify distinctive seep-carbonate facies and structures (breccias, non-systematic fractures, spongy fabric, drusy-like structures) referable to clathrites, described in present day settings [18,25,27]. We report the carbon and oxygen isotopic composition of Apennine seep-carbonates (new data and previous papers, Table 1, Supplementary Materials S1). We investigated the spatial and stratigraphic relationships between seep-carbonates and sedimentary instability structures related to fluid expulsion processes, such as mud volcanoes and diapirs, neptunian dikes, soft-sediment deformations, slide block, chaotic and mass transport deposits cemented by depleted micrites. The nannofossil biostratigraphy of host sediments allowed us to constrain the age of the stratigraphic interval containing seep-carbonates. We analysed outcrops that do not show any evidences of reworking. Smear slides from unprocessed sediment were prepared and analysed under a polarising light microscope at a magnification of 1250X. A semiquantitative analysis was performed for each sample by observing 100 fields of view in random traverses. In Table 1 we also included age data from previous papers [50,57]. Petrographic observations on thin sections were conducted using an optical microscope in order to identify the main carbonate phases and microfacies. Stable carbon and oxygen isotopic analyses were conducted on matrix micrite and on calcite cements filling veins and cavities. Carbonate phases were isolated by microdrilling in thin-section counterparts. Analyses were carried out at the ISO4 Stable Isotope Laboratory of the University of Turin, using a Finnigan MAT 251 mass spectrometer (Thermo Electron Corp., Waltham, MA, USA). Data are reported as range of values in ‰ notation relative to VPDB. Analytical error was better than 0.1‰ for both carbon and oxygen. Full isotopic data are reported in Supplementary Materials S1.

Table 1. Field and geochemical data of the examined seep-carbonate outcrops.

Outcrop	Geosetting	Dimension (Length × Thickness)		Samples (n)	^{13}C (‰ V-PDB)	^{18}O (‰ V-PDB)	Clathrite Facies	SSD	References
		Outcrop	MDAC						
MSS	Wedge-top	400 × 70	10–250 × 5–30	16	−39.1 to −18.2	+0.3 to +5.5	X	X	[49]
CM		n.m.	35 × 25	7	−30.0 to −11.0	+1.1 to +2.9	X	n.o.	[49]
SV		200 × 40	10–70 × 5–20	11	−33.2 to −27.2	+0.1 to +3.6	X	X	This work, [58]
BU		n.m.	50 × 15	2	−36.4 to −35.1	−0.3 to +1.5	X	n.o.	This work
PC	Slope	70 × 30	10–20 × 5–8	1	−32.2	+2.2	X	X	This work
SAB		250 × 35	10–25 × 5–8	18	−41.4 to −8.7	+0.2 to + 2.9	X	X	[59]
MP		100–150	10–40 × 5–10	20	−52.7 to −19,1	+0.7 to +6.0	X	X	[53]
LL		350 × 40	10–150 × 5–30	18	−51,7 to −27.4	−1.6 to + 5.0	X	n.o.	[58]
COR		1000 × 50	20–300 × 10–25	16	−42.3 to −26.6	−5.7 to +1.2	X	X	[51]
CA	Intrabasinal highs	90 × 50	12–30 × −5–10	30	−41.3 to −15.0	+0.9 to +1.2	X	X	[60]
COL		100 × 30	5–10 × 2–5	7	−56.2 to −38.9	+0.6 to +3.5	X	n.o.	This work
DE		150 × 40	10–80 × 5–20	14	−46.0 to −11.0	−4.7 to +2.2	X	X	[61]

Dimensions are reported in meters. SSD, soft-sediment deformation; n.m., not measured; n.o., not observed. Biostratigraphic data for each outcrop are reported in Section 4.1.

4. Results

We report the results from field-work, facies analysis, petrography, geochemistry and biostratigraphy of 12 seep-carbonate outcrops of the northern Apennines that show the peculiar features (i.e., dimensions, clathrite-like structures, $\delta^{18}O$ isotope and sediment instability in enclosing sediments) suggesting relationships with gas hydrates.

Seep-carbonate dimensions—The examined carbonates mainly consist of large lenticular stratiform or pinnacle-like bodies, from 10 to 300 m wide and 8 to 30 m thick (Table 1). They are arranged in horizons that extend for 700–1000 m conformable with the stratification of the enclosing fine-grained turbidites and hemipelagites. Carbonate bodies are vertically repeated and seep-impacted sediment can reach 150 m in thickness (MP outcrop).

Clathrite-like structures—Examined seep-carbonates show peculiar structures (Figure 3) such as thin layered structures consisting of an alternation of micrite and sparry calcite (Figure 3a), radial pattern of fractures (Figure 3b), spongy and vuggy-like fabrics resembling drusy crystals similar to those reported in present-day gas hydrate-associated carbonates (Figure 3c,d). Cavities are irregular in shape, circular to ellipsoidal in cross-section, empty or filled with carbonate cements and/or coarser sediments (calcarenites, peloidal sediments, microbreccias, coquina debris). They resemble voids previously occupied by solid substances (gas hydrates) that successively disappeared. The examination of cavities in thin section has allowed the identification of non-gravitational cementation fabrics and pinch-out terminations (Figure 4), ascribable to irregularly oriented dissociation of gas hydrates as proposed by [48,62] for fossil deposits in the Tertiary Piedmont Basin (north-western Apennines).

Clathrite-like structures also include fluid-induced breccias (Figure 5a–c). Monogenic and polygenic breccias are abundant, generally restricted but not exclusive, to the basal portion of carbonate bodies. Monogenic breccias are constituted by heterometric angular clasts, ranging in size from few millimetres to 5–10 cm, composed of the authigenic micrite from the seep-carbonates. Clasts are chaotically dispersed in a micritic matrix or in a fine to medium-grained sandy matrix. Larger clasts derive from the coalescence of heterometric smaller clasts, testifying various cycles of cementation and fragmentation. In many cases, monogenic breccias pass gradually to a dense and intricate network of non-systematic carbonate-filled veins and microfractures, irregularly connected to a larger vein network and conduits (Figure 5d). Polygenic breccias contain carbonate, arenitic and pelitic clasts of various stratigraphic provenance and dimensions, floating in the micritic matrix. Clasts are heterometric (from some millimetres to 50 cm in diameter), with sharp edges; clast size decreases from the base to the top of carbonate bodies. Polygenic breccias form units ranging in thickness from centimetres to a few meters, often interdigitated with fine-grained carbonate cemented sediments.

Figure 3. Examples of various clathrite-like structures. (**a**) Thin layered structures surrounding a carbonate breccia with shell fragments (MP outcrop in Figure 1). (**b**) Irregular network (mainly radial to concentric) of carbonate-filled veins (DE outcrop in Figure 1). (**c,d**) Vacuolar, spongy and vuggy-like fabrics: cavities have various shape, empty or filled with carbonate cements and/or coarser sediments and coquina debris. PC outcrop (**c**) and SS outcrop (**d**) of Figure 1.

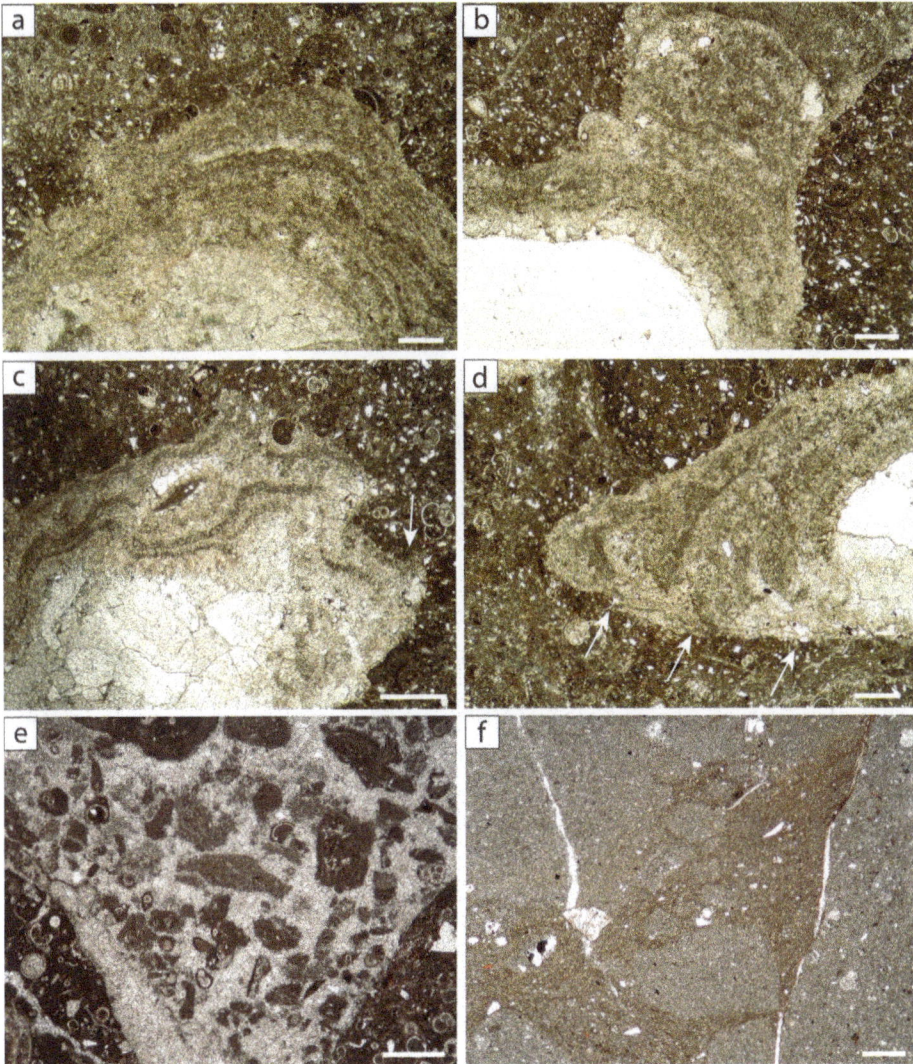

Figure 4. Micrographs displaying cavity infillings in the vuggy carbonate facies (**a–d**) and microbreccias (**e,f**). (**a,b**) Alternating laminae of microspatite and micrite. The thickness of the laminae varies from few microns to tens of microns. (**b**) Voids can be connected by conduits showing laminated fabric. Laminae do not line all the cavity wall but stop abruptly, in some cases producing pinch out terminations (white arrows; **c,d**). Calcite cements and laminae may include some tests of foraminifera or siliciclastic grains, indicating precipitation in an early diagenetic stage within semi-consolidated material. (**e**) Microbreccias composed by very angular and poorly sorted micritic clasts indicative of autoclastic fragmentation; clasts are floating in microsparitic cement. (**f**) Matrix-supported microbreccias composed by subangular clasts of micrite; clasts seem to fit to each other. Thin sections from samples of PC and COR outcrops of Figure 1. Scale bar = 1 mm.

Figure 5. Examples of fluid-induced breccias: (**a**) Polygenic breccias with abundant disarticulated shells (COL outcrop in Figure 1). (**b**) Polygenic breccias with clasts floating in a micritic matrix (SV outcrop in Figure 1). (**c**) Carbonate monogenic breccia (CO outcrop in Figure 1). (**d**) Network of veins and a large conduit (arrows) filled with coarse sediments and carbonate cement (DE outcrop in Figure 1).

Stable C and O isotopic composition—Table 1 reports isotopic data for 160 carbonate samples collected in 12 outcrops. The examined seep-carbonates are mainly constituted of low-Mg calcite and minor aragonite and dolomite. They are typically depleted in $\delta^{13}C$ with values around −30‰ (Table 1). Values more negative than −50‰ occur in three outcrops (MP, LL, COL), with the most negative value is −56.2‰ (COL). Most samples display positive $\delta^{18}O$ values (up to 6.0‰), enriched respect to the carbonate fraction of surrounding sediments, that is generally around −1‰ [58,60]. The $\delta^{13}C$ depletion and ^{18}O enrichment reach the maximum values in clathrite-like textures, in particular from brecciated structures and conduit-rich facies.

Sediment instability in enclosing sediments related to fluid expulsion processes—Seep-carbonates are hosted in fine-grained deposits that in many cases show soft-sediment deformation structures, related to fluid expulsion (Figure 6a,b). In wedge-top outcrops (MSS), seep-carbonates are associated to chaotic deposits formed by the ascent of mud diapirs. In the foredeep, fluid-expulsion structures in host sediment adjacent to seep-carbonates include neptunian dikes and injection structures, chimneys and conduits filled by coarse deposits, cylindrical and ellipsoidal concretions, brecciated and septaria-like concretions made up of ^{13}C-depleted carbonate (Figure 6d). In the slope setting, slumps and slides have been frequently observed (MP). In many outcrops, fluidized resedimented arenites and debris flow are cemented by $\delta^{13}C$ depleted micrite (Figure 6c).

Figure 6. Examples of fluid expulsion structures: (**a,b**) Soft-sediment deformation structures and disrupted strata in host sediment in proximity of the seep carbonate body (COR outcrop in Figure 1). (**c**) Debris flow deposits cemented by ^{13}C-depleted micrite (DE outcrop of Figure 1). (**d**) Septaria-like concretion made up of depleted methane-derived carbonates (BU outcrop of Figure 1).

4.1. Dating of Seep Carbonate Outcrops

The biostratigraphy of the examined outcrops results from new analyses on nannofossil assemblages and from previous studies based on both nannofossils and foraminifera (Figure 7), carried out on the fine-grained hosting sediment. The examined sediments are distributed from the Langhian up to the early Messinian. In detail, two outcrops in the foredeep (CA, COR) are referred to the Langhian (nannofossil subzone MNN5a) in agreement with previous studies [51,60]. Based on nannofossil biostratigraphy, most of the seep outcrops (MMS, PC, COL, DE) indicate a Serravallian age. The CM and SAB outcrops provided similar ages, confirming previous analyses obtained with foraminifera [59]. Other seep-carbonate outcrops are referred to the early Tortonian; more in detail, SV correspond to the zone MNN8 and CB outcrop indicates subzone MNN8b and zone MNN9. Nannofossil assemblage in LL outcrop indicates the late Tortonian, whereas MP outcrop points to an age comprised between the late Tortonian and the early Messinian based on previous foraminifera biostratigraphy [53].

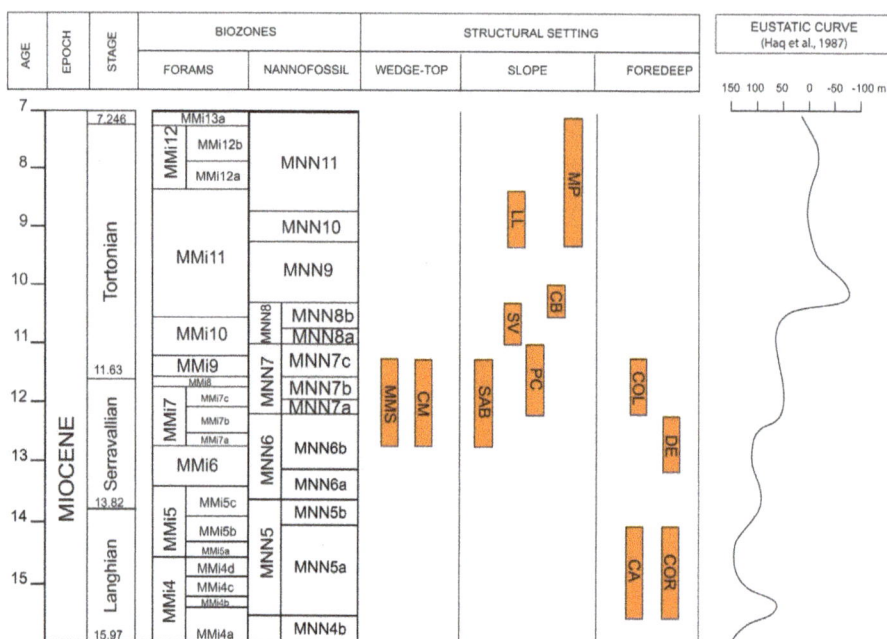

Figure 7. Biostratigraphic scheme, based on nannofossil and foraminifera assemblages, indicating the Miocene biozones and the stratigraphic distribution of the examined outcrops. For the full name of the outcrops refer to Figure 1. The third-order eustatic curve is reported to show a possible correlation between the development of seepage systems and low stands.

5. Discussion

We report several features in fossil seep-carbonates of the northern Apennines that can be related to the occurrence of paleo-gas hydrates. Some of these are similar to those described from present-day settings where gas hydrates have been mapped or directly observed on the seafloor, therefore their identification in the sedimentary record provides a quite robust evidence. The vacuolar, spongy and vuggy-like fabrics in the examined Apenninic carbonates are interpreted as voids and pores previously occupied by solid substances (possibly crystallization of gas hydrates at the rim of bubbles trapped in the sediment) [18] preserved by the precipitation of methane-derived carbonates. Monogenic breccias are made up of very angular micritic clasts of the previously precipitated carbonate crusts and may represent collapse breccias resulting from the rapid destabilization of gas hydrates within the sediment pore space [27]. Similar fabrics have been reported in seep-carbonates of the Hydrate Ridge [18] and in the Oligocene seep-carbonates of the Carpathians [42]. Polygenic breccias include exotic material derived from the underlying formations and could be related to rapid ascent of fluids released by gas hydrate destabilization: Similar structures are also described in carbonates associated to gas hydrates from the Hydrate Ridge (intraformational breccias) [18]. Fluid overpressures, as expected from the rapid decomposition of hydrates, can also result in the opening of fractures and injection of fluidized sediments [63–65] as shown in the dense and intricate network of non-systematic carbonate-filled veins and microfractures, irregularly connected to larger conduits, in the studied seep-carbonates. Layered carbonate facies show a zebra-like appearance due to the alternation of thin micrite and sparry calcite layers. This texture likely resulted from carbonate precipitation in contact with pure gas hydrate layers roughly oriented parallel to the bedding surfaces of the host sediments, as reported from present-day sediments at Hydrate Ridge, offshore Oregon [18].

Other markers are more uncertain, as the large dimension and volumes of carbonate bodies hardly explainable as originated by local anaerobic degradation of organic matter and more likely are indicative of abundant methane supply by destabilization of gas hydrates [56]. Moreover, according to various studies, fluid circulation for gas hydrate destabilization may trigger soft-sediment deformation and mass-wasting processes responsible for the emplacement of chaotic deposits [5,13]. Gas hydrates may act as cement, particularly in fine-grained deposits, improving cohesion and stability of paleo-slopes; conversely, their rapid decomposition increases sediment pore-pressure and favours sediment mobilization at all scales. Examined seep carbonate in the northern Apennines are sometimes associated with slumps, slides and resedimented arenites and debris flow, cemented or encrusted by micrite with depleted $\delta^{13}C$ and positive $\delta^{18}O$. In wedge-top outcrops, seep-carbonates are associated to chaotic deposits formed by the ascent of a mud diapir [49].

The oxygen isotopic composition of methane-derived carbonates provides important information on fluid sources and processes occurring within the sediments [27,56]. The ultimate $\delta^{18}O$ value recorded by authigenic carbonates is controlled by several factors, that is, the temperature of formation, the $\delta^{18}O$ composition of the precipitating fluid, the mineralogy and the pH of the solution [66–70]. Heavy $\delta^{18}O$ values (>+3.4‰ up to 14.8‰) [18,71] in methane-derived carbonates have been reported in gas hydrate-bearing sediments worldwide (Hydrate Ridge, Gulf of Mexico). Heaviest oxygen isotopic values in the examined seep-carbonates are around +5‰ and +6‰ (Table 1). Based on these $\delta^{18}O$ values, the temperature of formation of examined carbonates has been calculated applying the equation by [72] and assuming an original aragonitic mineralogy and pore fluid composition of 0‰ versus SMOW (average modern seawater). The resulting temperatures are below 0 °C which are clearly unrealistic. Therefore a contribution of isotopically heavier fluids has to be assumed: a possible influence of gas hydrate decomposition [27] or a contribution from deep-sourced fluids recording the dehydration of clay minerals [73]. We exclude the clay dehydration process, as the depth to the detachment surface under the foredeep deposits during the Miocene did not probably exceed 3 km [53], which is not sufficient to make the clay dehydration process a significant controlling factor on the oxygen isotopic composition of the carbonates [73]. We suggest that gas hydrate decomposition could be the main source at least for the heaviest $\delta^{18}O$ values in our study. In order to validate this hypothesis, we discuss the gas hydrate stability in the Apennines during the Miocene.

Gas Hydrate Stability along the Northern Apennine Margin during the Miocene

The stability of gas hydrates within the sediments depends on many factors such as temperature, pressure, gas composition and saturation, as well as pore-water composition (e.g., salinity) [27]. All these parameters are used to construct the stability curve for a specific gas hydrate structure. To calculate the vertical extent of the gas hydrate stability zone (GHSZ) we need to know the temperature profile in the water column and local geothermal gradient. The true depth interval in which hydrate potentially form within the sediments, called gas hydrate occurrence zone (GHOZ), is defined by the interceptions of the phase boundary curve with the seafloor and with the geothermal gradient (Figure 8) [74]. In order to reconstruct the gas hydrates stability in a fossil environment at a specific depth, we need to consider the bottom water temperature to calculate the GHOZ. In this way, we avoid introducing errors related to uncertainties on the oceanographic conditions due to paucity of data.

Paleobathimetric estimates of middle Miocene seep-carbonates in the inner foredeep by [75] and more recently by [57], place the seepage systems at the average depth of ~1000 m and we used this value as representative of the seafloor depth in our reconstruction (Figure 8). We calculated the stability field of pure methane hydrates (crystal structure I) [27] in the Miocene along the northern Apennine margin, since this methane hydrate structure is the most frequently observed along modern continental margins. We considered two different hypotheses of temperatures of bottom water, as reported by various authors for the Mediterranean region during the Miocene and a normal geothermal gradient of 30 °C/km. The value of 4 °C has been utilized by [48] for calculating the gas hydrate stability in the lower Messinian sediments of the Tertiary Piedmont Basin (northern Italy). The authors extrapolated that value by applying an exponential curve decreasing from the Messinian sea surface temperature of 18 °C [76] until reaching the depth of 1000 m. By using 4 °C as minimum bottom water temperature and assuming modern-day Mediterranean salinity (3.8%) [77] we obtained a GHOZ of ~400 m for the Apennine setting (Figure 8). We then considered the bottom temperature of 11 °C as the highest value allowing a potential occurrence of gas hydrates within the sediments and consistent with a GHOZ of few tens of meters (Figure 8). A similar value of bottom water temperature (12–13 °C) has been also used by [41] to demonstrate that gas hydrates were stable at 1000 m water depth in the Western Mediterranean during the upper Miocene. Considering the paleobathimetry of our seepage systems and the expected depth of the thermocline located between 500 and 1000 m, the bottom temperature reported by [78] during the Middle Miocene Climatic Optimum (~15 °C at 500 m depth) would support the value of about 10 °C obtained in our model. Therefore, it is conceivable to use this latter value for the calculation of paleo-gas hydrates stability zones in the Miocene, indicating the potential occurrence of shallow gas hydrates.

The chronostratigraphic distribution of examined outcrops (Figure 7) shows that they are roughly concentrated in three main intervals: in the Langhian (MNN5a), in the late Serravallian-early Tortonian (MNN6b-MNN7) and the late Tortonian-early Messinian (MNN10-MNN11). When comparing seep distributions with 3rd order eustatic curves of [79], they approximately seem to match phases of sea-level lowering. However, in the studied cases, we are aware that the biostratigraphic resolution is conditioned by the duration of the biozone, that could be longer than the seepage activity. Some authors investigated the correlation between seep-carbonate precipitation and sea-level changes and proposed the correspondence of seep-carbonates with sea-level lowering [9,15,80]. A drop in the hydraulic pressure on the plumbing system during sea-level lowering would shift the bottom of the gas hydrate stability zone into shallower depths, inducing gas-hydrate destabilization [27]. Specific paleoclimatic reconstructions for the Miocene of the western Mediterranean will provide more accurate estimations.

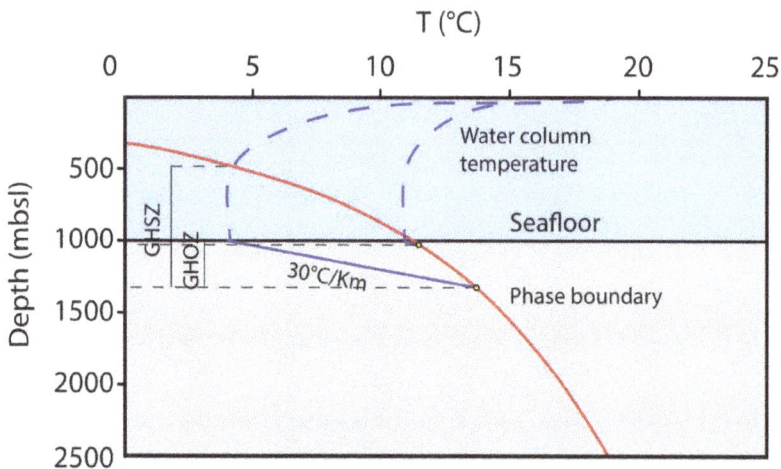

Figure 8. Evaluation of the gas hydrate stability along the northern Apennines in the Miocene, based on the assumptions reported in the text. For bottom water temperature of 4 °C, gas hydrates would be stable for 400 m below the seafloor. The value of 11 °C represents the highest temperature allowing the occurrence of gas hydrates in the upper tens of meters of sediments. Water column profiles were drawn following [48].

6. Conclusions

We document the occurrence of clathrite-like features in Miocene seep-carbonates from different structural settings of the Apennine wedge-foredeep system and discuss their possible relationships with gas hydrate dissociation.

Vuggy fabrics are interpreted as related to voids and pores previously occupied by gas hydrates dispersed in the sediment, preserved by the precipitation of methane-derived carbonates. It is suggested that monogenic and polygenic breccias could be related to rapid ascent of bubbles and growing gas hydrate layers that caused sediment brecciation. Fluid overpressures could result in the opening of fractures and injection of fluidized sediments. We also considered dimension and volumes of carbonate bodies, as well as sedimentary instability in host deposits, as possible indicator of abundant and prolonged methane supply by destabilization of gas hydrates.

We report heavy oxygen isotopic values up to +6‰ in the examined seep-carbonates and propose a contribution of isotopically heavier fluids released by gas hydrate decomposition.

Based on the calculation of gas hydrate stability zone for the selected settings, we suggest a potential occurrence of shallow gas hydrates within the sediments in the upper few tens of meters.

Supplementary Materials: The following are available online at http://www.mdpi.com/2076-3263/9/3/134/s1, S1: Full isotopic data.

Author Contributions: The authors contributed equally to this work.

Funding: This research was funded by the International Association of Sedimentologists, IAS Post-Graduate Grant 2nd session 2017, and by the University of Modena and Reggio Emilia PhD student research grant.

Acknowledgments: We are indebted to two anonymous Reviewers for helpful comments and suggestions.

Conflicts of Interest: The authors declare no conflict of interest.

References

1. Archer, D. Methane hydrate stability and anthropogenic climate change. *Biogeosciences* **2007**, *4*, 993–1057. [CrossRef]
2. Maslin, M.; Owen, M.; Betts, R.; Simon, D.; Dunkley Jones, T.; Ridgwell, A. Gas hydrates: Past and future geohazard? *Philos. Trans. R. Soc. A* **2010**, *368*, 2369–2393. [CrossRef] [PubMed]
3. Plaza-Faverola, A.; Klaeschen, D.; Barnes, P.; Pecher, I.; Henrys, S.; Mountjoy, J. Evolution of fluid expulsion and concentrated hydrate zones across the southern Hikurangi subduction margin, New Zealand: An analysis from depth migrated seismic data. *Geochem. Geophys. Geosyst.* **2012**, *13*. [CrossRef]
4. Ruppel, C.D.; Kessler, J.D. The interaction of climate change and methane hydrates. *Rev. Geophys.* **2017**, *55*, 126–168. [CrossRef]
5. Paull, C.K.; Buelow, W.J.; Ussler, W., III; Borowski, W.S. Increased continental-margin slumping frequency during sea-level lowstands above gas hydrate-bearing sediments. *Geology* **1996**, *24*, 143–146. [CrossRef]
6. Collett, T.S. Energy resource potential of natural gas hydrates. *AAPG Bull.* **2002**, *86*, 1971–1992.
7. Matsumoto, R.; Freire, A.F.M.; Machiyama, H.; Satoh, M.; Hiruta, A. Low velocity anomaly of gas hydrate bearing silty and clayey sediments, Joetsu Basin, Eastern Margin of Japan Sea. *AOGS* **2009**, *4*, 529–543.
8. Freire, A.F.M.; De Matos Maia Lete, C.; De Oliveira, F.M.; Guimaraes, M.F.; Da Silva Milhomen, P.; Pietzsch, R.; D'Avila, R.S.F. Fluid escape structures as possible indicators of past gas hydrate dissociation during the deposition of the Barremian sediments in the Reconcavo Basin, NE, Brasil. *Braz. J. Geol.* **2017**, *47*, 79–93. [CrossRef]
9. Haq, B.U. Natural gas hydrates: Searching for the long-term climatic and slope-stability records. In *Gas Hydrates: Relevance to World Margin Stability and Climate Change*; Henriet, J.P., Mienert, J., Eds.; Geological Society: London, UK, 1998; Volume 137, pp. 303–318.
10. Henriet, J.P.; Mienert, J. *Gas Hydrates—Relevance to World Margin Stability and Climatic Change*; Geological Society: London, UK, 1998; Volume 137, pp. 1–338.
11. Vogt, P.R.; Jung, W.Y. Holocene mass wasting on upper non-Polar continental slopes- due to post Glacial ocean warming and hydrate disssociation? *Geophys. Res. Lett.* **2002**, *29*, 1341–1348. [CrossRef]

12. Sultan, N.; Cochonat, P.; Foucher, J.P.; Mienert, J. Effect of gas hydrates melting on seafloor slope instability. *Mar. Geol.* **2004**, *213*, 379–401. [CrossRef]

13. Handwerger, A.L.; Rempel, A.W.; Skarbek, R.M. Submarine landslides triggered by destabilization of high-saturation hydrate anomalies. *Geochem. Geophys. Geosyst.* **2017**, *18*, 2429–2445. [CrossRef]

14. Pierre, C.; Fouquet, Y. Authigenic carbonates from methane seeps of the Congo deep-sea fan. *Geo-Mar. Lett.* **2007**, *27*, 249–257. [CrossRef]

15. Pierre, C.; Blanc-Valleron, M.; Demange, J.; Boudouma, O.; Foucher, J.P.; Pape, T.; Himnler, T.; Fekete, N.; Spiess, P. Authigenic carbonates from active methane seeps offshore southwest Africa. *Geo-Mar. Lett.* **2012**, *32*, 501–513. [CrossRef]

16. Kennett, J.P.; Fackler-Adams, B.N. Relationship of clathrate instability to sediment deformation in the upper Neogene of California. *Geology* **2000**, *28*, 215–218. [CrossRef]

17. Aloisi, G.; Pierre, C.; Rouchy, J.M.; Foucher, J.P.; Woodside, J.; Medinaut Scientific-Party. Methane-related authigenic carbonates of eastern Mediterranean Sea mud volcanoes and their possible relation to gas hydrate destabilisation. *Earth Planet. Sci. Lett.* **2000**, *184*, 321–338. [CrossRef]

18. Greinert, J.; Bohrmann, G.; Suess, E. Gas hydrate associated carbonates and methane venting at Hydrate Ridge: Classification, distribution and origin of authigenic lithologies. In *Natural Gas Hydrates: Occurrence, Distribution and Detection*; Geophysical Monograph Series; Paull, C.K., Dillon, W.P., Eds.; American Geophysical Union: Washington, DC, USA, 2001; Volume 124, pp. 99–113.

19. Sassen, R.; Roberts, H.H.; Carney, R.; Milkov, A.; DeFreitas, D.A.; Lanoil, B.; Zhang, C.L. Free hydrocarbon gas, gas hydrate and authigenic minerals in chemosynthetic communities of the northern Gulf of Mexico continental slope: Relation to microbial process. *Chem. Geol.* **2004**, *205*, 195–217. [CrossRef]

20. Mazzini, A.; Svensen, H.; Hovland, M.; Planke, S. Comparison and implications from strikingly different authigenic carbonates in a Nyegga complex pockmark, G11, Norwegian Sea. *Mar. Geol.* **2006**, *231*, 89–102. [CrossRef]

21. Ho, S.A.; Cartwright, J.A.; Imbert, P. Vertical evolution of fluid venting structures in relation to gas flux, in the Neogene-Quaternary of the Lower Congo Basin, Offshore Angola. *Mar. Geol.* **2012**, *332–334*, 40–55. [CrossRef]

22. Suess, E. Marine cold seeps and their manifestations: Geological control, biogeochemical criteria and environmental conditions. *Int. J. Earth Sci.* **2014**, *103*, 1889–1916. [CrossRef]

23. Smith, J.P.; Coffin, R.B. Methane-flux and authigenic carbonate in shallow sediments overlying methane hydrate bearing strata in Alaminos Canyon, Gulf of Mexico. *Energies* **2014**, *7*, 6118–6141. [CrossRef]

24. Bohrmann, G.; Heeschen, K.; Jung, C.; Weinrebe, W.; Baranov, B.; Cailleau, B.; Heath, R.; Huhnerbach, V.; Hort, M.; Masson, D.; et al. Widespread fluid expulsion along the seafloor of the Costa Rica convergent margin. *Terra Nova* **2002**, *14*, 69–79. [CrossRef]

25. Teichert, B.M.A.; Gussone, N.; Eisenhauer, A.; Bohrmann, G. Clathrites: Archives of near-seafloor pore-fluid evolution ($\delta^{44/40}$Ca, δ^{13}C, δ^{18}O) in gas hydrate environments. *Geology* **2005**, *33*, 213–216. [CrossRef]

26. Abegg, F.; Bohrmann, G.; Freitag, J.; Kuhs, W. Fabric of gas hydrate in sediments from Hydrate Ridge-results from ODP Leg 204 samples. *Geo-Mar. Lett.* **2007**, *27*, 269–277. [CrossRef]

27. Bohrmann, G.; Torres, M.E. Gas hydrates in marine sediments. In *Marine Geochemistry*; Schulz, H.D., Zabel, M., Eds.; Springer: Berlin, Germany, 2006; pp. 481–512.

28. Tong, H.; Feng, D.; Cheng, H.; Yang, S.; Wang, H.; Min, A.G.; Edwards, R.L.; Chen, Z.; Chen, D. Authigenic carbonates from seeps on the northern continental slope of the South China Sea: New insights into fluid sources and geochronology. *Mar. Pet. Geol.* **2013**, *13*, 260–271. [CrossRef]

29. Han, X.; Suess, E.; Liebetrau, V.; Eisenhauer, A.; Huang, Y. Past methane release events and environmental conditions at the upper continental slope of the South China Sea: Constraints by seep carbonates. *Int. J. Earth Sci.* **2014**, *103*, 1873–1887. [CrossRef]

30. Loyd, S.J.; Sample, J.; Tripati, R.E.; Defliese, W.F.; Brooks, K.; Hovland, M.; Torres, M.; Marlow, J.; Hancock, L.G.; Martin, R.; et al. Methane seep carbonates yield clumped isotope signatures out of equilibrium with formation temperatures. *Nat. Commun.* **2016**, *7*, 12274. [CrossRef]

31. Naehr, T.H.; Rodriguez, N.M.; Bohrmann, G.; Paull, C.K.; Botz, R. Methane derived authigenic carbonates associated with gas hydrate decomposition and fluid venting above the Blake Ridge Diapir. *Proc. ODP Sci. Results* **2000**, *164*, 285–300.

32. Matsumoto, R.; Borowski, W.S. Gas-hydrate estimate from newly determined oxygen isotopic fractionaction (α_{GH-IW}) and ^{18}O anomalies of the interstitial waters. Leg 164, Blake Ridge. *Proc. ODP Sci. Results* **2000**, *164*, 59–66.

33. Heeschen, K.U.; Haeckel, M.; Klaucke, I.; Ivanov, M.K.; Bohrmann, G. Quantifying in situ gas hydrates at active seep sites in the eastern Blake Sea using pressure coring technique. *Biogeosciences* **2011**, *8*, 3555–3565. [CrossRef]

34. Larrasoaña, J.C.; Roberts, A.P.; Musgrave, R.J.; Gràcia, E.; Piñero, E.; Vega, M.; Martínez-Ruiz, F. Diagenetic formation of greigite and pyrrhotite in gas hydrate marine sedimentary systems. *Earth Planet. Sci. Lett.* **2007**, *261*, 350–366. [CrossRef]

35. Wendel, J. Ancient methane seeps tell tale of sudden warming. *EOS* **2017**, *98*. [CrossRef]

36. Hesselbo, S.P.; Grocke, D.R.; Jenkins, H.C.; Bjerrum, C.J.; Farrimond, P.; Bell, H.S.M.; Green, O.R. Massive dissociation of gas hydrate during a Jurassic oceanic anoxic event. *Nature* **2000**, *406*, 392–395. [CrossRef] [PubMed]

37. Padden, M.; Weissert, H.; De Rafelis, M. Evidence for Late Jurassic release of methane from gas hydrate. *Geology* **2001**, *29*, 223–226. [CrossRef]

38. Kennedy, M.J.; Christie-Blick, N.; Sohl, L.E. Are Proterozoic cap carbonates and isotopic excursions a record of gas hydrate destabilization following Earth's coldest intervals? *Geology* **2001**, *29*, 443–446. [CrossRef]

39. Krause, F.F. Genesis and geometry of the Meiklejohn Peak lime mud-mound, Bare Mountain Quadrangle, Nevada, USA. Ordovician limestone with submarine frost heave structures—A possible response to gas clathrate hydrate evolution. *Sediment. Geol.* **2001**, *145*, 189–213. [CrossRef]

40. Pierre, C.; Rouchy, G.M.; Blanc-Valleron, M.M. Gas hydrate dissociation in the Lorca Basin (SE Spain) during the Mediterranean Messinian salinity crisis. *Sediment. Geol.* **2002**, *147*, 247–252. [CrossRef]

41. Pierre, C.; Rouchy, J.M. Isotopic compositions of diagenetic dolomites in the Tortonian marls of the western Mediterranean margins: Evidence of past gas hydrate formation and dissociation. *Chem. Geol.* **2004**, *205*, 469–484. [CrossRef]

42. Bojanowski, M.J.; Bagiński, B.; Guillermier, C.; Franchi, I.A. Carbon and oxygen isotope analysis of hydrate-associated Oligocene authigenic carbonates using NanoSIMS and IRMS. *Chem. Geol.* **2015**, *416*, 51–64. [CrossRef]

43. Himmler, T.; Freiwald, A.; Stollhofen, H.; Peckmann, J. Late Carboniferous hydrocarbon-seep carbonates from the glaciomarine Dwyka Group, southern Namibia. *Palaeogeogr. Palaeoclimatol. Palaeoecol.* **2008**, *257*, 185–197. [CrossRef]

44. Wang, J.; Jiang, G.; Xiao, S.; Li, Q.; Wei, Q. Carbon isotope evidence for widespread methane seeps in the ca.635 Ma Doushantuo cap carbonate in south China. *Geology* **2008**, *36*, 347–350. [CrossRef]

45. Nyman, S.L.; Nelson, C.S.; Campbell, K.A. Miocene tubular concretions in East Coast Basin, New Zealand: Analogue for subsurface plumbing of cold seeps. *Mar. Geol.* **2010**, *272*, 319–336. [CrossRef]

46. Iadanza, A.; Sanpalmieri, G.; Cipollari, P.; Mola, M.; Cosentino, D. The "Brecciated Limestones" of Maiella, Italy: Rheological implications of hydrocarbon-charged fluid migration in the Messinian Mediterranean Basin. *Palaeogeogr. Palaeoclimatol. Palaeoecol.* **2013**, *390*, 130–147. [CrossRef]

47. Peckmann, J.; Thiel, V. Carbon cycling at ancient methane-seeps. *Chem. Geol.* **2004**, *205*, 443–467. [CrossRef]

48. Dela Pierre, F.; Martire, L.; Natalicchio, M.; Clari, P.; Petrea, C. Authigenic carbonates in Upper Miocene sediments of the Tertiary Piedmont Basin (NW Italy): Vestiges of an ancient gas hydrate stability zone? *GSA Bull.* **2010**, *122*, 994–1010. [CrossRef]

49. Conti, S.; Fontana, D.; Lucente, C.C.; Pini, G.A. Relationships between seep-carbonates, mud volcanism and basin geometry in the Late Miocene of the northern Apennines of Italy: The Montardone mélange. *Int. J. Earth Sci.* **2014**, *103*, 281–295. [CrossRef]

50. Conti, S.; Fioroni, C.; Fontana, D. Correlating shelf carbonate evolutive phases with fluid expulsion episodes in the foredeep (Miocene, northern Apennines, Italy). *Mar. Pet. Geol.* **2017**, *79*, 351–359. [CrossRef]

51. Argentino, C.; Conti, S.; Crutchley, G.J.; Fioroni, C.; Fontana, D.; Johnson, J.E. Methane-derived authigenic carbonates on accretionary ridges: Miocene case studies in the northern Apennines (Italy) compared with modern submarine counterparts. *Mar. Pet. Geol.* **2019**, *102*, 860–872. [CrossRef]

52. Conti, S.; Fioroni, C.; Fontana, D.; Grillenzoni, C. Depositional history of the Epiligurian wedge-top basin in the Val Marecchia area (northern Apennines, Italy): A revision of the Burdigalian-Tortonian succession. *Ital. J. Geosci.* **2016**, *135*, 324–335. [CrossRef]

53. Conti, S.; Fontana, D.; Mecozzi, S.; Panieri, G.; Pini, G.A. Late Miocene seep-carbonates and fluid migration on top of the Montepetra intrabasinal high (Northern Apennines, Italy): Relations with synsedimentary folding. *Sediment. Geol.* **2010**, *231*, 41–54. [CrossRef]

54. Tinterri, R.; Magalhaes, P.M. Synsedimentary structural control on foredeep turbidites: An example from Miocene Marnoso-arenacea Formation, Northern Apennines, Italy. *Mar. Pet. Geol.* **2011**, *28*, 629–657. [CrossRef]

55. Ricci Lucchi, F. The Oligocene to Recent Foreland basins of the Northern Apennines. In *Foreland Basins*; Allen, P.A., Homewood, P., Eds.; The International Association of Sedimentologists: Gent, Belgium, 1986; Volume 8, pp. 103–139.

56. Campbell, K.A. Hydrocarbon seep and hydrothermal vent paleoenvironments and paleontology: Past developments and future research directions. *Paleogeogr. Paleoclimatol. Paleoecol.* **2006**, *232*, 362–407. [CrossRef]

57. Grillenzoni, C.; Monegatti, P.; Turco, E.; Conti, S.; Fioroni, C.; Fontana, D.; Salocchi, A.C. Paleoenvironmental evolution in a high-stressed cold-seep system (Vicchio Marls, Miocene, northern Apennines, Italy). *Palaeogeogr. Palaeoclimatol. Palaeoecol.* **2017**, *487*, 37–50. [CrossRef]

58. Terzi, C.; Lucchi, F.R.; Vai, G.B.; Aharon, P. Petrography and stable isotope aspects of cold-vent activity imprinted on Miocene-age "calcari aLucina" from Tuscan and Romagna Apennines, Italy. *Geo-Mar. Lett.* **1994**, *14*, 177–184. [CrossRef]

59. Conti, S.; Fontana, D.; Gubertini, A.; Sighinolfi, G.; Tateo, F.; Fioroni, C.; Fregni, P. A multidisciplinary study of middle Miocene seep-carbonates from the northern Apennine foredeep (Italy). *Sediment. Geol.* **2004**, *169*, 1–19. [CrossRef]

60. Conti, S.; Fontana, D.; Lucente, C.C. Authigenic seep-carbonates cementing coarse-grained deposits in a fan-delta depositional system (middle Miocene, Marnoso-arenacea Formation, central Italy). *Sedimentology* **2008**, *55*, 471–486. [CrossRef]

61. Artoni, A.; Conti, S.; Turco, E.; Iaccarino, S. Tectonic and climatic control on deposition of seep-carbonates: The case of middle-late Miocene Salsomaggiore Ridge (Northern Apennines, Italy). *Rivista Italiana di Paleontologia e Stratigrafia* **2014**, *120*, 317–335.

62. Martire, L.; Natalicchio, M.; Petrea, C.C.; Cavagna, S.; Clari, P.; Pierre, F.D. Petrographic evidence of the past occurrence of gas hydrates in the Tertiary Piedmont Basin (NW Italy). *Geo-Mar. Lett.* **2010**, *30*, 461–476. [CrossRef]

63. Shanmugam, G. Global case studies of soft-sediment deformation structures (SSDS): Definitions, classifications, advances, origins and problems. *J. Palaeogeogr.* **2017**, *6*, 251–320. [CrossRef]

64. Schwartz, H.; Sample, J.; Weberling, K.D.; Minisini, D.; Moore, J.C. An ancient linked fluid migration system: Cold-seep deposits and sandstone intrusions in the Panoche Hills, California, USA. *Geo-Mar. Lett.* **2003**, *23*, 340–350. [CrossRef]

65. Mazzini, A. Mud volcanism: Processes and implications. *Mar. Pet. Geol.* **2009**, *26*, 1677–1680. [CrossRef]

66. Friedman, I.; O'Neil, J.R. Compilation of stable isotope fractionation: Factors of geochemical interest. In *Data of Geochemistry*, 6th ed.; Geological Survey Professional Paper; Fleischer, M., Ed.; USGS: Washington, DC, USA, 1977; pp. 1–12.

67. Tremaine, D.M.; Froelich, P.N.; Wang, Y. Speleothem calcite farmed in situ: Modern calibration of ^{18}O and ^{13}C paleoclimate proxies in a continuously-monitored natural cave system. *Geochim. Cosmochim. Acta* **2011**, *75*, 4929–4950. [CrossRef]

68. Tarutani, T.; Clayton, R.N.; Mayeda, T.K. The effect of polymorphism and Mg substitution on oxygen isotope fractionation between calcium carbonate and water. *Geochim. Cosmochim. Acta* **1969**, *33*, 987–996. [CrossRef]

69. Emrich, K.; Ehhalt, D.H.; Vogel, J.C. Carbon isotope fractionation during the precipitation of calcium carbonate. *Earth Planet. Sci. Lett.* **1970**, *8*, 363–371. [CrossRef]

70. Zeebe, R.E.; Wolf-Gladrow, D. *CO$_2$ in Seawater: Equilibrium, Kinetics, Isotopes*; Elsevier: Amsterdam, The Netherlands, 2001; 346p.

71. Feng, D.; Birgel, D.; Peckmann, J.; Roberts, H.H.; Joye, S.B.; Sassen, R.; Liu, X.L.; Hinrichs, K.U.; Chen, D. Time integrated variation of sources of fluids and seepage dynamics archived in authigenic carbonates from Gulf of Mexico Gas Hydrate Seafloor Observatory. *Chem. Geol.* **2014**, *385*, 129–139. [CrossRef]

72. Hudson, J.D.; Anderson, T.F. Ocean temperatures and isotopic compositions through time. *Earth Environ. Sci. Trans. R. Soc. Edinb.* **1989**, *80*, 183–192. [CrossRef]

73. Dählmann, A.; De Lange, G.J. Fluid–sediment interactions at Eastern Mediterranean mud volcanoes: A stable isotope study from ODP Leg 160. *Earth Planet. Sci. Lett.* **2003**, *212*, 377–391. [CrossRef]

74. Tréhu, A.M. Gas hydrates in marine sediments: Lessons from scientific ocean drilling. *Oceanography* **2006**, *19*, 124–142. [CrossRef]

75. Aharon, P.; Sen Gupta, B.K. Bathymetric reconstructions of the Miocene age "calcari a Lucina" (Northern Apennines, Italy) from oxigen isotopes and benthic foraminifera. *Geo-Mar. Lett.* **1994**, *14*, 219–230. [CrossRef]

76. Bosellini, F.R.; Perrin, C. Estimating Mediterranean Oligocene-Miocene sea-surface temperatures: An approach based on coral taxonomic richness. *Paleogeogr. Paleoclimatol. Paleoecol.* **2008**, *258*, 71–88. [CrossRef]

77. Wenzhöfer, F. *Short Cruise Report MERIAN MSM 13/4 HOMER Limassol-Limassol 21.11. 2009 14.12. 2009*; Max Planck Institut für Marine Mikrobiologie: Bremen, Germany, 2009.

78. Scheiner, F.; Holcová, K.; Milovský, R.; Kuhnert, H. Temperature and isotopic composition of seawater in the epicontinental sea (Central Paratethys) during the Middle Miocene Climate Transition based on Mg/Ca, δ^{18}O and δ^{13}C from foraminiferal tests. *Palaeogeogr. Palaeoclimatol. Palaeoecol.* **2018**, *495*, 60–71. [CrossRef]

79. Haq, B.U.; Hardenbol, J.A.N.; Vail, P.R. Chronology of fluctuating sea levels since the Triassic. *Science* **1987**, *235*, 1156–1167. [CrossRef] [PubMed]

80. Fontana, D.; Conti, S.; Grillenzoni, C.; Mecozzi, S.; Petrucci, F.; Turco, E. Evidence of climatic control on hydrocarbon seepage in the Miocene of the northern Apennines: The case study of the Vicchio Marls. *Mar. Pet. Geol.* **2013**, *48*, 90–99. [CrossRef]

MDPI

St. Alban-Anlage 66

4052 Basel

Switzerland

Tel. +41 61 683 77 34

Fax +41 61 302 89 18

www.mdpi.com

Geosciences Editorial Office

E-mail: geosciences@mdpi.com

www.mdpi.com/journal/geosciences